BIODEGRADABLE HYDROGELS FOR DRUG DELIVERY

HOW TO ORDER THIS BOOK

BY PHONE: 800-233-9936 or 717-291-5609, 8AM–5PM Eastern Time

BY FAX: 717-295-4538

BY MAIL: Order Department
Technomic Publishing Company, Inc.
851 New Holland Avenue, Box 3535
Lancaster, PA 17604, U.S.A.

BY CREDIT CARD: American Express, VISA, MasterCard

BIODEGRADABLE HYDROGELS FOR DRUG DELIVERY

Kinam Park
Waleed S. W. Shalaby
Haesun Park

Purdue University, School of Pharmacy

LANCASTER · BASEL

Biodegradable Hydrogels for Drug Delivery
a **TECHNOMIC**® publication

Published in the Western Hemisphere by
Technomic Publishing Company, Inc.
851 New Holland Avenue, Box 3535
Lancaster, Pennsylvania 17604 U.S.A.

Distributed in the Rest of the World by
Technomic Publishing AG
Missionsstrasse 44
CH-4055 Basel, Switzerland

Printed in the United States of America
10 9 8 7 6 5 4 3 2 1

Main entry under title:
Biodegradable Hydrogels for Drug Delivery

A Technomic Publishing Company book
Bibliography:
Includes index p. 241

Library of Congress Catalog Card No. 93-60193
ISBN No. 1-56676-004-6

To our parents
and
Elaine

TABLE OF CONTENTS

PREFACE

The advances made in the area of controlled drug delivery during the last two decades are remarkable. Drug delivery in the future will require much more sophisticated delivery devices for the delivery of protein drugs, self-regulated drug delivery, and the release of drugs at target sites. Of the many polymeric materials, biodegradable hydrogels present unique advantages and opportunities in the development of the future delivery devices. Many biodegradable hydrogel systems have been described in the literature and the number of publications on the subject is increasing steadily. We have undertaken the challenge of putting together information relevant to biodegradable hydrogels in one place. This book covers the mechanisms of biodegradation, types of biodegradable hydrogels, chemical and physical gels, chemical and enzymatic degradation, and examples of biodegradable drug delivery systems. Due to the variety of subtopics covered in this book, it was virtually impossible to cover all the references. There is no doubt that we inadvertently missed many important articles and we send our sincere apologies to those whose articles relevant to biodegradable hydrogels are not cited in this book. We hope that the information contained in this book is helpful to those who wish to explore the potential of biodegradable hydrogels in the development of new drug delivery systems.

We wish to express our sincere gratitude to the staff of Technomic Publishing Company, Inc., especially Joseph L. Eckenrode who suggested that we write a book on biodegradable hydrogels and Susan G. Farmer who helped us in many ways during the preparation of the manuscript. Finally, we would like to pay our respects to those who contributed to the development of biodegradable hydrogels, especially Dr. Jorge Heller and Dr. Jindrich Kopeček.

CHAPTER 1

Introduction

1.1 CONTROLLED RELEASE DRUG DELIVERY SYSTEMS

During the last two decades, significant advances have been made in the controlled release drug delivery of therapeutic agents [1 – 17]. In the early stages of research on controlled release drug delivery, major emphasis was focused on the development of zero-order release devices. Current technology has improved to such a level that delivery of some drugs at a constant rate for a certain period of time ranging from days to years is not a major issue anymore. Transdermal patches deliver drugs at a constant rate for 24 hours or longer [18] and the Norplant System (Wyeth-Ayerst Labs) releases progestin levonorgestel from silicone rubber tubular capsules for several years [19].

The premise of the zero-order release is to maintain a constant drug concentration in blood for an extended period of time. The zero-order release of a drug, however, does not necessarily result in a constant drug concentration in blood. The absorption of the drug by the body usually does not follow the zero-order kinetics, except when the drug is delivered directly into the blood stream by an infusion pump. The nonzero-order drug absorption is still effective in most cases as long as the drug concentration is maintained between the minimum effective and the maximum safe concentrations. Cumulative evidence suggests that circadian rhythms need to be considered in drug delivery [20,21]. Many drugs show rhythmic changes in bioavailability, absorption, distribution, and excretion [22]. Furthermore, the zero-order drug release is not always necessary and sometimes it is not desirable. The drug delivery needs to be feedback controlled depending on the drug concentration in blood and pharmacologic effect. In many situations, a drug needs to be released only when the body requires it [23]. For example, insulin is required only when the glucose concentration in blood is increased. Once the glucose level is decreased, no further insulin is required. Chronotherapy will require new, improved controlled release dosage forms, such as self-regulated drug delivery systems [24 – 28].

There are additional important issues that have to be dealt with for further development of controlled release dosage forms. Controlled release devices should be able to deliver large molecular weight drugs, such as peptides and proteins, without altering the bioactivities. The recombinant DNA technology has allowed production of many protein drugs in large quantities which were not readily available previously [29,30]. The stability, bioavailability, and bioactivity of the protein drugs can be modified by protein engineering [31]. Most protein drugs have exquisite specificity. Controlled delivery, however, is a major obstacle in the effective use of the protein drugs [32]. Controlled delivery of bioactive peptides and proteins is not as easy as the delivery of small molecular weight drugs. The loading of protein drugs into, and their release from, polymeric controlled release devices without altering the bioactivity still remains an unsolved problem.

Another important issue in controlled release drug delivery is drug targeting, i.e., delivery of a drug to specific cells and organs in the body and release of the drug at the target sites [33–35]. Delivery of therapeutic agents to a particular cell population also minimizes exposure of normal cells to the agents and thereby reduces toxic side effects. For drug targeting one needs to establish distinct molecular differences between target and normal cells, and develop a carrier that will show a high degree of selectivity for target cells [36]. Drug targeting usually has been achieved using water-soluble polymer chains or polymeric microparticles containing drugs and specific homing devices such as antibodies [37,38]. The use of biodegradable polymers is essential to avoid accumulation of polymers in the body during prolonged applications.

Hydrogels have played a vital role in the development of controlled release drug delivery systems [15]. Many novel drug delivery systems are based on hydrogels. Hydrogels are also used in the preparation of molecularly recognizable synthetic membranes for biosensors [39,40], which may become a critical component of signal-responsive drug delivery systems. Continued advances in hydrogel technology will undoubtedly contribute to the further development of controlled release drug delivery systems.

1.2 HYDROGELS

1.2.1 Definition of Hydrogel

Certain materials, when placed in excess water, are able to swell rapidly and retain large volumes of water in their swollen structures. The materials do not dissolve in water and maintain three-dimensional networks. Such aqueous gel networks are called hydrogels (also called aquagels). Hydrogels are usually made of hydrophilic polymer molecules which are crosslinked either by chemical bonds or other cohesion forces such as ionic interaction, hydrogen bonding, or hydrophobic interaction. Hydrogels are elastic solids in the sense that there exists a remembered reference configuration to which the system returns even after being deformed for a very long time [41].

A hydrogel swells for the same reason that an analogous linear polymer dissolves in water to form an ordinary polymer solution [42]. If a hydrogel, for any reason, dissolves in aqueous solvent, then the gel has become a hydrosol, which is a dispersion of colloidal particles in water [43]. From a general physicochemical point of view, a hydrosol is simply an aqueous solution. Many polymers can undergo a reversible transformation between hydrogel and hydrosol. Chemical crosslinking of dispersed particles in hydrosols will result in an irreversible hydrogel. It is noted that a gel is an infinitely large macromolecule, or supermacromolecule, which forms a network extending from one end to the other and occupying the whole reaction vessel [44]. The polymer networks of small particles with diameters smaller than 1 μm (typically in the range of 100 nm) are called microgels [45]. Microgels have crosslinked networks just like the supermacromolecules. Microgels, however, dissolve in water like linear or branched macromolecules due to their molecular nature [46].

The term hydrogel implies that the material is already swollen in water. The dried hydrogel is called a xerogel or dry gel. During the drying process water evaporates from the gel and the surface tension causes collapse of the gel body. Thus the gel shrinks to only a small fraction of its swollen size. If water is removed without disturbing the polymer network, either by lyophilization (i.e., freeze drying) or by extraction with organic solvents, then the remaining material is extremely light with a porosity as high as 98 percent. Such a dehydrated hydrogel is called an aerogel [47] or sponge [48,49]. Aerogels are known to have lower thermal conductivities than all other thermal insulants at ambient conditions [50]. Not all dried hydrogels (xerogels) maintain the ability to swell in water. Ceramics prepared by the sol-gel process are also called xerogels [51]. Xerogels such as silica gel will not swell again. To indicate the swelling ability of the dry material, a term ''xerogellant'' was proposed [52]. There is no lower limit defining how much water a material has to absorb to be called a hydrogel. In general, however, hydrogels are expected to absorb at least 10–20 percent of their own weight in water [53,54]. If a dried hydrogel imbibes at least 20 times its own weight of aqueous fluid while retaining its original shape, it is also called a superabsorbent [52]. Due to their high water content, hydrogels usually have low mechanical strength [55].

Hydrogels have been a topic of extensive research because of their unique bulk and surface properties. Although the research on hydrogels is more than three decades old, the research interest in hydrogels is still growing.

Since the first report on the biomedical use of poly(2-hydroxyethyl methacrylate) hydrogel [56], hydrogels with various properties have been prepared. Hydrogels can be made to respond (i.e., either shrink or expand) to changes in environmental conditions and the extent of the response can be controlled. Sometimes the volume change in response to the alteration in environmental conditions is so drastic that the phenomenon is called volume collapse or phase transition [57–59]. The environmental conditions include pH [60–62], temperature [63–67], electric field [68–72], ionic strength [73], salt type [74,75], solvent [57,76,77], external stress [69,78], light [79–81], and combinations of these. It is these unique properties that have made hydrogels find numerous applications in pharmaceutical, agricultural, biomedical, and consumer-oriented fields.

1.2.2 Hydrogel as a Biomaterial

Biomaterials are any materials which are designed to restore, augment, or replace the natural functions of the living tissues or organs in the body [82,83]. Simply speaking, biomaterials are those which become a part of the body either temporarily or permanently. Biomaterials are used not only for prosthetic applications but also for diagnostic and therapeutic applications [84]. Biomaterials should perform with an appropriate host response in a specific application without toxic, inflammatory, carcinogenic, and immunogenic responses [85,86]. An appropriate host response ranges from inertness and no interaction to one of positive interaction [87]. In general, the body's reaction to implants is to extrude them from the body or form sheath-like capsules around the implants if they cannot be removed [82]. The injury created by the implantation procedure usually results in inflammation which can be defined as the local reaction of vascularized tissue to injury [88]. The implanted biosensors can maintain the required sensitivity only for a few days mainly due to inflammation and immune reactions which decrease the sensor's sensitivity [89].

Success in the application of biomaterials relies heavily on the biocompatibility of biomaterials. Biocompatibility is the appropriate biological performance, both local and systemic, of a given polymer in a specific application [90]. A clear, specific, and absolute definition of biocompatibility does not exist at this time. Numerous interdisciplinary factors must be used to describe the biocompatibility of a given polymer in a given application for a given duration. The importance of biocompatibility cannot be overemphasized. The recent highly publicized controversy on silicone gel-filled breast implants is a case in point [91]. The typical tissue reaction around the implanted biomaterial is the formation of a thin fibrous capsule similar to scar tissue. The fibrous capsule often contracts and causes pain and deformity [92]. The formation of a fibrous membrane capsule around the implant is an attempt by the body to extrude the implant [90]. Some silicone implants were coated with polyurethane foam featuring micro-pillars which were designed to disrupt the fibrous capsule architecture and prevent the formation of scar tissue [93,94]. The polyurethane foam, however, was slowly degraded in the body [95]. Furthermore, the silicone shell of the implant ruptured in many cases releasing silicone gel into the body. Such leakage was blamed for harmful immune reactions and for causing cancer. The issue of biocompatibility becomes even more important if biomaterials are in contact with blood. Interactions between blood and biomaterials may result in thrombus formation, destruction or sensitization of cellular elements in blood, infection, and adverse immune responses [84].

Cumulative evidence shows that hydrogels are highly biocompatible. Hydrogels possess a few unique properties that make them biocompatible. First, hydrogels have low interfacial tension with surrounding biological fluids and tissues and that minimizes the driving force for protein adsorption and cell adhesion [56,96]. Because of its very high water content, the hydrogel surface is called a superhydrophilic diffuse surface [97]. The superhydrophilic diffuse surface is known to be highly biocompatible [98]. This is probably due to the fact that the hydrogel surface makes the actual interface more vague [99]. Second, hydrogels simulate some hydrodynamic properties of natural biological gels, cells, and tissues in many ways [55,84,100]. The high mobility of polymer chains at the hydrogel surface contributes to the prevention of protein adsorption and cell adhesion [101,102]. This is mainly due to the steric repulsion exerted by the polymer chains [103]. The protein adsorption and cell adhesion is prevented by the nonspecific repulsion resulting from "entropic" and "mixing" interactions between the polymer chains and proteins or cell membranes [104]. This phenomenon is the same as the well-known steric stabilization of protein molecules [105] and colloidal particles [106–108]. Third, the soft, rubbery nature of hydrogels minimizes mechanical and frictional irritation to the surrounding tissue [109]. Low friction surfaces cause no pain and no damage to mucous membranes or to the intima of the blood vessels, and thus no infections and no mural thrombus formation [110].

A major disadvantage of using hydrogels is that they have poor mechanical strength and toughness after swelling. This disadvantage can be overcome by grafting a hydrogel with good mechanical properties onto the biomaterial. The grafting of hydrogels onto biomaterial surfaces changes only the surface properties while the bulk properties remain unchanged. Hydrogels can be grafted onto biomaterials by physical adsorption, physical entrapment, graft coupling, and polymerization [97,111,112].

The principal markets for biomaterials are in the areas of cardiovascular implants, orthopedic implants, intravascular and urinary tract catheters, soft tissue replacements, intraocular lenses, wound dressings, biosensors, and controlled release devices. All of these biomaterials will improve their biocompatibility through coating with hydrogels [113].

1.2.3 Biodegradable Hydrogels

Biodegradable polymeric systems have been used frequently in the development of advanced drug delivery systems. The use of biodegradable polymeric systems in controlled release drug delivery is desirable, since the dosage forms will be degraded and eliminated from the body. This will avoid removal of the device from the body by surgery or other means when the device is no longer needed. Biodegradable polymeric systems also provide flexibility in the design of delivery systems for large molecular weight drugs, such as peptides and proteins, which are not suitable for diffusion-controlled release through nondegradable polymeric matrices. Our impression of biodegradable polymeric systems has changed significantly in recent years. When the rate of drug release was mainly controlled by diffusion through the polymer matrix, the degradation of the polymer was considered to be a less well defined and unnecessary variable [114]. Currently, the degradation of polymers is regarded as highly desirable and is frequently used to control the drug release rate.

Most of the research on biodegradable drug delivery systems has employed water-insoluble polymers such as poly(glycolic acid) or poly(lactic acid) [115,116]. Not much work has been done, however, on biodegradable hydrogel systems. Because of the unique properties of hydrogels, the biodegradable hydrogels are expected to find wide applications in the improvement of existing dosage forms and the development of new and better drug delivery systems [117].

1.3 REFERENCES

1. Paul, D. R. and F. W. Harris, eds. 1976. *Controlled Release Polymeric Formulations*. Washington, DC: American Chemical Society.
2. Robinson, J. R., ed. 1978. *Sustained and Controlled Release Drug Delivery Systems*. New York, NY: Marcel Dekker, Inc.
3. Kydonieus, A. F., ed. 1980. *Controlled Release Technologies: Methods, Theory, and Applications, Vols. I and II*. Boca Raton, FL: CRC Press.
4. Goldberg, E. P. and A. Nakajima, eds. 1980. *Biomedical Polymers. Polymeric Materials and Pharmaceuticals for Biomedical Use*. New York, NY: Academic Press.
5. Lewis, D. H., ed. 1981. *Controlled Release of Pesticides and Pharmaceuticals*. New York, NY: Plenum Press.
6. Bundgaard, H., A. B. Hansen and H. Kofod, eds. 1982. *Optimization of Drug Delivery*. Copenhagen, Denmark: Munksgaard.

7. Das, K. G., ed. 1983. *Controlled-Release Technology. Bioengineering Aspects*. New York, NY: John Wiley & Sons.
8. Roseman, T. J. and S. Z. Mansdirf, eds. 1983. *Controlled Release Delivery Systems.* New York, NY: Marcel Dekker, Inc.
9. Langer, R. S. and D. L. Wise, eds. 1984. *Medical Applications of Controlled Release, Vols. I and II.* Boca Raton, FL: CRC Press.
10. Tirrell, D. A., L. G. Donaruma and A. B. Turek, eds. 1985. *Macromolecules as Drugs and as Carriers for Biologically Active Materials, Annals of the New York Academy of Sciences, Vol. 446.*
11. Juliano, R. L., ed. 1987. *Biological Approaches to the Controlled Delivery of Drugs, Annals of the New York Academy of Sciences, Vol. 507.*
12. Baker, R. 1987. *Controlled Release of Biologically Active Agents*. New York, NY: John Wiley & Sons.
13. Lee, P. I. and W. R. Good, eds. 1987. *Controlled-Release Technology. Pharmaceutical Applications, ACS Symposium Series 348.* Washington, DC: American Chemical Society.
14. Robinson, J. R. and V. H. L. Lee. 1987. *Controlled Drug Delivery. Fundamentals and Applications*. New York, NY: Marcel Dekker, Inc.
15. Peppas, N. A., ed. 1987. *Hydrogels in Medicine and Pharmacy, Vols. I–III.* Boca Raton, FL: CRC Press.
16. Lenaerts, V. and R. Gurny, eds. 1990. *Bioadhesive Drug Delivery Systems*. Boca Raton, FL: CRC Press.
17. Chien, Y. W. 1992. *Novel Drug Delivery Systems*. New York, NY: Marcel Dekker, Inc.
18. Higuchi, W. I. and D. Sharma, eds. 1990. *Transdermal Delivery of Drugs, Proceedings of the Workshop on Current Status and Future Directions*. Washington, DC: U.S. Department of Health and Human Services, National Institute of Health.
19. Flattum-Riemer, J. 1991. "Norplant: A New Contraceptive," *Am. Fam. Physician*, 44:103–108.
20. Hrushesky, W. J. M. 1991. "Temporally Optimizable Delivery Systems–*sine qua non* for Molecular Medicine," *Ann. N.Y. Acad. Sci.*, 618:xi–xvii.
21. Lemmer, B. 1991. "Circadian Rhythms and Drug Delivery," *J. Controlled Rel.*, 16:63–74.
22. Labrecque, G. and P.-M. Belanger. 1985. "Time-Dependency in the Pharmacokinetics and Disposition of Drugs," in *Topics in Pharmaceutical Sciences 1985*, D. D. Breimer and P. Speiser, eds., New York, NY: Elsevier Science Publishers B.V., pp. 167–178.
23. Hrushesky, W. J. M., R. Langer and F. Theeuwes, eds. 1991. *Temporal Control of Drug Delivery, Annals of the New York Academy of Sciinces, Vol. 618.*
24. Kost, J., ed. 1990. *Pulsed and Self-Regulated Drug Delivery*. Boca Raton, FL: CRC Press.
25. Heller, J. 1988. "Chemically Self-Regulated Drug Delivery Systems," *J. Controlled Rel.*, 8:111–125.
26. Klumb, L. A. and T. A. Horbett. 1992. "Design of Insulin Delivery Devices Based on Glucose Sensitive Membranes," *J. Controlled Rel.*, 18:59–80.

27. Goldner, H. J. 1991. "Self-Regulating Drug Delivery Makes Treatment Automatic," *R&D Magazine* (November):64–68.

28. Pitt, C. G., Z. W. Gu, R. W. Hendren, J. Thompson and M. C. Wani. 1985. "Triggered Drug Delivery Systems," *J. Controlled Rel.*, 2:363–374.

29. Check, W. A. 1984. "New Drugs and Drug Delivery Systems in the Year 2000," *Amer. Pharm.*, NS24:44–56.

30. Ovellette, R. and P. Cheremisinoff. 1985. *Application of Biotechnology*. Lancaster, PA: Technomic Publishing Co., Inc.

31. McPherson, J. M. and D. J. Livingston. 1989. "Protein Engineering: New Approaches to Improved Therapeutic Proteins, Part I," *Pharm. Technol.*, 13:22–32.

32. Racker, E. 1987. "Structure, Function, and Assembly of Membrane Proteins," *Science*, 235:959–961.

33. Widder, K. J. and R. Green, eds. 1985. *Drug and Enzyme Targeting, Part A. Methods in Enzymology, Vol. 112.* New York, NY: Academic Press.

34. Tomlinson, E. and S. S. Davis, eds. 1986. *Site-Specific Drug Delivery*. New York, NY: John Wiley & Sons.

35. Kopeček, J. 1991. "Targetable Polymeric Anticancer Drugs: Temporal Control of Drug Activity," *Ann. N.Y. Acad. Sci.*, 618:335–344.

36. Poznansky, M. J. and R. L. Juliano. 1984. "Biological Approaches to the Controlled Delivery of Drugs: A Critical Review," *Pharmacol. Rev.*, 36:277–336.

37. Duncan, R. and J. Kopeček. 1984. "Soluble Synthetic Polymers as Potential Drug Carriers," *Adv. Polymer Sci.*, 57:51–101.

38. Brich, E., S. Ravel, T. Kissel, J. Fritsch and A. Schoffmann. 1992. "Preparation and Characterization of a Water Soluble Dextran Immunoconjugate of Doxorubicin and the Monoclonal Antibody (ABL 364)," *J. Controlled Rel.*, 19:245–258.

39. Aizawa, M. 1985. "Biofunctional Synthetic Membranes," *ACS Symposium Ser.*, 269:447–480.

40. Oliver, B. N., L. A. Coury, J. O. Egekeze, C. S. Sosnoff, Y. Zhang, R. W. Murray, C. Keller and M. X. Umana. 1990. "Electrochemical Reactions, Enzyme Electrocatalysis, and Immunoassay Reactions in Hydrogels," in *Biosensor Technology. Fundamentals and Applications*, R. P. Buck, W. E. Hatfield, M. Umana and E. F. Bowden, eds., New York, NY: Marcel Dekker, Inc., pp. 117–135.

41. Silberberg, A. 1989. "Network Deformation in Flow," in *Molecular Basis of Polymer Networks*, A. Baumgärtner and C. E. Picot, eds., Berlin: Spring-Verlag, pp. 147–151.

42. Flory, P. J. 1953. *Principles of Polymer Chemistry*. Ithaca, NY: Cornell University Press, pp. 576–589.

43. Napper, D. H. and R. J. Hunter. 1972. "Hydrosols," Chapter 8 in *Surface Chemistry and Colloids, Vol. 7,* M. Kerker, ed., Baltimore, MD: University Park Press.

44. Stauffer, D., A. Coniglio and M. Adam. 1982. "Gelation and Critical Phenomena," *Adv. Polymer Sci.*, 44:103–158.

45. Antonietti, M., W. Bremser and M. Schmidt. 1990. "Microgels: Model Polymers for the Cross-Linked State," *Macromolecules*, 23:3796–3805.

46. Funke, W. 1989. "Reactive Microgels–Polymers Intermediate in Size between Single Molecules and Particles," *Brit. Polymer J.*, 21:107–115.

47. Fricke, J. 1988. "Aerogel," *Sci. Amer.*, 258:92–97.

48. Gorham, S. D. 1991. "Collagen," in *Biomaterials. Novel Materials from Biological Sources*, D. Byrom, ed., New York, NY: Stockton Press, pp. 55–122.
49. Chvapil, M. 1977. "Collagen Sponge: Theory and Practice of Medical Applications," *J. Biomed. Mater. Res.*, 11:721–741.
50. Lu, X., M. C. Arduini-Schuster, J. Kuhn, O. Nilsson, J. Fricke and R. W. Pekala. 1992. "Thermal Conductivity of Monolithic Organic Aerogels," *Science*, 255: 971–972.
51. Ellerby, L. M., C. R. Nishida, F. Nishida, S. A. Yamanaka, B. Dunn, J. S. Valentine and J. I. Zink. 1992. "Encapsulation of Proteins in Transparent Porous Silicate Glasses Prepared by the Sol-Gel Method," *Science*, 255:1113–1115.
52. Gross, J. R. 1990. "The Evolution of Absorbent Materials," in *Absorbent Polymer Technology*, L. Brannon-Peppas and R. S. Harland, eds., New York, NY: Elsevier, pp. 3–22.
53. Ratner, B. D. and A. S. Hoffman. 1976. "Synthetic Hydrogels for Biomedical Applications," *ACS Symposium Ser.*, 31:1–36.
54. Piirma, I. 1992. *Polymeric Surfactants*. New York, NY: Marcel Dekker, Inc., Chapter 10.
55. Hoffman, A. S. 1975. "Hydrogels–A Broad Class of Biomaterials," in *Polymers in Medicine and Surgery*, R. L. Kronenthal, Z. Oser and E. Martin, eds., New York, NY: Plenum Press, pp. 33–44.
56. Wichterle, O. and D. Lim. 1960. "Hydrophilic Gels for Biological Use," *Nature*, 185:117–118.
57. Tanaka, T. 1981. "Gels," *Sci. Amer.*, 244:124–138.
58. Tanaka, T. 1992. "Phase Transition of Gels," *ACS Symp. Ser.*, 480:1–21.
59. Grosberg, A. Y. and Nechaev, S. K. 1991. "Topological Constraints in Polymer Network Strong Collapse," *Macromolecules*, 24:2789–2793.
60. Siegel, R. A., M. Falamarzian, B. A. Firestone and B. C. Moxley. 1988. "pH-Controlled Release from Hydrophobic/Polyelectrolyte Copolymer Hydrogels," *J. Controlled Rel.*, 8:179–182.
61. Pradny, M. and J. Kopeček. 1990. "Poly[(acrylic Acid)-*co*-(butyl Acrylate)] Crosslinked with 4,4′-bis(Methacryloylamino)azobenzene," *Makromol. Chem.*, 191:1887–1897.
62. Brannon-Peppas, L. and N. A. Peppas. 1991. "Equilibrium Swelling Behavior of Dilute Ionic Hydrogels in Electrolyte Solutions," *J. Controlled Rel.*, 16:319–330.
63. Taylor, L. D. and L. D. Cerankowski. 1975. "Preparation of Films Exhibiting a Balanced Temperature Dependence to Permeation by Aqueous Solutions–A Study of Lower Consolute Behavior," *J. Polym. Sci., Polym. Chem.*, 13:2551–2570.
64. Hirose, Y., T. Amiya, Y. Hirokawa and Y. Tanaka. 1987. "Phase Transition of Submicron Gel Beads," *Macromolecules*, 20:1342–1344.
65. Otake, K., H. Inomata, M. Konno and S. Saito. 1990. "Thermal Analysis of the Volume Phase Transition with *N*-Isopropylacrylamide Gels," *Macromolecules*, 23:283–289.
66. Hoffman, A. S. 1987. "Application of Thermally Reversible Polymers and Hydrogels in Therapeutics and Diagnostics," *J. Controlled Rel.*, 6:297–305.
67. Bae, Y. H., T. Okano and S. W. Kim. 1991. "On-Off Thermocontrol of Solute Transport. II. Solute Release from Thermosensitive Hydrogels," *Pharm. Res.*, 8:624–628.

68. Tanaka, T., I. Nishio, S.-T. Sun and S. Ueno-Nishio. 1982. "Collapse of Gels in an Electric Field," *Science*, 218:467–469.

69. Sawahata, K., M. Hara, H. Yasunaga and Y. Osada. 1990. "Electrically Controlled Drug Delivery System Using Polyelectrolyte Gels," *J. Controlled Rel.*, 14:253–262.

70. Kurauchi, T., T. Shiga, Y. Hirose and A. Okada. 1991. "Deformation Behaviors of Polymer Gels in Electric Field," in *Polymer Gels. Fundamentals and Biomedical Applications*, D. DeRossi, K. Kajiwara, Y. Osada and A. Yamauchi., eds., New York, NY: Plenum Press, pp. 237–246.

71. Osada, Y., H. Okuzaki and H. Hori. 1992. "A Polymer Gel with Electrically Driven Motility," *Nature*, 355:242–244.

72. Kishi, R., M. Hasebe, M. Hara and Y. Osada. 1989. "Mechanism and Process of Chemomechanical Contraction of Polyelectrolyte Gels under Electric Field," *Polym. Adv. Technol.*, 1:19–25.

73. Hooper, H. H., J. P. Baker, H. W. Blanch and J. M. Prausnitz. 1990. "Swelling Equilibria for Positively Ionized Polyacrylamide Hyrogels," *Macromolecules*, 23:1096–1104.

74. Hughlin, M. B. and J. M. Rego. 1991. "Influence of a Salt on Some Properties of Hydrophilic Methacrylate Hydrogels," *Macromolecules*, 24:2556–2563.

75. Ohmine, I. and T. Tanaka. 1982. "Salt Effects on the Phase Transition of Ionic Gels," *J. Chem. Phys.*, 77:5725–5729.

76. Ilavsky, M. 1982. "Phase Transition in Swollen Gels. 2. Effect of Charge Concentration on the Collapse and Mechanical Behavior of Polyacrylamide Networks," *Macromolecules*, 15:782–788.

77. Amiya, T. and T. Tanaka. 1987. "Phase Transition in Cross-Linked Gels of Natural Polymers," *Macromolecules*, 20:1162–1164.

78. Dusek, K. and D. Patterson. 1968. "Transition in Swollen Networks Induced by Intramolecular Condensation," *J. Polym. Sci., Part A-2*, 6:1209–1216.

79. Suzuki, A. and T. Tanaka. 1990. "Phase Transition in Polymer Gels Induced by Visible Light," *Nature*, 346:345–347.

80. Van der Veen, G. and W. Prins. 1974. "Photoregulation of Polymer Conformation," in *Polyelectrolytes*, E. Sélégny, ed., Dordrecht, Holland: D. Reidel Publishing Co., pp. 483–505.

81. Mamada, A., T. Tanaka, D. Kungwatchakun and M. Irie. 1990. "Photoinduced Phase Transition of Gels," *Macromolecules*, 23:1517–1519.

82. Park, J. B. 1984. *Biomaterials Science and Engineering*. New York, NY: Plenum Press, Chapters 1 and 7.

83. Duncan, E. 1990. "Biomaterials. What is a Biomaterial?" *Med. Dev. Diag. Ind.*, 12:138–142.

84. Brook, S. D. 1980. *Properties of Biomaterials in Physiological Environment*. Boca Raton, FL: CRC Press, Chapter 4.

85. Williams, D. F. 1991. "The Significance of Surfaces in Biocompatibility Phenomena," *Surfaces in Biomaterials Symposium*, pp. 1–4.

86. Anderson, J. M. 1986. "*In vivo* Biocompatibility Studies: Perspectives on the Evaluation of Biomedical Polymer Biocompatibility," in *Polymeric Biomaterials*, E. Piskin and A. S. Hoffman, eds., Boston, MA: Martinus Nijhoff Publishers, pp. 29–39.

87. Braybrook, J. H. and L. D. Hall. 1990. "Organic Polymer Surfaces for Use in Medicine: Their Formation, Modification, Characterization and Application," *Prog. Polymer Sci.*, 15:715–734.

88. Anderson, J. M. 1988. ''Inflammatory Response to Implants,'' *Trans. Am. Soc. Artif. Intern. Organs*, 34:101–107.

89. Reach, G. and G. S. Wilson. 1992. ''Can Continuous Glucose Monitoring Be Used for the Treatment of Diabetes?'' *Anal. Chem.*, 64:381A–386A.

90. Coleman, D. L., R. N. King and J. D. Andrade. 1974. ''The Foreign Body Reaction: A Chronic Inflammatory Response,'' *J. Biomed. Mater. Res.*, 8:199–211.

91. Weiss, R. 1991. ''Breast Implant Fears Put Focus on Biomaterials,'' *Science*, 252:1059–1060.

92. Ersek, R. A. 1991. ''Rate and Incidence of Capsular Contracture: A Comparison of Smooth and Textured Silicone Double-Lumen Breast Prostheses,'' *Plast. Reconstr. Surg.*, 87:879–884.

93. 1991. *Materials News*. Dow Corning (March/April):4–6.

94. Caffee, H. H. and C. Hathaway. 1990. ''Polyurethane Foam-Covered Implants and Capsular Contracture: A Laboratory Investigation,'' *Plast. Reconst. Surg.*, 86: 708–710.

95. Capozzi, A. 1991. ''Long-Term Complications of Polyurethane-Covered Breast Implants,'' *Plast. Reconstr. Surg.*, 88:458–461.

96. Jhon, M. S. and J. D. Andrade. 1973. ''Water and Hydrogels,'' *J. Biomed. Mater. Res.*, 7:509–516.

97. Ikada, Y. 1984. ''Blood-Compatible Polymers,'' *Adv. Polymer Sci.*, 57:104–140.

98. Ikada, Y. 1991. ''Membranes as Biomaterials,'' *Polymer J.*, 23:551–560.

99. Van Blitterwijk, C. A., D. Bakker, S. C. Hesseling and H. K. Koerten. 1991. ''Reactions of Cells at Implant Surfaces,'' *Biomaterials*, 12:187–193.

100. Andrade, J. D., ed. 1976. *Hydrogels for Medical and Related Applications*. Washington, DC: American Chemical Society, pp. xi–xiii.

101. Ratner, B. D. 1986. ''Hydrogel Surfaces,'' Chapter 4 in *Hydrogels in Medicine and Pharmacy, Vol. I,* N. A. Peppas, ed., Boca Raton, FL: CRC Press.

102. Horbett, T. A. 1986. ''Protein Adsorption to Hydrogels,'' in *Hydrogels in Medicine and Pharmacy, Vol. I,* N. A. Peppas, ed., Boca Raton, FL: CRC Press.

103. Amiji, M. A. and K. Park. 1992. ''Prevention of Protein Adsorption and Platelet Adhesion on Surfaces by PEO/PPO/PEO Triblock Copolymers,'' *Biomaterials*, 13:682–692.

104. Tadros, Th. F. 1982. ''Polymer Adsorption and Dispersion Stability,'' in *The Effect of Polymers on Dispersion Properties*, Th. F. Tadros, ed., New York, NY: Academic Press, pp. 1–38.

105. Tomlinson, E. 1990. ''Control of the Biological Dispersion of Therapeutic Proteins,'' Chapter 16 in *Protein Design and the Development of New Therapeutics and Vaccines*, J. B. Hook and G. Poste, eds., New York, NY: Plenum Press.

106. Vincent, B. and S. G. Whittington. 1982. ''Polymers at Interfaces and in Disperse Systems,'' *Surface Colloid Sci.*, 12:1–117.

107. Napper, D. H. and R. J. Hunter. 1972. ''Hydrosols,'' Chapter 8 in *Surface Chemistry and Colloids, Vol. 7,* M. Kerker, ed., Baltimore, MD: University Park Press.

108. Hunter, R. J. 1987. *Foundations of Colloid Science, Vol. I.* New York, NY: Oxford University Press, Chapter 8.

109. Ratner, B. D. 1981. ''Biomedical Applications of Hydrogels: Review and Critical Appraisal,'' Chapter 7 in *Biocompatibility of Clinical Implant Materials, Vol. II,* D. F. Williams, ed., Boca Raton, FL: CRC Press.

110. Nagaoka, S. and R. Akashi. 1990. "Low Friction Hydrophilic Surface for Medical Devices," *J. Bioact. Compat. Polymers*, 5:212–226.

111. Auroy, P., L. Auvray and L. Leger. 1991. "Building of a Grafted Layer. 1. Role of the Concentration of Free Polymers in the Reaction Bath," *Macromolecules*, 24:5158–5166.

112. Kronick, P. 1980. "Fabrication and Characterization of Grafted Hydrogel Coatings as Blood Contacting Surfaces," in *Synthetic Biomedical Polymers. Concepts and Applications*, M. Szycher and W. J. Robinson, eds., Lancaster, PA: Technomic Publishing Co., Inc., pp. 153–169.

113. Barenberg, S. A. 1988. "Abridged Report of the Committee to Survey the Needs and Opportunities for the Biomaterials Industry," *J. Biomed. Mater. Res.*, 22: 1267–1291.

114. Pitt, C. G. and A. Schindler. 1983. "Biodegradation of Polymers," Chapter 3 in *Controlled Drug Delivery, Vol. I. Basic Concepts*, S. D. Bruck, ed., Boca Raton, FL: CRC Press.

115. Sanders, L. M., G. I. McRae, K. M. Vitale and B. A. Kell. 1985. "Controlled Delivery of an LHRH Analogue from Biodegradable Injectable Microspheres," *J. Controlled Rel.*, 2:187–195.

116. Müller, R. H. 1991. *Colloidal Carriers for Controlled Drug Delivery and Targeting. Modification, Characterization and* in vivo *Distribution*. Boca Raton, FL: CRC Press, Chapter 5.

117. Baker, R. W., M. E. Tuttle and R. Helwing. 1984. "Novel Erodible Polymers for the Delivery of Macromolecules," *Pharm. Technol.*, 8:26–30.

CHAPTER 2

Biodegradation

Biodegradable materials have been studied extensively for their biomedical applications such as temporary scaffold [1], temporary barrier for surgical adhesions [2], and matrices for drug delivery systems [3,4]. Recently, the term degradation or biodegradation, has become a key word in the development of new technologies not only for pharmaceutical and medical applications but also for agricultural and environmental applications. Degradability of polymers is relatively new, and no adequate definitions or test protocols exist. In reality everything degrades. The question is how fast it degrades. Polyanhydrides or polyorthoesters can be hydrolyzed in a matter of hours, while polycarbonates or polyurethanes take thousands of years to hydrolyze [5]. Significant degradation in human time scale is important. The application of biodegradable polymers depends on the properties of the polymers. For drug delivery, polymers need to have the appropriate water permeability, biocompatibility, and tensile strength [6]. The most desirable biodegradable system would be the one which leaves no residual polymer following the release of the drug and the degradation of the polymer. It is not always easy, however, to find such a completely degradable system which meets the criteria for the controlled delivery of a wide variety of bioactive agents.

2.1 DEFINITIONS OF BIODEGRADATION

Polymer degradation can be classified into photo- and photo-oxidative, thermo- and thermo-oxidative, mechanochemical, ozone-induced, radiolytic, ionic, and biodegradation [7]. Biodegradation of polymers is the classification of greatest importance for biomedical, pharmaceutical, and environmental applications. Biodegradation has been defined in various ways by different investigators. It has been defined as changes in surface properties or loss of mechanical strength [8], assimilation by microorganisms [9], degradation by enzymes [10,11], backbone chain breakage and subsequent reduction in the average molecular weight of the polymer [12,13], or extraction of low molecular weight material leading to surface defects [12]. Degradation can occur by each of the above mechanisms alone or in combination with one another.

While no consensus is found on the definition of biodegradation, it is clear that biodegradation can occur on many different structural levels, i.e., molecular, macromolecular, microscopic, and macroscopic, depending on the mechanism [14]. It was argued that the phenomenon of degradation *in vivo* may not be equated with the term biodegradation, since biodegradation implies the active participation of biological entities such as enzymes or organisms in the degradation process [15,16]. It is difficult, however, to identify the involvement and the role of biological species in the *in vivo* degradation. Both hydrolytic and enzymatic processes may contribute to the degradation to different extents during the different stages of the degradation process. Degradation may begin by hydrolysis, but as the polymer breaks up and surface area and accessibility increase, enzymatic degradation may dominate [16]. Thus, at least for now, the definition of biodegradation should be broad enough to include all types of degradation occurring *in vivo* whether the degradation is due to hydrolysis or metabolic processes [6,17–20].

In addition to biodegradation, the phenomenon of *in vivo* degradation of polymeric materials has been described as bioabsorption, bioresorption, bioerosion, or biodeterioration. Although no clear differences have been established in the definitions of these terms, attempts have been made to distinguish them [21,22]. The following is a summary of the definitions found in the literature.

2.1.1 Biodegradation

Biodegradation is defined as the conversion of materials into less complex intermediates or end products by solubilization, simple hydrolysis, or the action of biologically formed entities which can be enzymes and other products of the organism. Polymer molecules may, but not necessarily, break down to produce fragments in this process, but the integrity of the material decreases as a result of this process [17,23]. The formed fragments can move away from their site of action but not necessarily from the body [22].

2.1.2 Bioresorption

Bioresorption describes the degradation of materials into low molecular weight compounds which can be eliminated from the body through natural pathways [21,22].

2.1.3 Bioabsorption

Bioabsorption means the disappearance of materials from their initial application site with or without degradation of the dispersed polymer molecules [24]. The clearance of the dispersed polymer molecules may require special transport mechanisms since polymer molecules are too large for clearance by simple diffusion [25]. If the dispersed polymer molecules are metabolized or excreted from the body, the process becomes bioresorption.

2.1.4 Bioerosion

Bioerosion indicates conversion of water-insoluble polymers to water-soluble polymers or smaller molecules [26]. Erosion occurring only at the surface of a material is called surface (or heterogeneous) erosion, while erosion occurring throughout the material is called bulk (or homogeneous) erosion [4].

2.1.5 Biodeterioration

The term biodeterioration is used for any undesirable changes, either mechanical, physical, chemical, or aesthetic, in the properties of materials [27]. In general, biodeterioration is considered as an unwelcome and destructive process, while biodegradation is regarded as desirable.

These definitions suggest that biodegradation has a broader meaning than either bioresorption, bioabsorption, or bioerosion. Since it is quite often difficult to distinguish various mechanisms of degradation occurring *in vivo*, the term biodegradation will be used throughout this book to describe biodegradation, bioresorption, bioabsorption, and bioerosion, unless distinctions among them are absolutely necessary.

2.2 MECHANISMS OF BIODEGRADATION

As defined above, materials becomes less complex products through biodegradation and this occurs through four different mechanisms: solubilization, charge formation followed by dissolution, hydrolysis, and enzyme-catalyzed degradation [28]. Biodegradation of polymers is expected to undergo four stages: hydration, strength loss, loss of mass integrity, and mass loss [29]. The hydration of polymers depends on the hydrophilicity of the polymer. The hydration results from disruption of secondary and tertiary structures stabilized by van der Waals forces and hydrogen bonds. If the interaction between polymer chains is great, it may not dissolve in water. For example, intermolecular hydrogen bonding between hydroxyl groups on the glucose residues of cellulose is so strong that cellulose is insoluble in water [30]. During and after hydration, the polymer chains may become water-soluble and/or the polymer backbone may be cleaved by chemical- or enzyme-catalyzed hydrolysis to result in the loss of polymer strength. For crosslinked polymers, the polymer strength may be reduced by cleavage of either the polymer backbone, crosslinker, or pendant chains. Further cleavage of the polymer chains leads to the loss of mass. In nonswellable polymer systems, the reduction in the polymer molecular weight may lead to the loss of coherence between polymer chains. Subsequently, low molecular weight degradation products will be dissolved in body fluids and removed from the body [29].

2.2.1 Solubilization

Most natural polymers and many synthetic polymers dissolve in water. Synthetic polyelectrolytes, such as poly(acrylic acid), poly(styrene sulfonate), and carboxymethylcellulose (CMC), readily dissolve in water by interacting with the partial charge on water molecules. Other polar polymers, such as poly(vinyl alcohol), poly(ethylene oxide) (PEO), polyvinylpyrrolidone (PVP), and dextran, also readily dissolve in water by forming hydrogen bonds with water molecules.

In contact with an aqueous environment, the hydrophilic polymer in solid form imbibes water and swells to form a hydrogel [28]. This hydrogel is more properly called a mucilaginous composition [31]. Once a gel is formed, water molecules diffuse freely through a rather loose network formed by swollen polymer molecules. Upon further addition of water, polymer-polymer contacts are broken and individual polymer molecules are dissolved in water. This process is generally referred to as erosion or attrition. Water continues to penetrate toward the core and finally the gel is converted to a viscous solution [30]. The dissolution rate of water-soluble polymers varies depending on the molecular weight and stereo-regularity.

The formation of a gel or viscous solution depends on the amount of water, or the concentration of polymer in water. As the concentration of polymer increases, a gel is likely to be formed. The concentration at which the gel is formed varies depending on the type of polymer. At a given polymer concentration, a gel can be converted to a sol and vice versa by varying the environmental conditions, such as temperature, pH, or type of salt.

2.2.2 Charge Formation Followed by Dissolution

2.2.2.1 IONIZATION OR PROTONATION

Some polymers are initially water-insoluble but become solubilized by ionization or protonation of a pendant group. Examples are shown here.

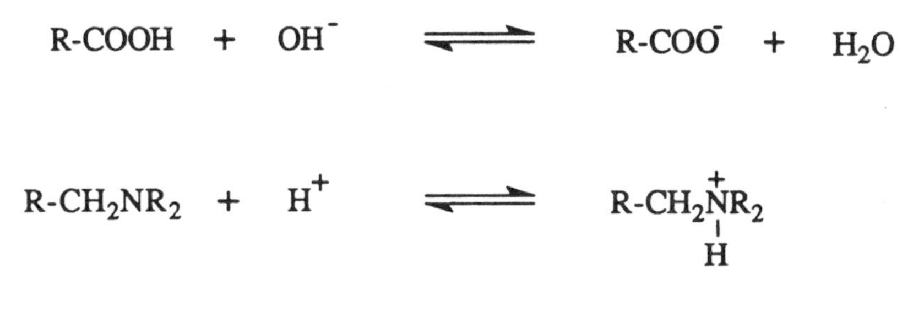

The solubility of polyacids is strongly pH-dependent. At low pH solution, polyacids are not water-soluble because the carboxyl groups of polyacids are protonated, i.e., not ionized. Upon an increase in the pH of the solution, the carboxyl groups release hydrogen atoms and become ionized. As the content of the ionized groups gradually increases, the polymer becomes more hydrophilic, absorbs water, swells, and finally dissolves in water. The solubility of polybases also strongly depends on pH, but the trend is opposite to that of polyacids. Polybases are water-soluble at low pH ranges.

Polyacids have been used widely as enteric coating materials for pharmaceuticals. They are poorly water-soluble in low pH environments, such as in the stomach, and dissolve in alkaline conditions such as those found in the intestines. The pH-sensitive polymers that have been used as enteric coating materials are shellac (esters of aleuritic acid), cellulose acetate phthalate, cellulose acetate succinate, polyvinyl acetate phthalate, hydroxypropylmethylcellulose phthalate, and poly(methacrylic acid-*co*-methyl methacrylate) [32]. Among these, the most widely used polymers for enteric coating are cellulose acetate phthalate, poly(vinyl acetate phthalate), and poly(methacrylic acid-*co*-methyl methacrylate) [33,34]. The structures of these enteric coating polymers are shown in Figure 2.1. Cellulose acetate phthalate becomes water-soluble at a pH greater than 6, while poly(vinyl acetate phthalate) and hydroxypropylmethylcellulose phthalate are ionized at a lower pH [35].

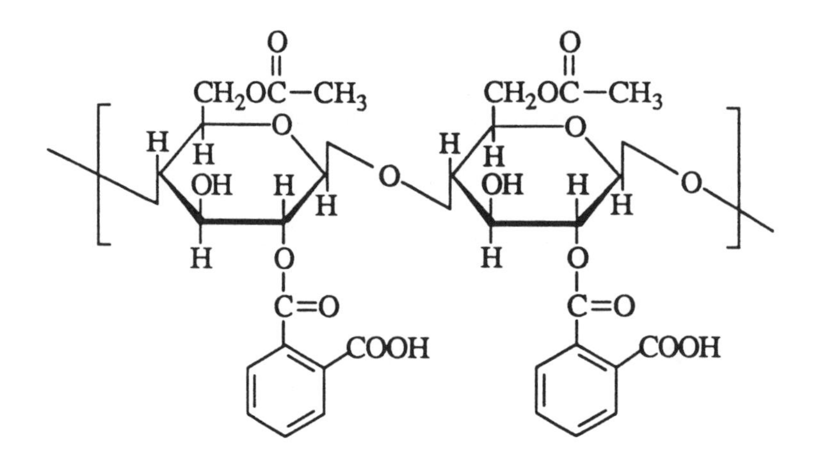

Cellulose acetate phthalate

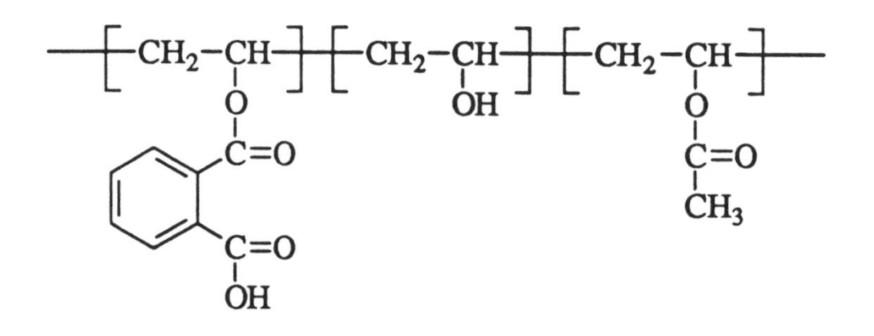

Poly(vinyl acetate phthalate)

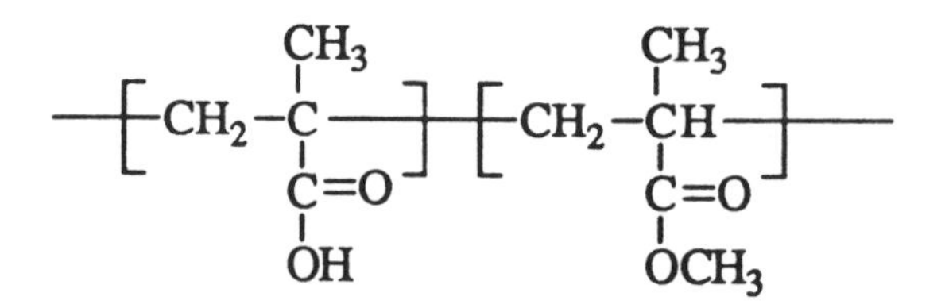

Poly(methacrylic acid-co-methyl methacrylate)

FIGURE 2.1. *Structures of enteric coating polymers.*

2.2.2.2 HYDROLYSIS FOLLOWED BY IONIZATION

Water-insoluble polymers containing pendent anhydrides or ester groups may be solubilized if anhydrides or esters hydrolyze to form ionized acids on the polymer chain. For example, poly(methyl acrylate) and poly(methyl methacrylate), which are esters derived from poly(acrylic acid) and poly(methacrylic acid), respectively, are not water-soluble, but become water-soluble upon hydrolysis of the pendent esters and subsequent ionization of the carboxyl groups.

Heller developed a chemically self-regulated drug delivery device using a pH-sensitive, bioerodible polymer. The pH sensitive, bioerodible polymer was surrounded by an albumin hydrogel containing an immobilized enzyme [26]. Methyl vinyl ether and maleic anhydride copolymer was esterified with 1-hexanol to prepare *n*-hexyl half ester of the copolymer which dissolves by ionization of the carboxyl groups (Figure 2.2). The dissolution of partially esterified copolymers is highly sensitive to the pH of the surrounding aqueous environment. They become abruptly water-soluble above a certain pH known as the dissolution pH. The dissolution pH increased from 4 to 8 as the size of the ester group in the copolymer increased. The dissolution pH is about 6 when the number of carbons in the alkyl substituent is 6 (hexyl) [36].

The labile esters of poly(glutamic acid) can be hydrolyzed *in vivo*. Copolymers of glutamic acid and ethyl glutamate have been used to prepare hydrogels with varying degrees of hydrophilicity [37–39].

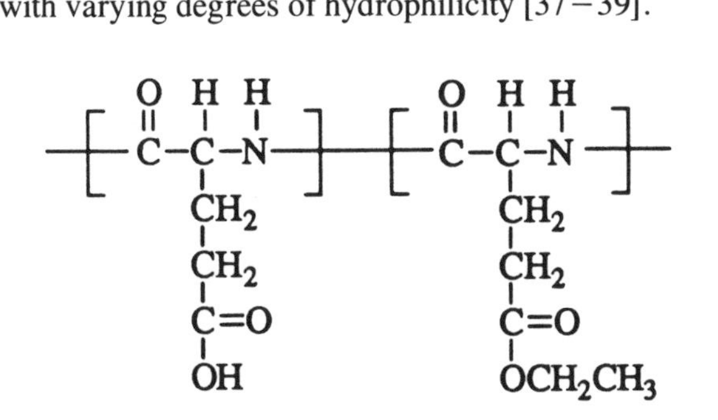

The copolymer gradually becomes hydrophilic as the ethyl glutamate ester groups undergo a slow, homogeneous hydrolysis. The biodegradation rate was dependent on the content of glutamic acid [40]. The copolymer becomes water-soluble at approximately 50 mol percent of glutamic acid. Copolymers of partially or totally esterified glutamic (or aspartic) acid and leucine were also known to undergo biodegradation [40]. Marck et al.

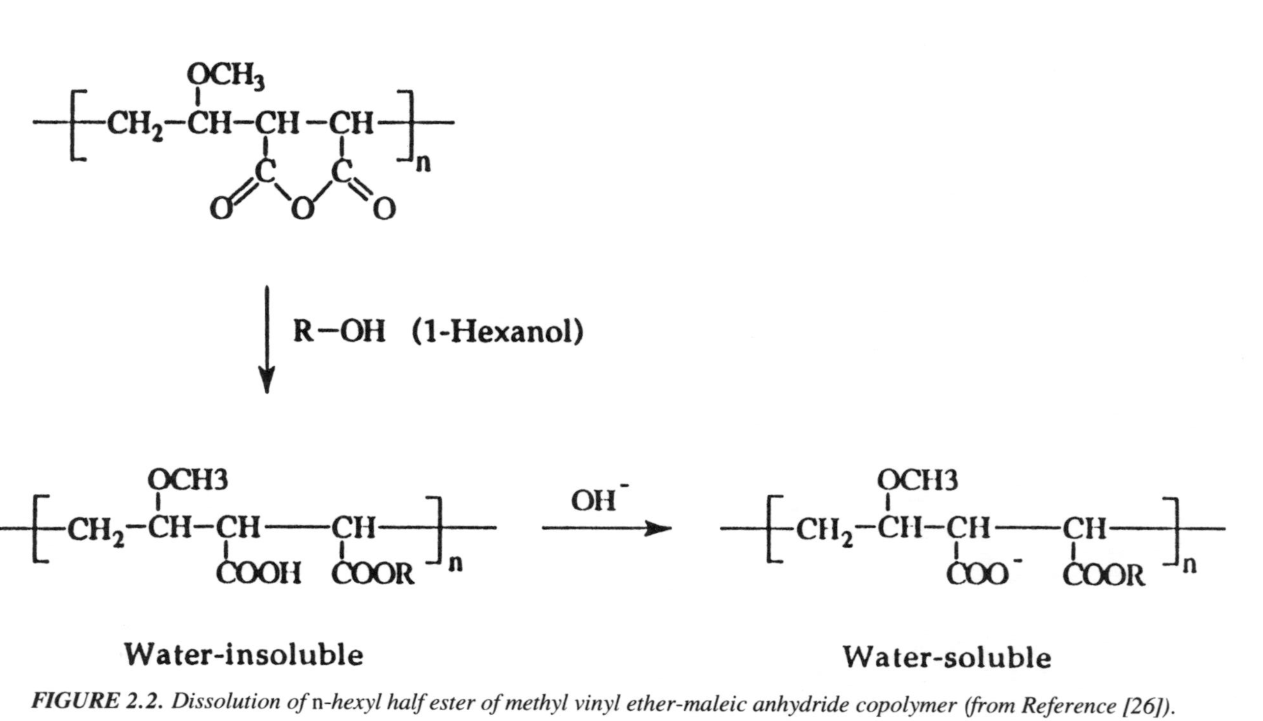

FIGURE 2.2. Dissolution of n-*hexyl half ester of methyl vinyl ether-maleic anhydride copolymer (from Reference [26]).*

prepared copolymers from a hydrophilic monomer L-aspartic acid and hydrophobic monomers such as L-leucine, β-methyl-L-aspartate, and β-benzyl-L-aspartate [41]. Poly(*tert*-butyloxycarbonylmethyl glutamate) (PBCMG) with various degrees of esterification was obtained by partial esterification of poly(glutamic acid) with *tert*-butylbromoacetate [39]. PBCMG with an ester content less than 50 percent showed visible swelling and fragmentation *in vivo* [38]. Poly(β-malate) was synthesized as the benzyl ester followed by reductive debenzylation. The solubility of this polymer was a function of pH, the degree of esterification, and the sequence distribution [42].

Crosslinked sequential polypeptides containing asparagine (Asn) or glutamine (Gln) [e.g., poly(Gly-Thr-Gln-Ala-Gly)] are known to undergo interfacial hydrolytic cleavage of a carboxamide to a carboxylate anion with resultant local swelling and subsequent degradation [43,44]. The rate of carboxamide hydrolysis can be controlled by the hydrophobicity of adjacent amino acid residues.

2.3 CHEMICAL HYDROLYSIS OF POLYMERS

Hydrolysis of the polymer backbone is most desirable since it will produce low molecular weight by-products. Table 2.1 lists examples of polymers which undergo degradation in the backbone chains by hydrolysis. It is well known that natural polymers such as proteins and polysaccharides undergo degradation by hydrolysis. Of the synthetic polymers listed in Table 2.1, only a few polymers are water-soluble. Water-insoluble polymers in the table are not very hydrophilic but contain functional groups capable of hydrogen bonding. They tend to be more crystalline, and this property accounts for their water-insolubility [5]. For hydrolysis to occur, the polymer has to contain hydrolytically unstable bonds which should be reasonably hydrophilic for the access of water. In the development of new biodegradable polymers, consideration of the toxicity of the degradation by-products has to be given the highest priority.

As listed in Table 2.1, an overwhelming number of biodegradable polymers are esters and ester-derivative polymers [20,45]. Polyesters include poly(glycolic acid), poly(lactic acid) [46], poly(ϵ-caprolactone) [47], poly(β-hydroxybutyric acid), poly(β-hydroxyvaleric acid) [48,49], polydioxanone, poly(ethylene terephthalate), poly(malic acid) [22], poly(tartronic acid) [20], and poly(ortho esters) [3,26]. Polyesters are degraded mainly by simple hydrolysis, although other mechanisms such as intracellular breakdown of small debris cannot be excluded [50].

Poly(glycolic acid) was the first synthetic polymer to be used in the preparation of bioabsorbable sutures. Water penetrates into the polymer

TABLE 2.1. Structures of Some Biodegradable Polymers.

Name	Structure
1. Polypeptides	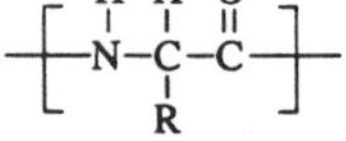 $\left[-NH-CH(R)-C(=O)-\right]$
2. Poly(lactic acid)	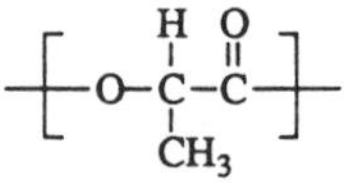 $\left[-O-CH(CH_3)-C(=O)-\right]$
3. Poly(glycolic acid)	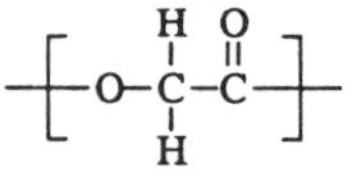 $\left[-O-CH(H)-C(=O)-\right]$
4. Poly(ϵ-caprolactone)	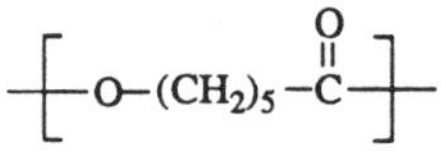 $\left[-O-(CH_2)_5-C(=O)-\right]$
5. Poly(β-hydroxybutyrate)	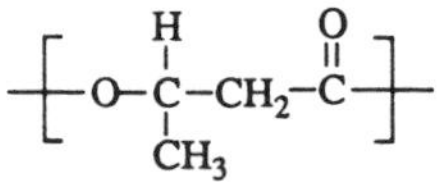 $\left[-O-CH(CH_3)-CH_2-C(=O)-\right]$
6. Poly(β-hydroxyvalerate)	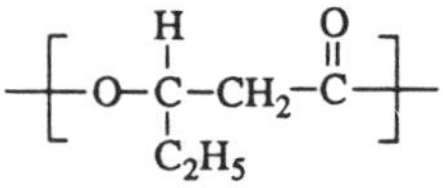 $\left[-O-CH(C_2H_5)-CH_2-C(=O)-\right]$
7. Polydioxanone	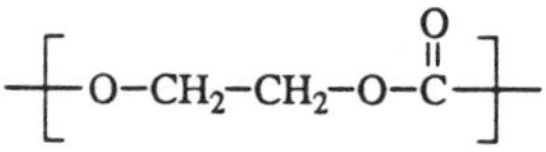 $\left[-O-CH_2-CH_2-O-C(=O)-\right]$
8. Poly(ethylene terephthalate)	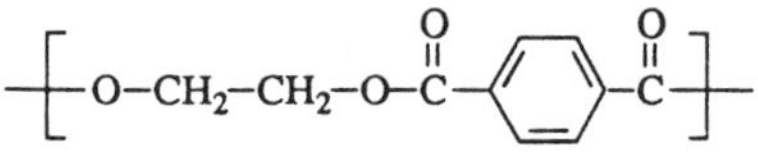 $\left[-O-CH_2-CH_2-O-C(=O)-C_6H_4-C(=O)-\right]$
9. Poly(malic acid)	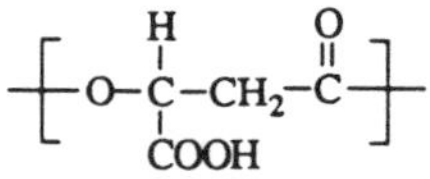 $\left[-O-CH(COOH)-CH_2-C(=O)-\right]$
10. Poly(tartronic acid)	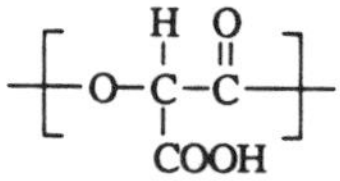 $\left[-O-CH(COOH)-C(=O)-\right]$
11. Poly(ortho esters)	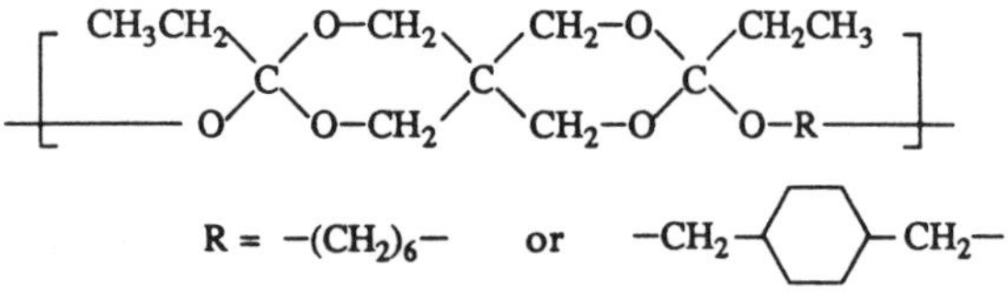 $\left[-O-C(CH_3CH_2)(-O-CH_2-)_2C(-CH_2-O-)_2C(CH_2CH_3)-O-R-\right]$; R = $-(CH_2)_6-$ or $-CH_2-C_6H_{10}-CH_2-$

TABLE 2.1. (continued).

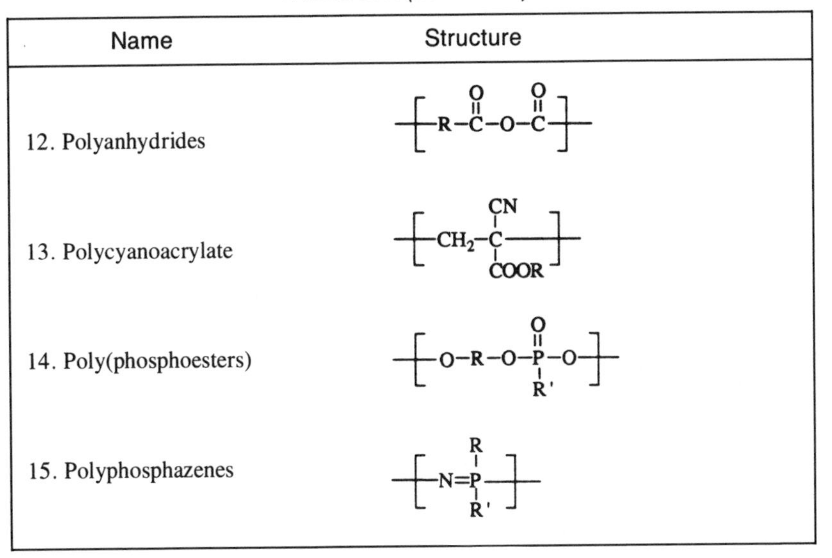

Name	Structure
12. Polyanhydrides	$-[R-C(=O)-O-C(=O)]-$
13. Polycyanoacrylate	$-[CH_2-C(CN)(COOR)]-$
14. Poly(phosphoesters)	$-[O-R-O-P(=O)(R')-O]-$
15. Polyphosphazenes	$-[N=P(R)(R')]-$

matrix and degradation occurs throughout the matrix [51]. For semi-crystalline polymers such as PGA, PLA, and poly(ϵ-caprolactone), degradation starts in the amorphous regions, followed by the crystalline domains [52]. The first stage of the degradation process involves nonenzymatic, random hydrolytic ester cleavage, and its duration is determined by the initial molecular weight of the polymer as well as its chemical structure [52]. Enzymes are thought to be involved in the degradation of hydrolysis products formed during the initial degradation stage. Since the diffusion of high molecular weight enzymes into the polymer bulk is kinetically prohibitive, enzyme participation in the first phase of polymer degradation is excluded [53,54]. Recently, amino acid was used to prepare copolymers with lactic acid, such as glycine/DL-lactic acid copolymers [55].

Poly(β-hydroxybutyric acid) (PHB) is known to be biocompatible with human tissue, resorbed by the body, and degradable to water and CO_2. Thus, it is suitable for use in sutures, blood-vessel grafts, and implantable controlled release drug delivery systems. PHB and poly(β-hydroxyvaleric acid) are synthesized by many species of bacteria maintained on a carbohydrate-rich, nitrogen-deficient diet [56]. Recently, PHB was produced in plants through genetic engineering [57]. Copolymers of β-hydroxybutyric acid and β-hydroxyvaleric acid are also widely used. The increase in β-hydroxyvalerate content led to a reduction in melting point and enabled milder processing conditions to be used. The processing conditions altered the crystallinity which in turn altered the hydrolytic degradation rate [56].

Unlike most polyesters which degrade by hydrolysis throughout the matrix, polyorthoesters and polyanhydrides undergo surface degradation, i.e., degradation is limited to the outer layer of the polymer. This property is ideal for the preparation of the "zero-order" release drug delivery systems. The hydrolysis rates of the latter two polymers are much higher than those of other polyesters [20]. The degradation of polyorthoesters is sensitive to pH and provides flexibility in controlling the drug release rate. Polyorthoesters and polyanhydrides have been developed by Jorge Heller [3,58] and Robert Langer [4,59], respectively, specifically for biodegradable controlled release drug delivery systems.

Poly(alkyl 2-cyanoacrylates) are also biodegradable. They are, however, unique in that they have a hydrolyzable all-carbon backbone. The C–C backbone in poly(butyl 2-cyanoacrylate) is relatively easily hydrolyzed [60]. The polymers degrade to alkyl cyanoacetate and formaldehyde which may be toxic [61]. The rate of degradation decreases as the size of the ester becomes larger [62]. Butyl 2-cyanoacrylate is used widely, since it spreads on biological fluids and polymerizes upon contact with water [63].

Polyphosphazenes are a new class of biodegradable polymers in which the hydrolytic stability is determined by changes in the side groups attached to an unconventional polymer backbone [64]. Many polyphosphazenes have already been used for drug delivery and cell encapsulation [65]. Polyphosphoesters are another class of new biodegradable polymers characterized by P(O)–O–C linkages in the polymer backbone. The pentavalent phosphorus atom allows for the attachment of a pendant component [66,67]. Cleavage of phosphoester bonds resulted in the swelling of the polymer. For example, poly(bisphenol A-phenylphosphate) absorbed more than 60 percent of the water in 10 days at 37°C [66].

Poly(ester urethanes) are also known to be biodegradable. The hydrolytic lability of poly(ester urethanes) increases as the ester content increases [45]. Poly(ether urethanes) are much less hydrolyzable than poly(ester urethanes). Periodate oxidized cellulose is also known to be hydrolyzed into small molecules such as glycolic and 2,4-dihydroxybutyric acids at physiological pH. Oxidized cellulose degraded slowly in rats and almost disappeared from the implantation sites in a few months [68].

It is possible to synthesize biodegradable copolymers from the polymers listed in Table 2.1. For example, Kobayashi et al. [69] copolymerized lactide and caprolactone using PEG as the initiator to obtain biodegradable prepolymers that can be cured upon contact with water in living tissues.

2.3.1 Enzyme-Catalyzed Hydrolysis

Enzymes in the body play vital roles in various reactions essential for daily activities. Enzymes are proteins which function as catalysts for a specific reaction or series of reactions, such as oxidation, reduction, hydrolysis, esterification, synthesis, and molecular interconversions [19,70]. Each enzyme is different in specificity, location, and concentration in the body. They act in concert to process and metabolize proteins and polysaccharides [71]. In each enzyme there is an active site which binds substrates and either forms or cleaves particular bonds in the substrates. The binding between enzymes and substrates depends on the three-dimensional structures of the active sites and substrate segments. Thus, the enzymatic degradation of substrates is substrate-specific and specific enzymes exist in the body for the degradation of particular proteins and polysaccharides.

Since natural polymers, such as proteins and polysaccharides, are well known for their degradability in the body, they were the first to be exploited to prepare biodegradable biomaterials. The use of catgut as an absorbable suture led to the development of synthetic, biodegradable sutures [20]. Reconstituted collagen has been used widely as a bioabsorbable suture, a wound dressing, and a drug delivery system [72]. From the enzymatic degradation of proteins, it is to be expected that enzymes will play a significant role in the degradation of synthetic poly(α-amino acids) such as poly-L-lysine, poly-L-arginine, poly-L-aspartic acid, or poly-L-glutamic acid [73]. The types of enzymes involved and the extent of degradation of poly(α-amino acids), however, are not easily predictable [45]. Of the many poly(α-amino acids), the enzymatic degradation of poly-L-lysine has been studied most extensively.

Hydrophobic polymers with no hydrolyzable bonds, such as polyethylene, polypropylene, poly(methyl methacrylate), or polystyrene, are expected to be the most stable. Cumulative evidence, however, suggests that even the most inert polymers suffer some degree of degradation after implantation into the body [74,75], or under physiological conditions in the presence of the proper enzymes [76]. The rate of hydrolysis of crosslinked elastomers of α-caprolactone and δ-valerolactone was measurably faster *in vivo* than *in vitro*. In addition, surface erosion was observed *in vivo* while bulk erosion was dominant *in vitro* [77]. These observations led to the suggestion that enzymes play an important role in the biodegradation process [50,74,75]. Although enzymatic degradation is known to be highly specific, there may be sufficient nonspecificity to make enzymes play a significant role in the degradation of many synthetic polymers [16]. Indeed, seemingly inert polymers, such as nylon, poly(ether urethane), poly(ethylene terephthalate), poly(hydroxybutyrate), poly(ester-urea), poly(ϵ-caprolactone), and poly(glycolic acid), are degradable by enzymes. Table 2.2 lists some synthetic polymers which are known to undergo significant enzymatic degradation *in vivo* as well as *in vitro*. Many of the synthetic polymers are degradable by esterases and this is not totally unexpected since they are known to be involved in the detoxification process in the body [5].

TABLE 2.2. Examples of Enzymatic Degradation of Water-Insoluble, Synthetic Polymers.

Polymers	Enzymes	Reference
Nylon 6,6	chymotrypsin	
	trypsin	
	papain	78
Poly(ϵ-caprolactone)	esterase	
	lipase	79
Polydioxanone	esterase	80
Poly(ester-urea)	chymotrypsin	10
	elastase	81
	substilisin	82
Poly(ether urethane)	liver homogenate	
	(acid phosphatase, esterase)	
	cathepsin	15
	leucine aminopeptidase	12
	papain	83
	urease	76
	trypsin	84
Poly(ethylene terephthalate)	esterase	78,85
	papain	12
	leucine aminopeptidase	86
Poly(glycolic acid)	esterase	87,88
	ficin	
	carboxypeptidase A	
	clostridiopeptidase A	
	bromelain	
	esterase	
	leucine aminopeptidase	86
Polyhydroxybutyrate	esterase	89
Poly(hydroxybutyrate-*co*-hydroxyvalerate)	esterase	90
Poly(2-hydroxyethyl-L-glutamine)	pronase	91
	papain	19
Poly(lactic acid)	pronase	
	proteinase K	
	bromelain	
	lipase	92,93
Polylysine	trypsin	
	chymotrypsin	
	carboxypeptidase B	
	elastase	
	papain	
	ficin	94

The ability of the enzymes to penetrate into synthetic polymers determines the extent of enzyme-catalyzed degradation and the mode of degradation, i.e., surface vs. bulk degradation. The degree of enzyme penetration into polymers depends on the size of the enzyme as well as the physical property of the polymer. Biomer, a segmented poly(ether urethane), underwent bulk degradation by papain, while urease caused surface degradation [76]. Papain which has a molecular weight of 20,700 daltons, was small enough to diffuse into the polymer. Urease, on the other hand, was localized at the surface due to its large size (a molecular weight of 473,000 daltons). In the study of enzyme-catalyzed degradation of low molecular weight copolymers of L-lactic acid and ϵ-caprolactone, Fukuzaki et al. observed that the extent of degradation was proportional to the degree of enzyme penetration into the copolymer [93,94]. Significant diffusion of enzymes (lipase and esterase) into the copolymer was observed when the content of ϵ-caprolactone was 30–70 mol percent and the resulting copolymer was an amorphous, paste-type. The degradation of semi-crystalline polymers may be significantly influenced by the proportion of amorphous regions in the polymer. As the polymer chains become more flexible, they have a greater possibility of interacting with active sites of enzymes [95]. Thus, amorphous regions of the polymer will degrade prior to the degradation of crystalline regions.

Nonbiodegradable polymers may be blended with enzyme-degradable polymers. For example, polyethylene-starch blends have been prepared by melt blending, and then used as biodegradable polymer films [96,97]. Apparently, only a portion of the materials occupied by starch will be degraded while the rest of the materials remain intact. Since enzymes or microorganisms have to diffuse into the polymer blend to digest starch, starch-filled channels have to be exposed to the surface. The connectivity of the starch-filled channels can be achieved at 8–16 percent (v/v) starch [98].

Other water-soluble, synthetic polymers are also known to undergo enzymatic degradation. Poly(ethylene glycol), poly(propylene glycol), and poly(tetramethylene glycol) are known to be biodegradable by various bacteria [33,99]. Poly(vinyl alcohol) is the only polymer which has the carbon-carbon backbone chains degradable by bacterial enzymes [33,100]. Polyvinylpyrrolidone (PVP) is also known to degrade in the body. The metabolism of intravenously administered PVP, however, was minimal (<0.3 percent) and involved only low molecular weight fractions [101].

2.4 BIODEGRADABLE HYDROGELS

The objective of modern materials synthesis is to find ways to generate new combinations of properties at three levels—at the molecular level, at the materials level, and at the surface level [102]. One advantage of modifying known polymers is that we can project the properties of the modified product fairly well from the properties of the mother polymers. Biodegradable hydrogels can be prepared from polymers with well-characterized properties.

Many of the polymers listed in Tables 2.1 and 2.2 are not water-soluble, and thus do not form hydrogels even in the presence of abundant water. The water-insoluble polymers, however, can be used to form hydrogels by preparing either block copolymers, polymer blends, or interpenetrating polymer networks, with water-soluble polymers. These preparations, containing substantial fractions of hydrophilic polymers, can absorb significant amounts of water and swell without dissolving in water to form hydrogels (see Chapter 3). Therefore, even water-insoluble polymers can be effectively used to prepare biodegradable hydrogels. It is advantageous to use the biodegradable polymers listed in Tables 2.1 and 2.2, since their properties are rather well characterized.

2.5 REFERENCES

1. Shieh, S.-J., M. C. Zimmerman and J. R. Parsons. 1990. "Preliminary Characterization of Bioresorbable and Nonresorbable Synthetic Fibers for the Repair of Soft Tissue Injuries," *J. Biomed. Mater. Res.*, 24:789–808.
2. Meislin, R. J., D. M. Wiseman, H. Alexander, T. Cunningham, C. Linksky, C. Carlstedt, M. Pitman and R. Casar. 1990. "A Biochemical Study of Tendon Adhesion Reduction Using a Biodegradable Barrier in a Rabbit Model," *J. Appl. Biomat.*, 1:13–19.
3. Heller, J. Y. Maa, F. Wuthrich, S. Y. Ng and R. Duncan. 1991. "Recent Developments in the Synthesis and Utilization of Poly(ortho Esters)," *J. Controlled Rel.*, 16:3–13.
4. Langer, R. 1990. "New Methods of Drug Delivery," *Science*, 249:1527–1533.
5. St. Pierre, T. and E. Chiellini. 1986. "Biodegradability of Synthetic Polymers Used for Medical and Pharmaceutical Applications: Part 1—Principles of Hydrolysis Mechanisms," *J. Bioact. Compat. Polymers*, 1:467–497.
6. Pitt, C. G. and A. Schindler. 1983. "Biodegradation of Polymers," Chapter 3 in *Controlled Drug Delivery, Vol. I. Basic Concepts*, S. D. Bruck, ed., Boca Raton, FL: CRC Press.
7. Grassie, N. and G. Scott. 1985. *Polymer Degradation & Stabilization*. New York, NY: Cambridge University Press.
8. Lemm, W., T. Krukenberg, G. Regier, K. Gerlach and E. S. Bucherl. 1981. "Biodegradation of Some Biomaterials after Subcutaneous Implantation," *Proc. Eur. Soc. Artif. Org.*, 8:71–75.
9. Potts, J. E., R. A. Clendinning, W. B. Ackart and W. D. Niegisch. 1973. "The Biodegradability of Synthetic Polymers," in *Polymers and Ecological Problems. Polymer Science and Technology Series, Vol. 3*, J. Guillet, ed., New York, NY: Plenum Press, pp. 61–79.

10. Huang, S. J., D. A. Bansleben and J. R. Knox. 1979. "Biodegradable Polymers: Chymotrypsin Degradation of Low Molecular Weight Poly(ester-urea) Containing Phenylalanine," *J. Appl. Polym. Sci.*, 23:429–437.

11. Swift, G. 1992. "Biodegradable Polymers in the Environment: Are They Really Biodegradable?" *Proc. ACS Div. Polym. Mat. Sci. Eng.*, 66:403–404.

12. Ratner, B. D., K. W. Gladhill and T. A. Horbett. 1988. "Analysis of *in vitro* Enzymatic and Oxidative Degradation of Polyurethanes," *J. Biomed. Mater. Res.*, 22:509–527.

13. Hergenrother, R. W., H. D. Wabers and S. L. Cooper. 1992. "The Effect of Chain Extenders and Stabilizers on the *in-vivo* Stability of Polyurethanes," *J. Appl. Biomater.*, 3:17–22.

14. Marchant, R. E., J. M. Anderson, K. Phua and A. Hiltner. 1984. "*In vivo* Biocompatibility Studies. II. Biomer: Preliminary Cell Adhesion and Surface Characterization Studies," *J. Biomed. Mater. Res.*, 18:309–315.

15. Smith, R., D. F. Williams and C. Oliver. 1987. "The Biodegradation of Poly(ether Urethanes)," *J. Biomed. Mater. Res.*, 21:1149–1166.

16. Kumar, G. S. 1987. *Biodegradable Polymers. Prospects and Progress.* New York, NY: Marcel Dekker, Inc., Chapters 1 and 3.

17. Albertsson, A.-C. and S. Karlsson. 1990. "Biodegradation and Test Methods for Environmental and Biomedical Applications of Polymers," in *Degradable Materials. Perspectives, Issues, and Opportunities*, S. A. Barenberg, J. L. Brash, R. Narayan and A. E. Redpath, eds., Boca Raton, FL: CRC Press, pp. 263–293.

18. Daniels, A. U., M. K. O. Chang, K. P. Andriano and J. Heller. 1990. "Mechanical Properties of Biodegradable Polymers and Composites Proposed for Internal Fixation of Bone," *J. Applied Biomaterials*, 1:57–78.

19. Kopeček, J. and P. Rejmanová. 1983. "Enzymatically Degradable Bonds in Synthetic Polymers," Chapter 4 in *Controlled Drug Delivery, Vol. I. Basic Concepts*, S. D. Bruck, ed., Boca Raton, FL: CRC Press.

20. Holland, S. J. and B. J. Tighe. 1992. "Biodegradable Polymers," in *Adv. Pharmaceutical Sci., Vol. 6,* D. Ganderton and T. Jones, eds., New York, NY: Academic Press, pp. 101–164.

21. Narayan, R. 1990. "Introduction," in *Degradable Materials. Perspectives, Issues, and Opportunities*, S. A. Barenberg, J. L. Brash, R. Narayan and A. E. Redpath, eds., Boca Raton, FL: CRC Press, pp. 1–37.

22. Vert, M. 1989. "Bioresorbable Polymers for Temporary Therapeutic Applications," *Die Angewandte Makromolekulare Chemie*, 166/167:155–168.

23. Anderson, J. P. E. 1989. "Principles of and Assay Systems for Biodegradation," in *Biotechnology and Biodegradation*, D. Kamely, A. Chakrabarty and G. S. Omenn, eds., Houston, TX: Gulf Publishing Co.

24. Shalaby, S. W. 1988. "Bioabsorbable Polymers," in *Encyclopedia of Pharmaceutical Technology, Vol. 1,* J. Swarbrick and J. C. Boylan, eds., New York, NY: Marcel Dekker, Inc., pp. 465–476.

25. Drobnik, J. and F. Rypàcek. 1984. "Soluble Synthetic Polymers in Biological Systems," *Adv. Polym. Sci.*, 57:1–50.

26. Heller, J., S. H. Pangburn and K. V. Roskos. 1990. "Development of Enzymatically Degradable Protective Coatings for Use in Triggered Drug Delivery Systems II. Derivatized Starch Hydrogels," *Biomaterials*, 11:345–350.

27. Brierley, C. L., D. P. Kelly, K. J. Seal and D. J. Best. 1985. "Materials and Biotechnology," Chapter 5 in *Biotechnology. Principles and Applications*, I. J. Higgins, D. J. Best and J. Jones, eds., Boston, MA: Blackwell Scientific Publications.

28. Gilding, D. K. 1981. "Biodegradable Polymers," Chapter 9 in *Biocompatibility of Clinical Implant Materials, Vol. II,* D. F. Williams, ed., Boca Raton, FL: CRC Press.

29. Kronenthal, R. L. 1975. "Biodegradable Polymers in Medicine and Surgery," in *Polymers in Medicine and Surgery*, R. L. Kronenthal, Z. Oser and E. Martin, eds., New York, NY: Plenum Press, pp. 119–137.

30. MacGreger, E. A and C. T. Greenwood. 1980. *Polymers in Nature*. New York, NY: John Wiley & Sons, Chapters 3 and 6.

31. Gross, J. R. 1990. "The Evolution of Absorbent Materials," in *Absorbent Polymer Technology*, L. Brannon-Peppas and R. S. Harland, eds., New York, NY: Elsevier, pp. 3–22.

32. Chambliss, W. G. 1983. "The Forgotten Dosage Form: Enteric-Coated Tablets," *Pharm. Technol.* (September):124–140.

33. Murphy, K. S., N. A. Enders, M. Mahjour and M. B. Fawzi. 1986. "A Comparative Evaluation of Aqueous Enteric Polymers in Capsule Coatings," *Pharm. Technol.* (October):36–45.

34. Lehmann, K. and D. Dreher. 1981. "Coating of Tablets and Small Particles with Acrylic Resins by Fluid Bed Technology," *Int. J. Pharm. Tech. & Prod. Mfr.*, 2:31–43.

35. Gennaro, A. R., ed. 1985. *Pharmaceutical Sciences, 17th Edition.* Easton, PA: Mack Publ. Co., pp. 1633–1643.

36. Heller, J., R. Bakeer, R. M. Gale and J. O. Rodin. 1978. "Controlled Drug Release by Polymer Dissolution I. Partial Esters of Maleic Anhydride Copolymers. Properties and Theory," *J. Appl. Polym. Sci.*, 22:1991–2009.

37. Sidman, K. R., W. D. Steber, A. D. Schwope, G. R. Schnaper and R. Gayle. 1983. "Controlled Release of Macromolecules and Pharmaceuticals from Synthetic Polypeptides Based on Glutamic Acid," *Biopolymers*, 22:547–556.

38. Lescure, F., R. Gurny, E. Doelker, M. L. Pelaprat, D. Bichon and J. M. Anderson. 1988. "Evaluation of New Poly(glutamic Acid Esters) for Implantable Drug Delivery Systems," *Proceed. Intern. Symp. Control. Rel. Bioact. Mater.*, 15:107–108.

39. Bichon, D. and B. Lamy. 1985. "New Polypeptidic Biomaterials," *Trans. 11th Ann. Meet. Soc. Biomater.*, 8:137.

40. Sidman, K. R., A. D. Schwope, W. D. Steber, S. E. Rudolph and S. B. Poulin. 1980. "Biodegradable, Implantable Sustained Release Systems Based on Glutamic Acid Copolymers," *J. Memb. Sci.*, 7:277–291.

41. Marck, K. W., C. H. Wildevuur, W. L. Sederel, A. Bantjes and J. Feijen. 1977. "Biodegradability and Tissue Reaction of Random Copolymers of L-Leucine, L-Aspartic Acid, and L-Aspartic Acid Esters," *J. Biomed. Mater. Res.*, 11:405–422.

42. Johns, D. B., R. W. Lenz and M. Vert. 1986. "Poly(malic Acid). Part I. Preparation and Polymerization of Benzyl Malolactonate," *J. Bioact. Comp. Polym.*, 1:47–60.

43. Urry, D. W. 1988. "Entropic Elastic Processes in Protein Mechanisms. II. Simple (Passive) and Coupled (Active) Development of Elastic Forces," *J. Protein Chem.*, 7:81–114.

44. Urry, D. W. 1982. "Characterization of Soluble Peptides of Elastin by Physical Techniques," *Methods in Enzymology*, 82:673–716.

45. St. Pierre, T. and E. Chiellini. 1986. "Biodegradability of Synthetic Polymers Used for Medical and Pharmaceutical Applications: Part 2—Backbone Hydrolysis," *J. Bioact. Compat. Polymers*, 2:4–30.

46. Lewis, D. H. 1990. "Controlled Release of Bioactive Agents from Lactide/Glycolide Polymers," Chapter 1 in *Biodegradable Polymers as Drug Delivery Systems*, M. Chasin and R. Langer, eds., New York, NY: Marcel Dekker, Inc.

47. Pitt, C. G. 1990. "Poly-ϵ-caprolactone and Its Copolymers," Chapter 3 in *Biodegradable Polymers as Drug Delivery Systems*, M. Chasin and R. Langer, eds., New York, NY: Marcel Dekker, Inc.

48. Steinbüchel, A. 1991. "Polyhydroxyalkanoic Acids," Chapter 3 in *Biomaterials. Novel Materials from Biological Sources*, D. Byrom, ed., New York, NY: Stockton Press.

49. Scandola, M., G. Ceccorulli, M. Pizzoli and M. Gazzano. 1992. "Study of the Crystal Phase and Crystallization Rate of Bacterial Poly(3-hydroxybutyrate-*co*-3-hydroxyvalerate)," *Macromolecules*, 25:1405–1410.

50. Schakenraad, J. M., M. J. Hardonk, J. Feijen, I. Molenaar and P. Nieuwenhuis. 1990. "Enzymatic Activity toward Poly(L-lactic Acid) Implants," *J. Biomed. Mater. Res.*, 24:529–545.

51. Chu, C. C. 1990. "Polyesters and Polyamides," in *Concise Encyclopedia of Medical & Dental Materials*, D. Williams, ed., New York, NY: Pergamon Press, pp. 261–271.

52. Pitt, C. G., M. M. Gratzl, G. L. Kimmel, J. Surles and A. Schindler. 1981. "Aliphatic Polyesters II. The Degradation of Poly(DL-lactide), Poly(ϵ-caprolactone), and Their Copolymers *in Vivo*," *Biomaterials*, 2:215–220.

53. Pitt, C. G., F. I. Chasalow, Y. M. Hibionada, D. M. Klimas and A. Schindler. 1981. "Aliphatic Polyesters. I. The Degradation of Poly(ϵ-caprolactone) *in Vivo*," *J. Appl. Polymer Sci.*, 26:3779–3787.

54. Woodward, S. C., P. S. Stannett, F. Moatamed, A. Schindler and C. G. Pitt. 1985. "The Intracellular Degradation of Poly(ϵ-caprolactone)," *J. Biomed. Mater. Res.*, 19:437–444.

55. Helder, J., P. J. Dijkstra and J. Feijen. 1990. "*In vitro* Degradation of Glycine/DL-Lactic Acid Copolymers," *J. Biomed. Mater. Res.*, 24:1005–1020.

56. Yasin, M., S. J. Holland and B. J. Tighe. 1990. "Polymers for Biodegradable Medical Devices. V. Hydroxybutyrate-Hydroxyvalerate Copolymers: Effects of Polymer Processing on Hydrolytic Degradation," *Biomaterials*, 11:451–454.

57. Poirier, Y., D. E. Dennis, K. Klomparens and C. Somerville. 1992. "Polyhydroxybutyrate, a Biodegradable Thermoplastic, Produced in Transgenic Plants," *Science*, 256:520–523.

58. Heller, J., S. Y. Ng, B. K. Fritzinger and K. V. Roskos. 1990. "Controlled Drug Release from Bioerodible Hydrophobic Ointments," *Biomaterials*, 11:235–237.

59. Leong, K. W., B. C. Brott and R. Ranger. 1985. "Bioerodible Polyanhydrides as Drug Carrier Matrices. I: Characterization, Degradation, and Release Characteristics," *J. Biomed. Mater. Res.*, 19:941–955.

60. Wood, D. A., T. L. Whateley and A. T. Florence. 1981. "Formation of Poly(butyl 2-Cyanoacrylate) Microcapsules and the Microencapsulation of Aqueous Solutions of [^{125}I]-Labelled Proteins," *Int. J. Pharm*, 8:35–43.

61. Tseng, Y-.C., S.-H. Hyon and Y. Ikada. 1990. "Modification, Synthesis and Investigation of Properties for 2-Cyanoacrylates," *Biomaterials*, 11:73–79.

62. Damge, C., C. Michel, M. Aprahamian, P. Couvreur and J. P. Devissaguet. 1990. "Nanocapsules as Carriers for Oral Peptide Delivery," *J. Controlled Rel.*, 13: 233–239.

63. Leonard, F. 1970. "Hemostatic Applications of Alpha Cyanoacrylates: Bonding Mechanism and Physiological Degradation of Bonds," Chapter 11 in *Adhesion in Biological Systems*, R. S. Manly, ed., London: Academic Press.

64. Allcock, H. R. 1990. "Polyphosphazenes as New Biomedical and Bioactive Materials," Chapter 5 in *Biodegradable Polymers as Drug Delivery Systems*, M. Chasin and R. Langer, eds., New York, NY: Marcel Dekker, Inc.

65. Bano, M. C., S. Cohen, K. B. Visscher, H. R. Allcock and R. Langer. 1991. "A Novel Synthetic Method for Hybridoma Cell Encapsulation," *Bio/Technology*, 9:468–471.

66. Richards, M., B. I. Dahiyat, D. M. Arm, P. R. Brown and K. W. Leong. 1991. "Evaluation of Polyphosphates and Polyphosphonates as Degradable Biomaterials," *J. Biomed. Mater. Res.*, 25:1151–1167.

67. Dahiyat, B., F. Shi, Z. Zhao and K. Leong. 1992. "Design of Degradable Elastomers for Medical Applications," *Proc. ACS Div. Polym. Mat. Sci. Eng.*, 66:87–88.

68. Singh, M., A. R. Ray and P. Vasudevan. 1982. "Biodegradation Studies on Periodate Oxidized Cellulose," *Biomaterials*, 3:16–20.

69. Kobayashi, H., S.-H. Hyon and Y. Ikada. 1991. "Water-Curable and Biodegradable Prepolymers," *J. Biomed. Mater. Res.*, 25:1481–1494.

70. Kaplan, A., L. L. Szabo and K. E. Opheim. 1988. "Enzymes," Chapter 6 in *Clinical Chemistry: Interpretation and Techniques*. Philadelphia, PA: Lea & Febiger.

71. Kenny, A. J. and N. M. Hooper. 1991. "Peptidase Involved in the Metabolism of Bioactive Peptides," Chapter 4 in *Degradation of Bioactive Substances: Physiology and Pathophysiology*, J. H. Henrikson, ed., Boca Raton, FL: CRC Press.

72. Gorham, S. D. 1991. "Collagen," Chapter 2 in *Biomaterials. Novel Materials from Biological Sources*, D. Byron, ed., New York, NY: Stockton Press.

73. Josefsson, J. and H. Sjostrom. 1966. "Intestinal Dipeptidases. IV. Studies on the Release and Subcellular Distribution of Intestinal Dipeptidases of the Mucosal Cells of the Pig," *Acta. Physiol. Scand.*, 67:27–33.

74. Oppenheimer, B. S., E. T. Oppenheimer, T. Danishefsky, A. P. Stout and E. F. Eirich. 1955. "Further Studies of Polymers as Carcinogenic Agents in Animals," *Cancer Res.*, 15:333–340.

75. Liebert, T. C., R. P. Chartoff, S. L. Cosgrove and R. S. McCluskey. 1976. "Subcutaneous Implants of Polypropylene Filaments," *J. Biomed. Mater. Res.*, 10: 939–945.

76. Phua, S. K., E. Castillo, J. M. Anderson and A. Hiltner. 1987. "Biodegradation of a Polyurethane *in Vitro*," *J. Biomed. Mater. Res.*, 21:231–246.

77. Schindler, A. and C. G. Pitt. 1982. "Biodegradable Elastometric Polyesters," *Polymer Preprints*, 23(2):111–112.

78. Smith, R., C. Oliver and D. F. Williams. 1987. "The Enzymatic Degradation of Polymers *in Vitro*," *J. Biomed. Mater. Res.*, 21:991–1003.

79. Tokiwa, Y. and T. Suzuki. 1977. "Hydrolysis of Polyesters by Lipases," *Nature*, 270:76–78.

80. Williams, D. F., C. C. Chu and J. Dwyer. 1984. "Effects of Enzymes and Gamma Irradiation on the Tensile Strength and Morphology of Poly(*p*-dioxanone) Fibers," *J. Appl. Poly. Sci.*, 29:1865–1877.

81. Britritto, M. M., J. P. Bell, G. M. Brenckle, S. J. Huang and J. R. Knox. 1979. "Synthesis and Biodegradation of Polymers Derived from α-Hydroxy Acids," *J. Appl. Polym. Sci., Appl. Polym. Symp.*, 35:405–414.

82. Huang, S. J. and K. W. Leong. 1989. "Biodegradable Polymers. Polymers Derived from Gelatin and Lysine Esters," *Polymer Preprints*, 20:552–554.

83. Takahara, A., R. W. Hergenrother, A. J. Coury and S. L. Cooper. 1992. "Effect of Soft Segment Chemistry on the Biostability of Segmented Polyurethanes. II. *In vitro* Hydrolytic Degradation and Lipid Sorption," *J. Biomed. Mater. Res.*, 26:801–818.

84. Bouvier, M., A. S. Chawla and I. Hinberg. 1991. "*In vitro* Degradation of a Poly(ether Urethane) by Trypsin," *J. Biomed. Mater. Res.*, 25:773–789.

85. Smith, R. and D. F. Williams. 1985. "The Degradation of a Synthetic Polyester by a Sysosomal Enzyme," *J. Mater. Sci. Letters*, 4:547–549.

86. Williams, D. F. 1990. "The Role of Active Species within Tissue in Degradation Processes", in *Degradable Materials. Perspectives, Issues, and Opportunities*, S. A. Barenberg, J. L. Brashm, R. Narayan and A. E. Redpathm, eds., CRC Press, Boca Raton, pp. 323–355.

87. Chu, C. C. and D. F. Williams. 1983. "The Effect of Gamma Irradiation on the Enzymatic Degradation of Polyglycolic Acid Absorbable Sutures," *J. Biomed. Mater. Res.*, 17:1029–1040.

88. Williams, D. F. and E. Mort. 1977. "Enzyme-Accelerated Hydrolysis of Polyglycolic Acid," *J. Bioeng.*, 1:231–238.

89. Kemnitzer, J. E., S. P. McCarthy and R. A. Gross. 1992. "Stereochemical and Morphological Effects on the Degradation Kinetics of Poly(β-hydroxybutyrate): A Model Study," *Proc. ACS Div. Polym. Mat. Sci. Eng.*, 66:405–407.

90. Parikh, M., R. A. Gross and S. P. McCarthy. 1992. "The Effect of Crystalline Morphology on Enzymatic Degradation Kinetics," *Proc. ACS Div. Polym. Mat. Sci. Eng.*, 66:408–409.

91. Dickinson, H. R. and A. Hiltner. 1981. "Biodegradation of a Poly(α-amino Acid) Hydrogel. II. *In Vitro*," *J. Biomed. Mater. Res.*, 15:591–603.

92. Williams, D. F. 1981. "Enzymatic Hydrolysis of Polylactic Acid," *Eng. in Med.*, 10:5–7.

93. Fukuzaki, H., M. Yoshida, M. Asano and M. Kumakura. 1989. "Synthesis of Copoly(D,L-lactic Acid) with Relatively Low Molecular Weight and *in vitro* Degradation," *Eur. Polym. J.*, 25:1019–1026.

94. Miller, G. W. 1964. "Degradation of Synthetic Polypeptides. III. Degradation of Poly-α-lysin by Proteolytic Enzymes in 0.20 M Sodium Chloride," *J. Am. Chem. Soc.*, 86:3918–3922.

95. Thombre, A. G. and J. R. Cardinal. 1990. "Biosynthesis of Drugs," in *Encyclopedia of Pharmaceutical Technology, Vol. 2*, J. Swarbrick and J. C. Boylan, eds., New York, NY: Marcel Dekker, Inc., pp. 61–88.

96. Peanasky, J. S., J. M. Long and R. P. Wool. 1991. "Percolation Effects in Degradable Polyethylene-Starch Blends," *J. Polymer Sci.: Part B: Polymer Physics*, 29:565–579.

97. Gonsalves, K. E., S. H. Patel and X. Chen. 1990. "Development of Potentially Degradable Materials for Marine Applications – I. Polyethylene-Starch Blends," *New Polymeric Mater.*, 2:175–189.

98. Glass , J. E. and G. Swift, eds. 1990. *Agricultural and Synthetic Polymers. Biodegradability and Utilization*. Washington, DC: American Chemical Society, Chapter 8.

99. Kawai, F. 1990. "Biodegradation of Polyethers," Chapter 10 in *Agricultural and*

Synthetic Polymers. Biodegradability and Utilization, J. E. Glass and G. Swift, eds., Washington, DC: American Chemical Society.

100. Suzuki, J., K. Hukushima and S. Suzuki. 1978. "Effect of Ozone Treatment upon Biodegradability of Water-Soluble Polymers," *Environ., Sci. Technol.*, 12: 1180–1183.
101. Robinson, B. V., F. M. Sullivan, J. F. Borzelleca and S. L. Schwartz. 1990. *PVP. A Critical Review of the Kinetics and Toxicology of Polyvinylpyrrolidone (Povidone)*. Chelsea, MI: Lewis Publishers, Inc., Chapter 5.
102. Allcock, H. R. 1992. "Rational Design and Synthesis of New Polymeric Materials," *Science*, 255:1106–1112.

CHAPTER 3

Types of Biodegradable Hydrogels

The degradation of hydrogels can be classified into at least three different categories as schematically described in Figure 3.1 [1,2]. First, the polymer backbone chain can be degraded by either hydrolysis or enzymatic degradation. Second, the crosslinking agent can be degraded while the polymer backbone may remain intact. Finally, the pendant groups attached to the polymer backbone may be cleaved, while the polymer backbone and the crosslinking agent remain unchanged. Of course, hydrogels can be degraded by more than one mechanism. The degradation of either backbone chains or crosslinkers or both will result in solubilization of the hydrogels. The cleavage of the pendant groups from the backbone chains, however, may or may not result in solubilization of the hydrogel. The hydrogel may become water-soluble upon release of the pendant groups, if the polymer chains constituting the hydrogel are not chemically crosslinked and the removal of the pendant groups increases the hydrophilicity of the polymer backbone chains.

3.1 HYDROGELS WITH DEGRADABLE POLYMER BACKBONE

In this type of hydrogel the bond cleavage occurs in the polymer backbone chains. Thus, the resultant degraded products are low molecular weight, water-soluble fragments. These types of hydrogels are most desirable, since they have a greater chance to be bioresorbed and eliminated from the body.

3.1.1 Crosslinked Hydrogels

3.1.1.1 HYDROGELS MADE OF NATURAL POLYMERS AND SYNTHETIC POLY(α-AMINO ACIDS)

It is well known that natural polymers, proteins and polysaccharides, are degraded by various enzymes as well as by hydrolysis. They break down

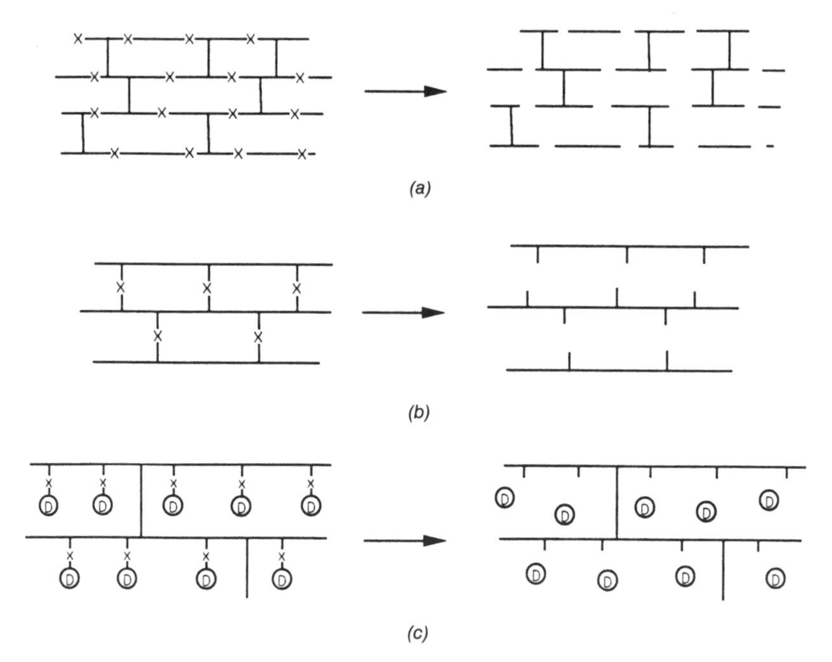

FIGURE 3.1. Hydrogel degradation by cleavage of polymer backbone (a), crosslinking agent (b), and pendant group (c) (from References [1] and [2]).

into smaller molecular weight products. Thus, hydrogels made of natural polymers undergo degradation by cleavage in the backbone chains. Water-soluble, natural polymers can form gels by crosslinking individual polymer molecules. The choice of crosslinking agents depends on the functional groups available on the polymer chains. The methods involved in preparing chemically crosslinked hydrogels from water-soluble polymers are described in Chapter 4.

Polysaccharides containing carboxyl groups, such as chondroitin sulfate and hyaluronic acid, were derivatized with cysteine methyl ester by substituting carboxyl groups of the polysaccharides. The cysteine-derivatized polysaccharides were crosslinked by mild oxidation to form hydrogels as shown in Figure 3.2. Hydrogels with various physical properties were prepared using this method [3].

In addition to naturally occurring proteins, synthetic poly(α-amino acids) have been used to prepare biodegradable hydrogels. Crosslinked poly(2-hydroxyethyl-L-glutamine) (PHEG) was prepared from poly(γ-benzyl-L-glutamate) by treating it with a mixture of ethanolamine and the crosslinker, 1,12-diaminododecane [4]. This reaction is described in Figure 3.3. The hydrogels degraded by a bulk degradation mechanism, when implanted subcutaneously in rats and in the peritoneal cavity of mice. There was an increase in the swelling ratio and a corresponding decrease in the overall weight of the implanted hydrogels. They were also degradable *in vitro* by general proteolytic enzymes such as pronase or papain. Since papain is similar in specificity to cathepsin B, it was thought that an enzyme present in the extracellular space during the inflammatory response contributed to the *in vivo* degradation of the implanted hydrogels [5].

The enzyme-catalyzed degradation of poly[N^5-(2-hydroxyethyl)-L-glutamine] (PHEG) and its copolymers was also studied by Pytela et al. [6]. While papain, pepsin, pronase, and cathepsin B were able to degrade PHEG, chymotrypsin and elastase could not. The latter two enzymes, however, degraded the polymer, if the hydrophobic amino acid was copolymerized. Phenylalanine was most effective in increasing the hydrolysis by chymotrypsin or elastase.

3.1.1.1.1 Interpenetrating Hydrogel Network (IPN)

An interpenetrating hydrogel network (IPN) is any material containing two polymers, each in network form as shown in Figure 3.4 [7]. Semi-IPN is an IPN which consists of one crosslinked and one uncrosslinked polymer. IPN is an excellent way to enhance the compatibility of polymer components and prevent phase separation. IPN hydrogels also allow access to properties that may be hybrids of those of the component macromolecules [8]. For example, pH-sensitive hydrogels can also be made temperature-sensitive

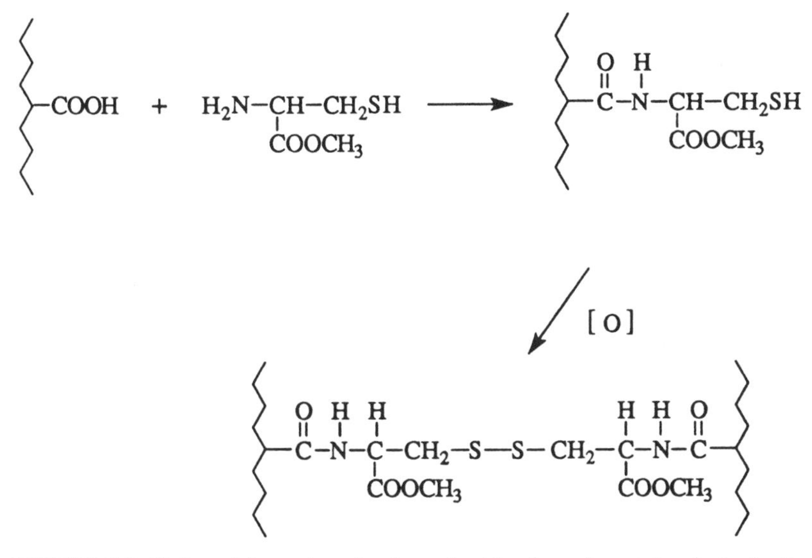

FIGURE 3.2. Hydrogel formation of polysaccharides through cysteine-derivatization (from Reference [3]).

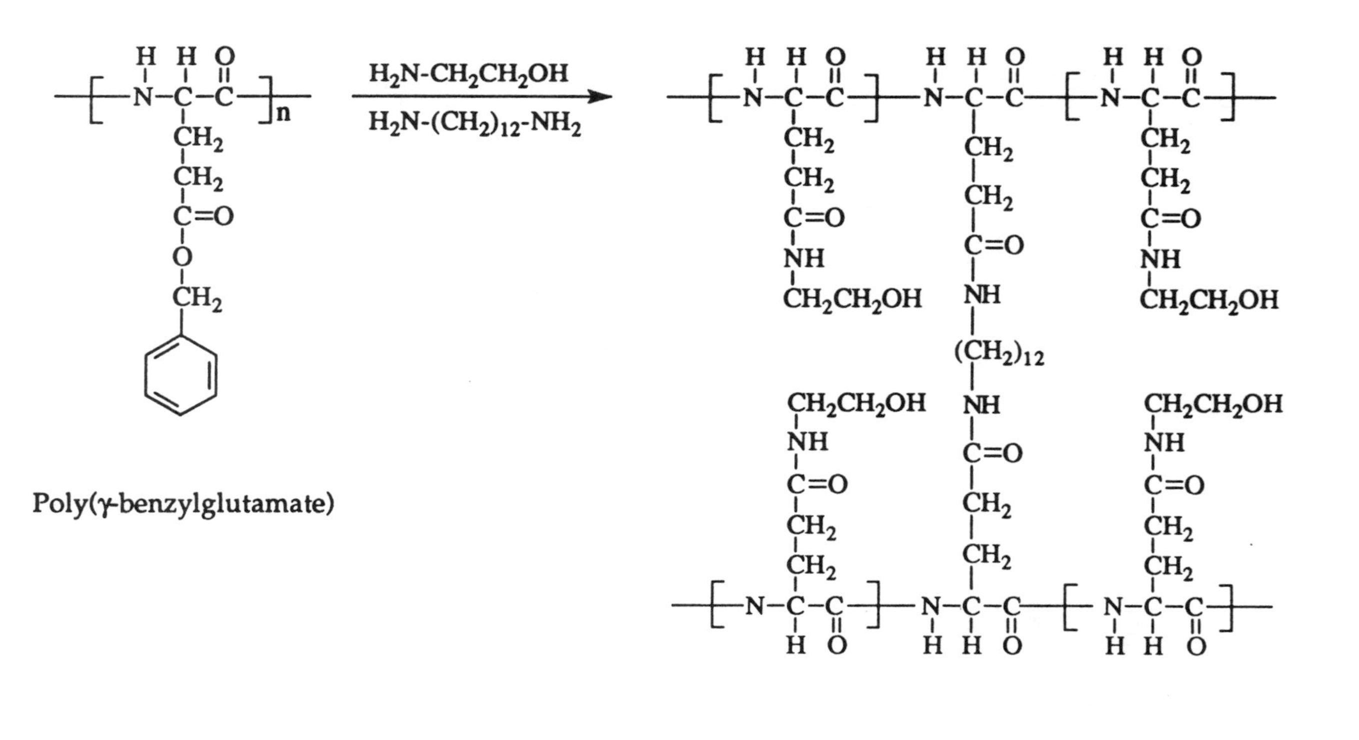

FIGURE 3.3. *Formation of poly(2-hydroxyethyl-L-glutamine) hydrogels.*

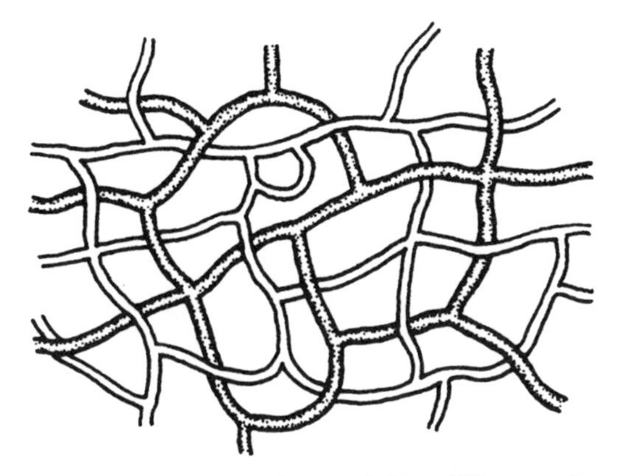

FIGURE 3.4. Interpenetrating hydrogel network. Two different polymer networks are interpenetrating each other.

by using a temperature-sensitive polymer as the second polymer. Mechanical strength of the hydrogel can be increased by using a relatively hydrophobic polymer as the second polymer. Furthermore, one polymer network or both networks of IPN can be made biodegradable.

Hydrogels of natural polymers can be crosslinked in the presence of a network of a synthetic polymer to form an interpenetrating hydrogel network. Acrylamide was copolymerized with *N,N'*-methylenebisacrylamide in the presence of dissolved gelatin. The gelatin was subsequently crosslinked with glutaraldehyde to form an interpenetrating hydrogel network [9,10]. Semi-IPN was prepared by polymerizing hydroxyethylmethacrylate (HEMA) monomers in the presence of 2 percent collagen dissolved in 1 mM HCl [11]. Alternatively, proteins or polysaccharides can be crosslinked in the presence of water-soluble, synthetic polymers to prepare semi-IPN. The advantage of a two component hydrogel matrix is that it can perform like synthetic polymers and yet degrade like natural polymers. This technique can be applied practically to all proteins and polysaccharides.

3.1.1.2 POLYESTER HYDROGELS

Heller and coworkers synthesized a series of biodegradable hydrogels from polyester prepolymers containing double bonds and the cleavable linkages in the backbone [12]. The prepolymers were prepared via condensation reactions between an unsaturated diacid, such as fumaric acid, itaconic acid, or allylmalonic acid, and a diol such as low molecular weight poly(ethylene glycol). The prepolymers were crosslinked by copolymerization with acrylamide or vinylpyrrolidone as shown in Figure 3.5.

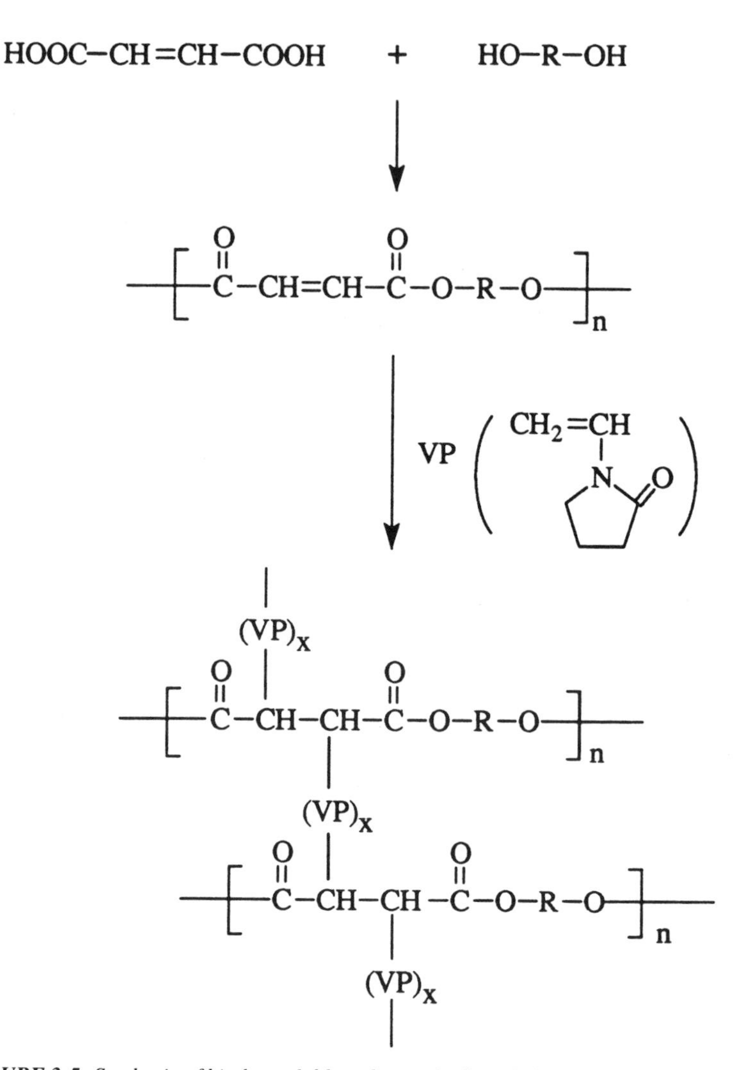

FIGURE 3.5. Synthesis of biodegradable polyester hydrogels by copolymerization with N-*vinylpyrrolidone (VP). Polyester prepolymers were synthesized from fumaric acid and poly(ethylene glycol) (from Reference [12]).*

To enhance the hydrolysis of aliphatic polyesters at pH 7.4 and 37°C, diacids, such as diglycolic acid, ketomalonic acid, or ketoglutaric acid, were added to the backbone structure. The degradation rate of the polyester hydrogels was controlled by varying the crosslinking density and the type of dicarboxylic acid. Copolymerization in the presence of bovine serum albumin resulted in hydrogels with physically-entrapped albumin. The diffusional release of albumin was minimal and the sustained release of albumin took place only after the matrix was eroded by hydrolysis of ester links. The release of albumin was faster when ketomalonic acid was used. Hydrolysis of ester links resulted in chain cleavage which generated poly(ethylene glycol) and PVP modified by vicinal carboxylic acid functions.

Heller and coworkers also synthesized biodegradable hydrogels from polyester prepolymers containing double bonds in the pendant groups [12]. Pendant unsaturation was provided by either itaconic acid or allylmalonic acid. These prepolymers were crosslinked by free radical initiation without the use of water-soluble vinyl monomers (Figure 3.6). Thus, degradation produced PEG and crosslinked diacids. The erosion rates of these hydrogels varied depending on the extent of the backbone unsaturation of the polyester prepolymers [13].

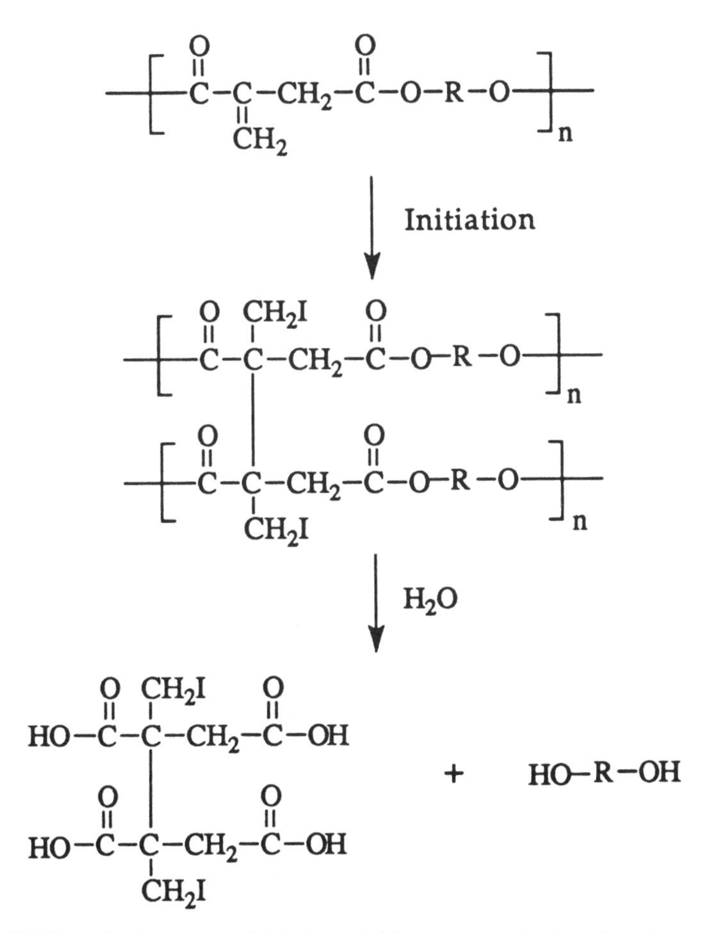

FIGURE 3.6. *Synthesis of biodegradable polyester hydrogels using polyester prepolymers containing double bonds in the pendant chain. Polyester prepolymers were synthesized from itaconic acid and poly(ethylene glycol) (from Reference [12]).*

3.1.1.3 POLYCAPROLACTONE IPN

Polycaprolactone was incorporated into poly(2-hydroxyethyl methacrylate) (PHEMA) hydrogels. Polycaprolactone in the hydrogel biodegrades to result in voids in the network where natural tissue can grow [14]. Typically, polycaprolactone was dissolved in HEMA monomers at 60−70°C and the HEMA monomers were polymerized to make semi-IPN [15]. For the preparation of IPN, entrapped polycaprolactone was extracted from the semi-IPN and modified polycaprolactone was incorporated into the PHEMA by swelling the dried PHEMA in the concentrated modified polycaprolactone. The end OH-groups of polycaprolactone reacted with itaconic anhydride to introduce double bonds. The itaconic acid-endcapped polycaprolactone in the PHEMA gel was polymerized to crosslink the polycaprolactone network [14]. The IPN of crosslinked PHEMA and polycaprolactone are expected to have improved temperature stability and mechanical properties. This approach should be useful in the preparation of hydrogels from water-insoluble biodegradable polymers such as those listed in Table 2.2.

3.1.2 Noncrosslinked Hydrogels

3.1.2.1 PHYSICAL GELS OF PROTEINS AND POLYSACCHARIDES

Some polymers form three-dimensional networks even in the absence of chemical crosslinking. These are called physical gels. Many proteins and polysaccharides are able to form physical gels, which are usually mechanically weak. To enhance the mechanical strength, physical gels can be further modified by chemical crosslinking. Chapter 5 describes the physical gels of the natural polymers in detail.

3.1.2.2 BLOCK COPOLYMER HYDROGELS

One way of improving the mechanical properties of hydrogels is to use segmented block copolymers. Copolymers offer flexibility in controlling physical and chemical properties including biodegradability [16]. The multicomponent nature of the system affords versatility in terms of both biodegradation and mechanical properties. A number of biodegradable, water-insoluble polymers can be copolymerized with hydrophilic polymers to form copolymers which absorb substantial amounts of water but do not dissolve in water.

Cohn and Younes synthesized a series of block copolymers comprising poly(ethylene oxide) (PEO) and poly(lactic acid) (PLA) segments through polyesterification of lactic acid in the presence of PEO chains [17]. The flexible PEO chains provide elasticity to the system while biodegradable PLA provides mechanical strength. The structure of the PEO-PLA segmented copolymer is shown in Figure 3.7. The uptake of water by PEO-PLA copolymer matrices depends on the relative concentration of ethylene oxide rather than the chain length of the PEO segment. The equilibrium water contents of PEO-PLA matrices containing 32 and 43 molar percentage of ethylene oxide were 31 percent and 55 percent, respectively. The poly(lactic acid) matrix attained the equilibrium water content of 6 percent. The substantial increase in water absorption is due to the hydrophilic nature of PEO. When PEO and PLA were physically mixed by dissolving and casting them from a chloroform solution, PEO chains were easily separated from PLA and extracted in water.

Zhu et al. [18] synthesized three- and four-arm star PEO-PLA copolymers (s-PEO-PLA) to make a molecular scale microcapsule termed a "super microcapsule." The phase separation behavior of s-PEO-PLA copolymers is known to be different from that of linear PEO-PLA copolymers with comparable block length. The s-PEO-PLA copolymers were prepared with an expectation that the star molecule may form a "capsule-like" structure under proper conditions in which hydrophilic blocks are aggregated as the inner core and hydrophobic segments as the outer shell. It is expected that the s-PEO-PLA copolymers can be used for a variety of applications.

Esterurethane prepolymer was prepared by polymerizing D,L-lactide or copolymerizing D,L-lactide-ϵ-caprolactone using poly(ethylene glycol) (PEG) as an initiator, and then reacting with excess diisocyanate [19]. This prepolymer can be cured in the presence of water and the cured polymer is biodegradable. A significant amount of water can be absorbed in the cured polymer depending on the amount of PEG in the polymer.

PEO-poly(ethylene terephthalate) (PEO-PET) copolymers were synthesized by polycondensation of diethylene glycol terephthalate and poly(ethylene glycol) [20–22]. The molecular weight of PEO segments was either 600 or 1,300 and the PEO composition in the copolymer ranged from 50 to 70 wt percent. As the PEO content increased from 50 to 70 wt percent, the water uptake increased from 69 percent to 137 percent of the dry weight and the mechanical integrity of the sample decreased due to the reduction in crystallinity of the PET segments. The ester linkage between PET and PEO was most vulnerable to hydrolysis. Thus, increasing the content of PEO also increases the rate of degradation both *in vitro* and *in vivo* [21,22]. The rate of hydrolysis increased dramatically as the pH increased from 7 to 9 [22]. Under slightly acidic conditions (pH 5), however, PET-PEO copolymers were more resistant to hydrolysis.

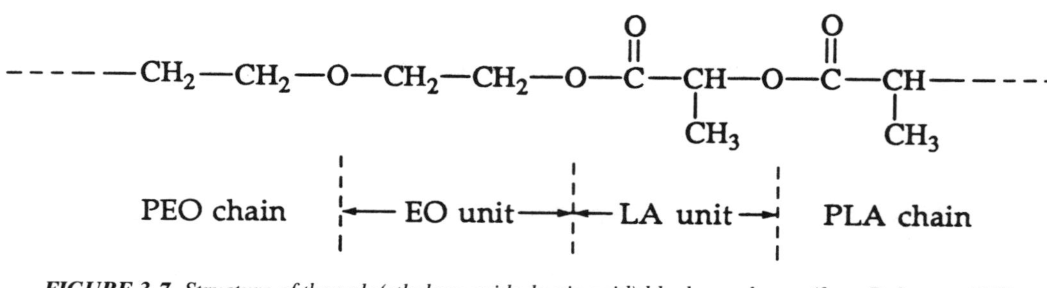

FIGURE 3.7. Structure of the poly(ethylene oxide-lactic acid) block copolymer (from Reference [17]).

3.1.2.3 COPOLY(AMINO ACID) HYDROGELS

As described in Chapter 2, Section 2.2.2, hydrogels can be prepared from copolymers of hydrophilic and hydrophobic amino acids. L-Glutamic acid or L-aspartic acid is commonly used as a hydrophilic amino acid, while hydrophobic amino acid can be L-leucine or the esters of hydrophilic amino acids, such as methyl glutamate, ethyl glutamate, methyl aspartate, or benzyl aspartate [23–26]. A wide range of water-uptake can be achieved by varying the relative proportions of the hydrophilic and hydrophobic components in the copolymer. Biodegradability usually increases with the content of hydrophilic amino acid which is related to the overall polymer hydrophilicity.

The enzymatic degradation of copolymers of *N*-hydroxyalkyl-L-glutamine and γ-methyl-L-glutamate was examined by Hayashi and Ikada [27]. As the content of hydroxyalkyl glutamine increased, the copolymers showed increased swelling and reduced tensile strength. The increase in swelling resulted in a faster rate of pronase E-catalyzed hydrolysis. The degradation of these copolymer fibers occurred through surface hydrolysis.

3.1.2.4 POLYELECTROLYTE COMPLEXES

Ionic interactions between two oppositely charged polyelectrolytes lead to the formation of a polysalt or polyelectrolyte complex which has properties quite different from the individual components [28]. Polyelectrolyte complexes are hydrogels which may be clear, rigid solids, or opaque, sponge materials [29]. Polyelectrolyte complexes are separated from solution either as flocculated precipitates which can form membranes or as complex coacervates [30]. The complex coacervate is a polymer-rich liquid phase which separates from a solution. The two polymers are collected together in one phase while the other phase consists almost entirely of solvent [31]. The simple coacervate is concerned with nonionized polymers. The factors which govern the formation of a polyelectrolyte complex are charge type and charge density of polyelectrolytes, ionic strength and the valency of ions, pH, and temperature. The polyelectrolyte complexes are different from ionomers and ion exchange resins. Ionomers contain only one type of polyelectrolyte with only a slight ionic functionality and ion exchange resins have covalent crosslinking which makes them water-insoluble and nonthermoformable [32].

While the individual polyelectrolytes are soluble in water, complexes of strong polyelectrolytes are usually not. They can maintain fairly high mechanical strength in the water-swollen state. Nonstoichiometric polyelectrolyte complexes can remain water-soluble at a certain pH and ionic strength [33]. The water-insoluble polyelectrolyte complexes can absorb a large amount of water and exhibit properties of hydrogels [28]. The polyelectrolyte complex with an equimolar composition is neutral while that with a nonequimolar composition is pH-sensitive. Thus, they may become soluble in an acid or alkaline aqueous solution. The morphology of polyelectrolyte complexes may be described by a ladder model or a "scrambled-egg" model [34]. The difference between polyelectrolyte complexes and other pH-sensitive hydrogels is that the degree of swelling of polyelectrolyte complexes increases as the ionic strength increases [35]. As the ionic strength increases further, the gel will dissolve due to the interruption of the electrostatic interactions of the complexes. The extent of water absorption depends on the excess charge in the complex.

Polyelectrolyte complexes can be formed by adding a dilute (<0.1 percent) aqueous solution of a polycation to an equally dilute solution of a polyanion or adding simultaneously two concentrated aqueous polyelectrolyte solutions to a large mass of water. Alternatively, each polyelectrolyte is dissolved in a ternary mixture for the complex, and the two solutions are combined. Then the solution is drowned in water to precipitate the polyelectrolytes and extract the remaining salts and other impurities [32]. The ternary mixtures that dissolve polyelectrolyte complexes consist of water, electrolyte, and an organic component [29]. Examples of ternary solvents are water-acetone-sulfuric acid, water-ethanol-hydrochloric acid, water-dioxane-calcium chloride, and water-acetone-sodium bromide [29]. A solution of the polyelectrolyte in a ternary solvent is used for fabrication of useful structures such as films or coatings. Films of polyelectrolyte complexes can be made by solution casting, followed by evaporation and leaching in water.

A classical example of a polyelectrolyte complex is the complex formed between gelatin and gum arabic [30]. Gum arabic is a polysaccharide which carries D-glucuronic acid residues, and gelatin is a protein which is positively charged below the isoelectric point, e.g., 4.8 [30]. The strongest coacervate will be formed when the charge densities of the two polyelectrolytes are equivalent. Thus, the coacervate formation is strongly dependent on pH and is readily reversible simply by changing the pH of the solution.

Recently, polyelectrolyte complexes have been used in the preparation of microcapsules [36,37]. Gelatin-acacia microcapsules were formed by the complex coacervation of acid gelatin and acacia at pH 4.1 [38]. Chitosan-alginate coacervate capsules were prepared by spraying sodium alginate solution into the chitosan solution. The formed chitosan-alginate capsules were mechanically strong and stable in a wide pH range [39]. Chitosan was also complexed with poly(acrylic acid) [40,41]. In this system, the pH had to be in the range of 2 to 6, since chitosan, a polycation, is insoluble at pH values higher than 6, and poly(acrylic acid) does not have a charge density sufficiently high to form a complex with chitosan at pH values below 2. The permeability of the polyelectrolyte complex membranes to liquid can be adjusted via the component polyelectrolytes and the conditions of complex preparation [34].

3.1.2.5 POLYMER BLENDS

Mixing two different types of polymers, either one or both of them are biodegradable and one of them is water-insoluble, will provide a polymer blend which has unique properties. Pitt et al. prepared blends of poly(vinyl alcohol) (PVA) and poly(glycolic acid-*co*-lactic acid) (PGLA) by dissolving different proportions of PVA and PGLA in hexafluoroisopropanol. Miscibility was observed when the PVA content was 70 (w/w) percent or higher. The water content of the blends increased from 30 to 80 percent as the PVA content increased from 25 to 95 percent [42]. The permeability of naltrexone, naltrexone HCl, cytochrome C, myoglobin, and somatotropin was proportional to the water content of the blends and it increased monotonically with the PVA content.

Many biodegradable, water-insoluble polymers can be blended with hydrophilic polymers to form hydrogels. The extent of water absorption will be determined mainly by the amount of the hydrophilic polymers in the blend. There will be an upper limit to which hydrophilic polymers can be added. Addition of too much hydrophilic polymer will not maintain gel structures. The degradation of the biodegradable component will reduce the mechanical structures of the polymer blends and cause them to fragment. One disadvantage of preparing polymer blends is using organic solvents which dissolve both hydrophilic and hydrophobic polymers.

3.2 HYDROGELS WITH DEGRADABLE CROSSLINKING AGENTS

In this type of hydrogel, water-soluble polymers are crosslinked with biodegradable crosslinking agents. Since the polymer backbone chains remain intact while the crosslinks are degraded, the final degradation products are high molecular weight, water-soluble polymers. The hydrogels swell in water and the space between polymer chains becomes large. Thus, the release of low molecular weight drugs from the hydrogel will not be hindered by the presence of polymer networks, but depends primarily on the solubility of the drugs. Drugs with appreciable water solubility will be released quite rapidly [43]. Thus, this type of hydrogel is useful in the delivery of drugs with low water solubility or high molecular weight drugs such as peptide and protein drugs [44].

3.2.1 Hydrogels Crosslinked with Small Molecules

3.2.1.1 HYDROGELS CROSSLINKED WITH N,N′*-METHYLENEBISACRYLAMIDE*

It was observed that the methylene group in *N,N′*-methylenebisacrylamide (BIS) underwent degradation by hydrolysis to produce formaldehyde as long as the concentration of the crosslinking agent was low (Figure 3.8) [45]. With 0.3 to 0.6 percent crosslinking agent, degradation occurred over two to ten days and the degradability was controlled by the crosslinking density. At concentrations higher than 1 percent, there was essentially no degradation, i.e., it took so long to degrade by our time scale that there appeared to be no degradation.

The BIS-crosslinked hydrogels were also used to deliver insulin in diabetic rats [46]. Ten milligrams of insulin was incorporated into the 40 percent acrylamide gel crosslinked with 2 percent BIS. The steady release of insulin for up to 20 days was observed. It was reported that the rats showed a normal growth rate and normal (nonglycosuria) urine. Since the concentration of BIS was 2 percent, hydrogels were not degraded during implantation in rats. Heller et al. also used this approach to deliver albumin from hydrogels crosslinked with lower concentrations of BIS [12].

The BIS crosslinker is also known to be degraded by chymotrypsin. In order to control the degradation of the BIS-crosslinked hydrogels, enzymes were either physically entangled in the hydrogel or chemically bound to the polymer backbone [45].

3.2.1.2 HYDROGELS CROSSLINKED WITH AZO REAGENTS

It is known that aromatic azo bonds are cleaved by azoreductase of bacteria which are present predominantly in the colon [47]. Thus, if hydrogels are crosslinked with the crosslinking agents containing aromatic azo groups, they are expected to degrade in the colon. Saffran et al. used this approach to deliver insulin by oral administration [48]. Insulin was coated with hydrogels which contain aromatic azo groups in crosslinks. The hydrogels crosslinked with azo bonds protected insulin from destruction in the stomach and released it in the lower bowel [49].

Recently, this technique was adapted by Kopeček et al. to develop colon-specific drug delivery systems [50,51]. They prepared fucosylamine-containing poly[*N*-(2-hydroxypropyl)methacrylamide] (PHPMA) hydrogels with azoaromatic crosslinks which are degradable by azoreductases. The enzymatic degradation of the crosslinks can be controlled by adjusting the hydrophilicity of the polymer backbone and the crosslinking agent. The increase in the hydrophilicity of the polymer chain will result in increased

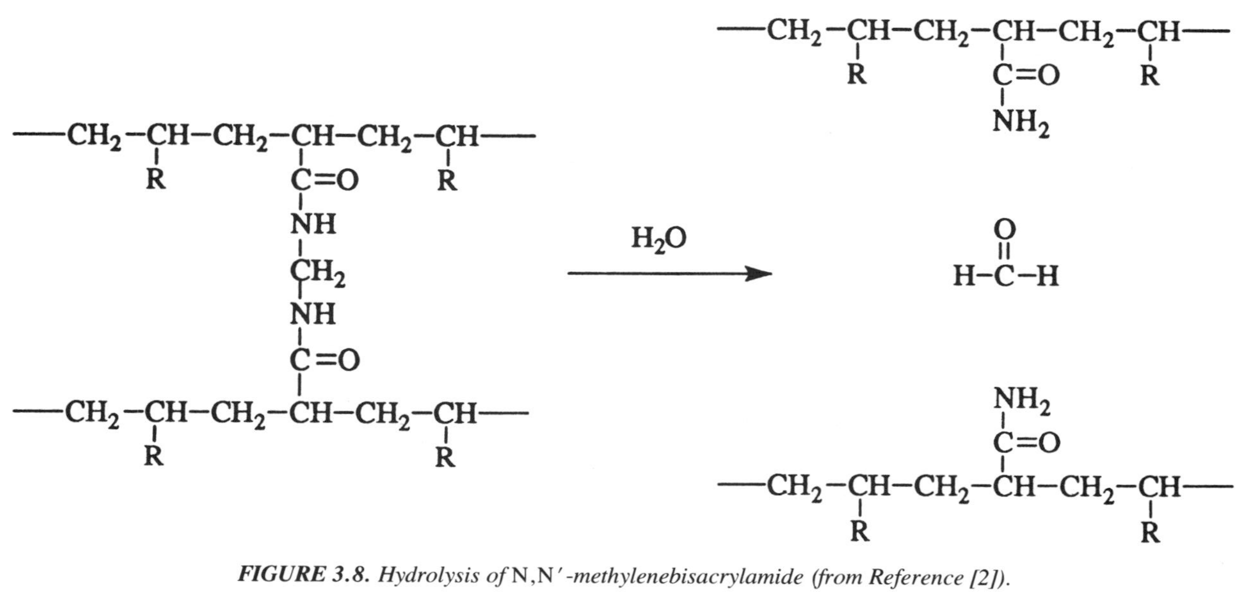

FIGURE 3.8. *Hydrolysis of* N,N′*-methylenebisacrylamide (from Reference [2]).*

swelling which in turn makes the crosslinks accessible to enzymes and mediators [51]. The increase in hydrophilicity of the crosslinking agent will enhance the reactivity of the azo bonds by bacterial enzymes in the aqueous environment. The hydrophilicity of the crosslinking agent can be increased by adding hydrophilic moiety such as azosulfonamide.

Some crosslinking agents with azo bonds are listed in Table 3.1. Crosslinking agents with divinyl groups (#1 – 4 in Table 3.1) can be used in the crosslinking copolymerization of vinyl monomers. On the other hand, crosslinking agents with amine groups at both ends (#5 in Table 3.1) can be used to chemically crosslink water-soluble polymers which have functional groups. The chemistry of forming crosslinking by diamines is described in Section 3.2.2.

3.2.1.3 HYDROGELS CROSSLINKED WITH SUCROSE

Gruber and Greber [52,53] explored the possibility of utilizing sucrose as a naturally renewable raw material. They synthesized various sucrose derivatives which have reactive functional groups other than hydroxyl groups. Polymerizable sucrose derivatives were prepared by reacting sucrose with methacrylic acid chloride, methacrylic acid anhydride, or methyl methacrylate. The average degree of substitution was regulated by the molar ratio of sucrose to the methacrylic acid derivative. The reaction always resulted in mixtures of monoesters and high substituted esters. Since the preparative separation of the components could be achieved only by tedious chromatographic methods, sucrose-methacrylate (SM) mixtures were used without purification.

Sucrose-crosslinked hydrogels were obtained by radical polymerization of the SM mixtures as described in Figure 3.9. The actual network of the formed sucrose gel may be much more complex than that shown in Figure 3.9, since some sucrose molecules may have three or more substituted esters. The resulting gels are claimed to be hydrolytically stable in the pH range of 1 – 10 and resistant to biochemical degradation. It is known, however, that lipase is able to hydrolyze esters of sucrose derivatives as long as the number of ester groups is three or less [54]. Thus, it is expected that sucrose with a low degree of substitution will undergo enzymatic degradation by either esterase and/or invertase (β-D-fructofuranosidase) as well as by chemical hydrolysis at a low pH range. Not only sucrose but also many other carbohydrates may be used as degradable crosslinking agents.

3.2.2 Hydrogels Crosslinked with Oligopeptides

It is quite rational to consider enzymatically degradable oligopeptides as crosslinking agents. Kopeček used this approach to prepare enzymatically

TABLE 3.1. Examples of Crosslinking Agents Containing Azo Bonds.

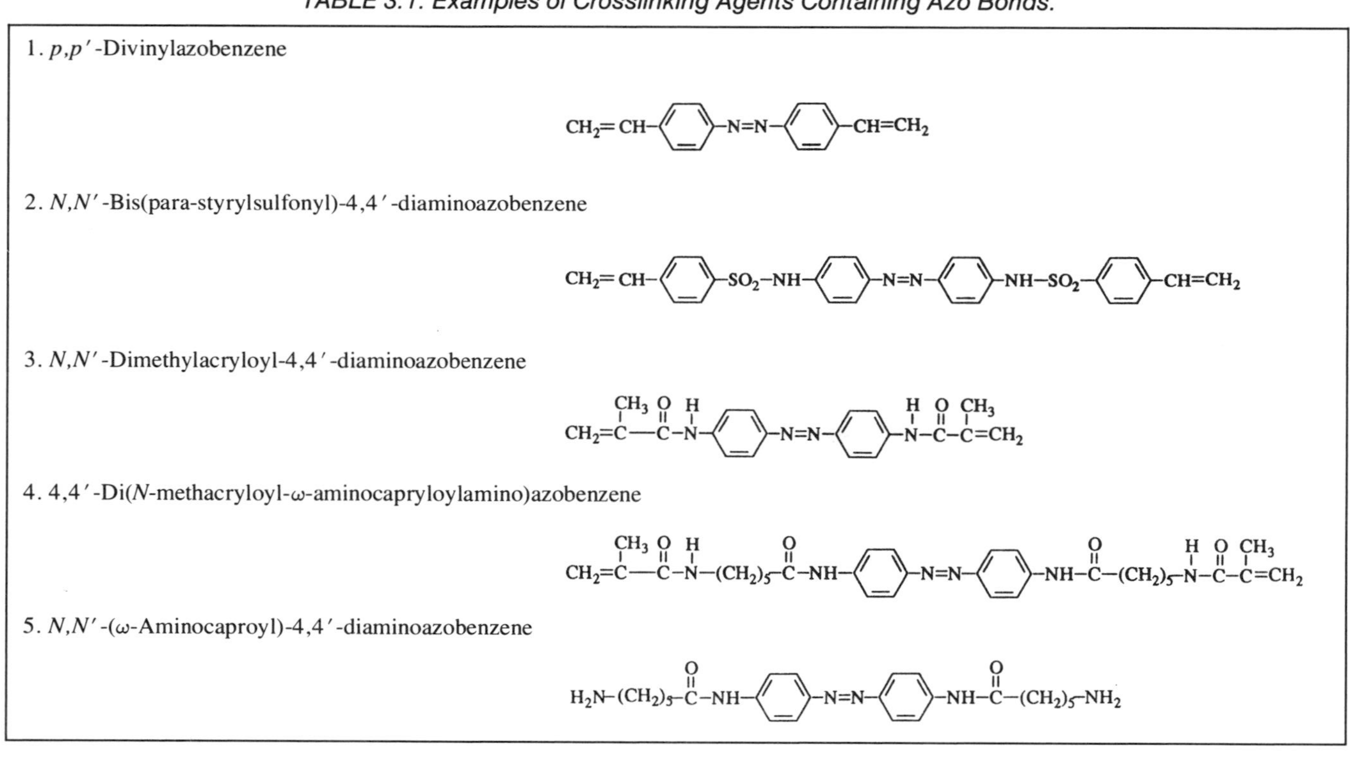

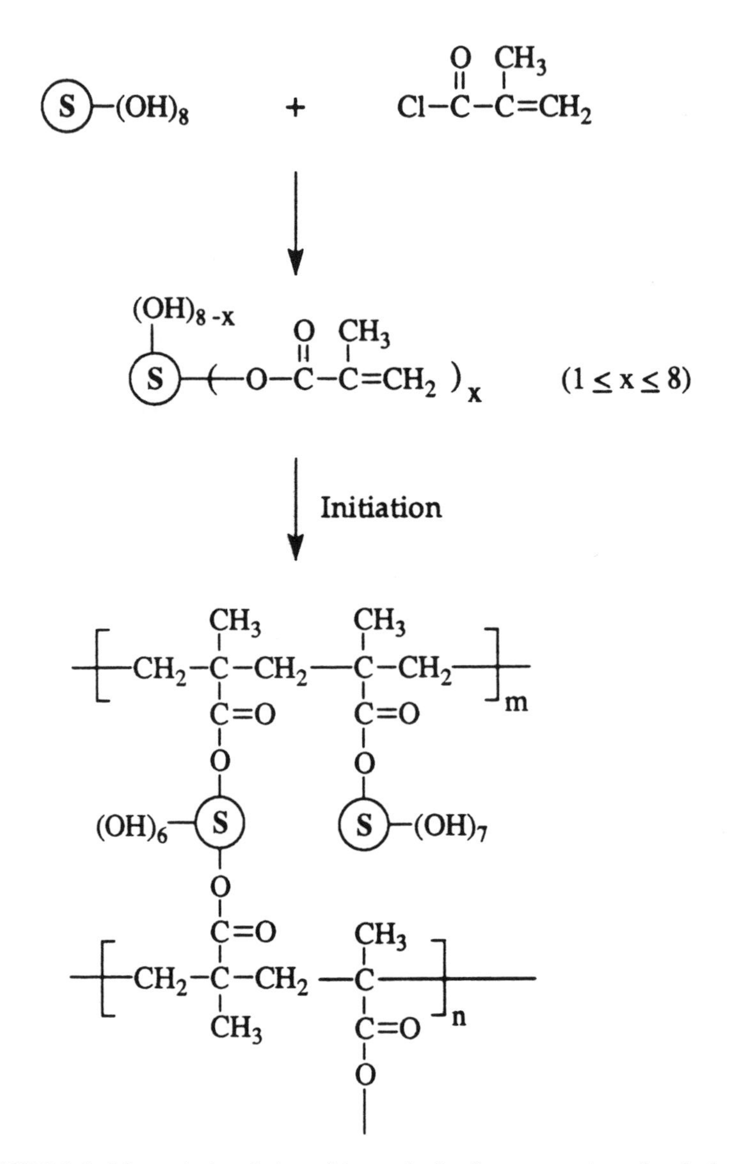

FIGURE 3.9. Schematic description of the synthesis of sucrose methacrylate hydrogels (from Reference [53]).

degradable hydrogels. Copolymers of *N*-(2-hydroxypropyl)methacrylamide (HPMA) and *p*-nitrophenyl esters of *N*-methacryloylated oligopeptides were crosslinked with diamines [55]. Figure 3.10 shows an example of the preparation of chymotrypsin-degradable hydrogels. A copolymer of HPMA and *N*-methacryloylglycylalanine *p*-nitrophenyl ester was crosslinked with *N,N'*-bis(H-D-Phe)hexamethylenediamine [56]. The *p*-nitrophenoxy group readily reacts with amine groups to form amide groups.

To study the relationship between the oligopeptide structure and its degradability with enzymes, Kopeček and coworkers varied the number and type of amino acids in the oligopeptide sequence of *N*-methacryloylated oligopeptides (R_1 in Figure 3.10) as well as in the diamines used for crosslinking (R_2 in Figure 3.10). Table 3.2 lists the degradability of hydrogels with different oligopeptide sequences [56]. Table 3.2 leads to a few important observations. First, the length of the oligopeptide sequence is most important in enzymatic degradation. Crosslinkers with only dipeptide sequences are not degraded by enzymes due to the high steric hindrance of the enzyme-substrate complex formation. Relatively bulky chymotrypsin molecules may not access the short oligopeptide chains linking the two polymer chains. The hydrolyzable linkage needs to have chain flexibility

TABLE 3.2. Structure of Oligopeptide Sequences and the Relative Degradability of Hydrogels as Expressed by the Percent of Cleaved Crosslinks after 48 Hours of Incubation with Chymotrypsin [54].

Oligopeptide Sequence		
R_1	R_2	% Degradation
Gly-Phe	H	0
Gly-Ala	Phe	16
Gly-Gly	Phe	22
Gly-Ile	Phe	37
Gly-Gly	Tyr	37
Gly-Val	Phe	45
Gly-Leu	Phe	56
Gly-Ile	Tyr	60
Gly-Phe	Phe	67
Gly-Val	Tyr	75
Gly-Phe-Tyr	H	82
Gly-D-Phe	Tyr	83
Ala-Val	Phe	87
Gly-Phe	Ala-Ala	44
Gly-Phe-Phe	Gly	60
Gly-Phe-Tyr	Gly	74
Gly-Phe-Phe	Gly	82
Gly-Gly-Phe	Phe	100
Gly-Gly-Phe	Tyr	100
Ala-Gly-Val	Phe	100

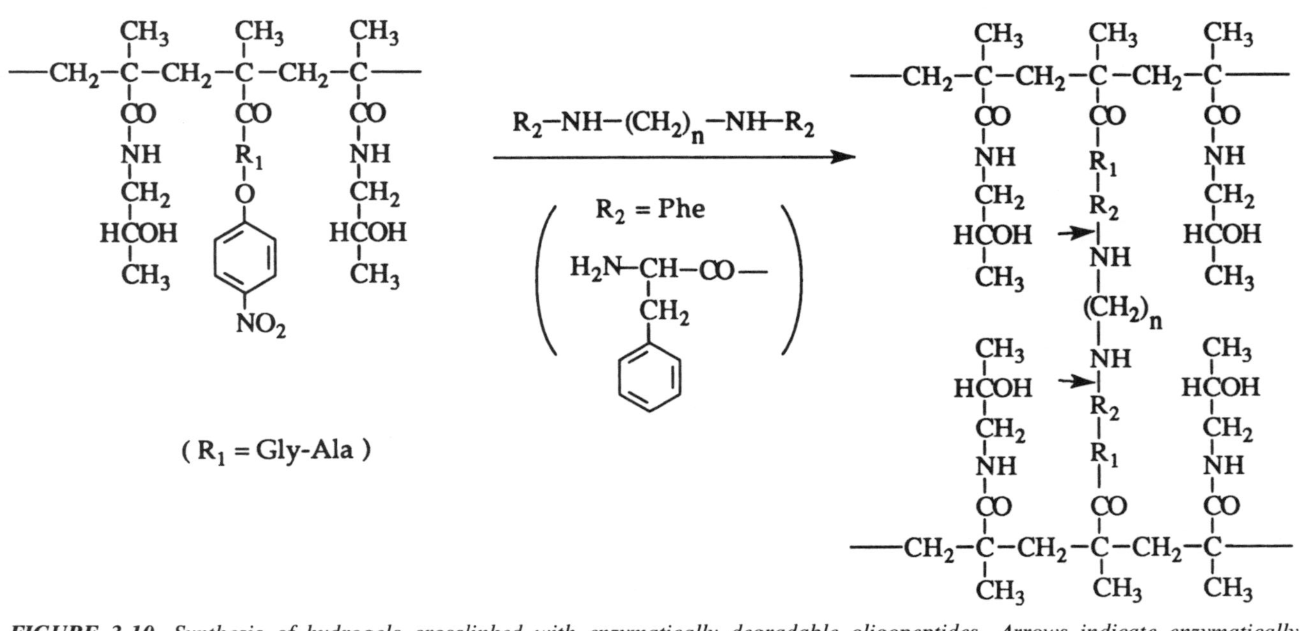

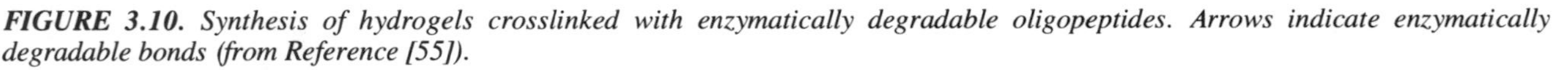

FIGURE 3.10. *Synthesis of hydrogels crosslinked with enzymatically degradable oligopeptides. Arrows indicate enzymatically degradable bonds (from Reference [55]).*

for enzymatic hydrolysis. Extension of the oligopeptide sequence from two to three or four amino acid residues resulted in a significant decrease in the steric effect of the polymer chain. In general, as the oligopeptide chain length increases, the steric inhibition is reduced and this results in a higher percentage of degradation. Second, the differences in degradability with the same oligopeptide chain length are due to structural differences in oligopeptide sequences. Tyr in position P_1 (i.e., the degradable site, see Chapter 7) increases the degradability compared to Phe in the same position. The presence of amino acids with a bulky side chain in position P_2 (i.e., second amino acid from the degradation site) increases the degradation by chymotrypsin. The structure of the oligopeptide sequence mainly determines the specificity toward cleavage by a certain enzyme.

The degradation of crosslinkers by chymotrypsin is also affected by the volume fraction of the polymers which is related to the density of the polymer chains or the crosslinking density [57]. Table 3.3 compares the degradability of hydrogels, with different volume fractions, by chymotrypsin [57]. Table 3.3 clearly indicates that an increase in network density leads to reduced formation of enzyme-substrate complexes by steric hindrances. Gels with high crosslinking density usually undergo surface degradation, since rather large chymotrypsin cannot penetrate the hydrogel easily.

By altering the structure of the oligopeptide sequences, the crosslinkers can be degraded by other enzymes such as papain [58], trypsin [59], or lysozymal enzymes (cathepsin) [60]. Table 3.4 lists the oligopeptide crosslinks which are degraded by trypsin. The data clearly show again that the length and the type of amino acids of oligopeptidic sequences are critical for enzymatic degradation. For the dipeptide crosslinkers, steric hindrances

TABLE 3.3. Time Necessary for the Chymotrypsin-Induced Dissolution of Hydrogels (τ) with Different Structure of Oligopeptide Sequences and/or Polymer Volume Fractions (ϕ) [55].

Oligopeptide Sequence			
R_1	R_2	ϕ	τ
Gly-Gly	Tyr	0.020	4
		0.040	5.5
		0.050	12
		0.100	150
Ala-Gly-Phe	Tyr	0.032	3.5
		0.102	8
		0.122	9
		0.150	11

TABLE 3.4. Structure of Oligopeptide Sequences and the Relative Degradability of Hydrogels as Expressed by the Percent of Cleaved Crosslinks after 24 Hours of Incubation with Trypsin [57].

Oligopeptide Sequence		
R_1	R_2	% Degradation
Gly	Lys	14
Gly-D-Phe	Lys	55
Gly-Phe	Lys	79
Gly-Val	Lys	88
Gly-Gly	Lys	90
Gly-Leu	Lys	93
Gly-Gly-Val	Lys	94
Gly-Gly-Phe	Lys	98
Ala-Gly-Val	Lys	100

by the polymer chains significantly impeded the formation of the trypsin-substrate complexes. The degradation of tripeptide crosslinkers was as effective as that of tetrapeptide crosslinkers [59].

3.2.3 Hydrogels Crosslinked with Macromolecules

Macromolecules, such as proteins and polysaccharides, can also be used as crosslinkers in the preparation of hydrogels at least two different ways. First, macromolecules, possessing various functional groups, can react directly with other functional groups on the polymer chains and form crosslinks. This depends solely on chemical reactions between the two functional groups. Second, the functional groups of macromolecules can be further modified to introduce new functional groups such as vinyl groups. The vinyl groups of macromolecules then participate in the polymerization of vinyl monomers. Since macromolecules can have two or more vinyl groups, they can effectively function as a crosslinking agent.

3.2.3.1 CROSSLINKING OF POLYMER CHAINS BY INSULIN MOLECULES

Kopeček and Rejmanová prepared poly-*N*-(2-hydroxypropyl)methacrylamide crosslinked by insulin molecules [61]. The preparation of the insulin-crosslinked gels was essentially the same as described in Section 3.2.2. Copolymers of *N*-(2-hydroxypropyl)methacrylamide (HPMA) and *p*-nitrophenyl esters of *N*-methacryloylated ϵ-aminocaproyl L-leucine were crosslinked with insulin which has three amine groups [55]. The gel became completely water-soluble after incubation with chymotrypsin. The insulin-crosslinked gels were degraded much faster than those crosslinked with oligopeptides, since the size of the insulin was much larger than that of the oligopeptide chains. The authors concluded that polymer backbone chains did not affect the enzymatic degradation of the crosslinkers if the size of the crosslinkers was large.

3.2.3.2 POLYMERIZATION OF VINYL MONOMERS IN THE PRESENCE OF FUNCTIONALIZED MACROMOLECULES

Double bonds can be introduced into macromolecules by various chemical reactions (see Chapter 4). Due to the presence of numerous chemically active groups on macromolecules, the introduction of only one double bond to each macromolecule is rather tricky. Quite often more than one double bond are introduced, and the modified macromolecules tend to become bi- or polyfunctional agents. Thus, they usually function as a crosslinking agent in the polymerization of vinyl monomers and as a result a gel is formed. The functionalized natural macromolecules retain their property of enzymatic degradation. Thus, the crosslinking agents of hydrogels can be made degradable by enzymes.

3.2.3.2.1 Hydrogels Crosslinked with Albumin

Albumin was modified with glycidyl acrylate to introduce double bonds. The number of double bonds introduced can be easily controlled by the reaction time and the concentration of glycidyl acrylate [62]. The epoxide group of glycidyl acrylate preferentially reacts with amine groups of albumin. The extent of reaction can be quantitated by measuring the free amine groups remaining before and after the reaction using 2,4,6-trinitrobenzenesulfonic acid [63].

The modified albumin was used as a crosslinking agent in the polymerization of vinyl monomers such as acrylic acid, acrylamide, and vinylpyrrolidone (Figure 3.11). The extent of albumin modification had a direct influence on the crosslinking ability of the modified albumin. When double bonds were introduced to less than 10 percent of the total amine groups in albumin, the modified albumin was not effective as a crosslinking agent, i.e., gel was not formed [62]. Ten percent of the total amine groups in

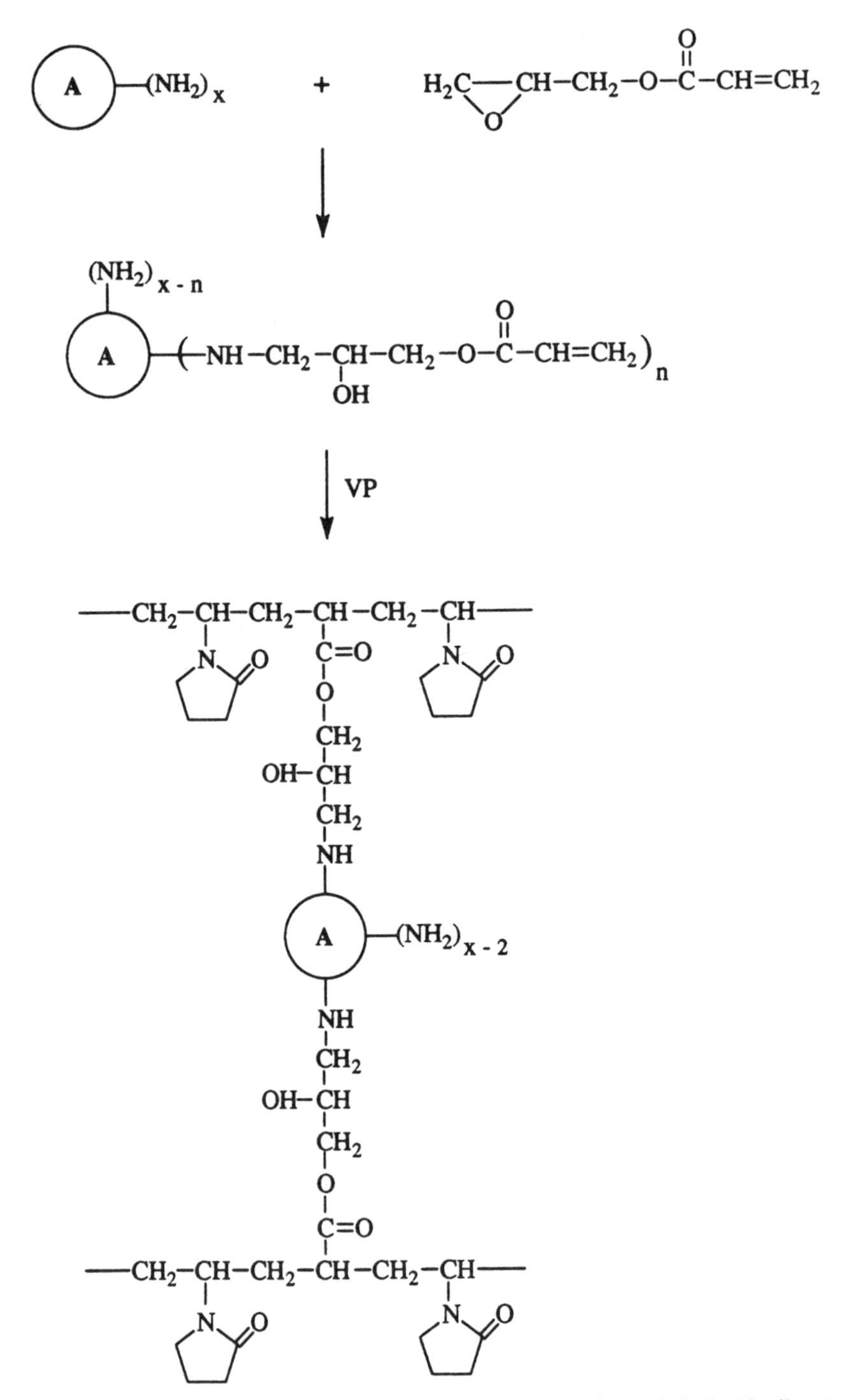

FIGURE 3.11. *Synthesis of a polyvinylpyrrolidone hydrogel crosslinked with albumin molecules.*

albumin represents about eight amine groups, since human serum albumin has more than 80 amine groups available for interaction with glycidyl acrylate [64]. Even with the presence of more than eight vinyl groups, the modified albumin could not function as a crosslinking agent. Many of the vinyl groups may not be able to participate in the polymerization of vinyl monomers. In any case, it is clear that many more than two vinyl groups should be present on each albumin molecule to form gels, although in theory only two vinyl groups are enough for forming gels.

3.2.3.2.2 Hydrogels Crosslinked with Polysaccharides

Polyacrylamide hydrogels were prepared using allyl carboxymethylcellulose, allyl hydroxyethylcellulose, or allyl dextran as crosslinking agents [65,66]. The allylic ethers of polysaccharides (AEP), being of high molecular weight and polyfunctional, can be effectively copolymerized with vinyl monomers to form a crosslinked polymer network. The efficiency of utilization of the allyl groups of cellulose ether is high in crosslinking reactions with high initial concentrations of the monomer and low concentrations of the ether. The macromolecules of the cellulose allyl ether are considered polyfunctional crosslinking "macronodes" with elastically efficient, synthetic polymer chains between them. Because of its flexibility, the polysaccharide crosslinked gel is used as a superabsorbent.

3.2.3.3 ADVANTAGES OF CROSSLINKING WITH MACROMOLECULES

The use of macromolecules, such as proteins or polysaccharides, as crosslinking agents, has a few advantages over using smaller molecular weight crosslinkers such as oligopeptides. First, the sizes of macromolecular crosslinkers are much larger than those of enzymes. Thus, steric hindrances for the formation of enzyme-substrate complexes is minimized. Polymer backbone chains are also known to exert minimum steric hindrance, if the size of crosslinker is large [61]. Second, macromolecules such as albumin can be degraded by a variety of enzymes such as pepsin, trypsin, or chymotrypsin. In certain applications such as in oral drug delivery using enzyme-degradable hydrogels [67], this nonspecific degradation by various enzymes is beneficial since it will ensure the complete degradation of the hydrogels. Third, the kinetics and extent of enzymatic degradation can be controlled by modifying the structure of the macromolecules. For example, the degree of introduction of double bonds to albumin molecules affects the degradation of the protein by pepsin [62]. Fourth, the modified macromolecules can be separated easily from small molecular weight modifying agents.

3.3 HYDROGELS WITH DEGRADABLE PENDANT CHAINS

High molecular weight drugs such as protein drugs can be entrapped in hydrogels, but low molecular weight drugs are rapidly released by simple diffusion. Thus, low molecular weight drugs have to be attached to the polymer backbone via a labile, easily degradable bond to prevent premature release. The drug molecules will be released only after degradation of the pendant chains either by chemical hydrolysis or by enzymes [68,69]. Many drugs were attached to the polymer backbone using this approach [70 – 72].

If the pendant chains are hydrophobic and contribute to the overall structure of the hydrogel, the degradation of the pendant chains will alter the overall architecture of the hydrogel. If the backbone chains also degrade as with the polypeptide or polysaccharide backbone, backbone hydrolysis will complicate the release profiles. It is desirable that the polymer backbone degrades at a much slower rate than the rate of cleavage of the labile bond.

Poly(L-glutamic acid) and poly(hydroxyalkylglutamines) have been used widely for attaching bioactive agents [70]. In most cases the drug component was linked to the polymer backbone via a spacer group. One approach was to use α,ω-hydroxyalkylamine with poly(γ-benzyl-L-glutamate) as described earlier in this chapter. In addition to poly(amino acids), polysaccharides have also been used frequently as drug carriers. The coupling of drugs to the polysaccharide backbone requires suitable functional groups.

Polysaccharides have also been used extensively to prepare polymer-drug complexes. Various types of functional groups, such as carboxyl, primary and secondary hydroxyl, and sulfate, are available on the polysaccharide backbone [3]. If suitable functional groups are not present, polysaccharides have to be activated to introduce more reactive groups. Schacht et al. [73] examined a variety of activation procedures and their effect on the degradation of the modified polysaccharides by enzymes. The biodegradability of dextran was significantly reduced upon chemical modification of the polymer backbone.

Kopeček prepared copolymers of *N*-(2-hydroxypropyl)methacrylamide (HPMA) and *p*-nitrophenyl esters of *N*-methacryloylated oligopeptides. The pendant oligopeptide sequences are degradable by enzymes. The *p*-nitrophenoxy group readily reacts with amine groups to form amide groups. Thus, drugs containing amine groups can be easily grafted to the polymer chains. The chemistry used in the preparation of hydrogels with degradable pendant chains is basically the same as that described in Section 3.2.2.

If the release of pendant drug molecules depends on enzymatic degradation, it is important to conjugate drug molecules to the polymer backbone through enzymatically degradable linkages. Studies by Kopeček and his coworkers identified a few important design features of the hydrogels with hydrolyzable pendant groups. First, the comonomer unit should be hydrophilic to promote water swellability. Second, the hydrolyzable bond must be away from the backbone chain for easy access by the enzymes. The pendent amide groups used by Kopeček et al. are resistant to chemical hydrolysis but have a potential for enzymatic hydrolysis. Third, specific oligopeptide sequences may be used to achieve substrate-specific, enzymatic hydrolysis. Kopeček and his coworkers have prepared the specific oligopeptide sequences which are degradable by chymotrypsin [74], trypsin, or lysosomal enzymes such as cathepsin [75–77].

Although in essence the enzyme-induced drug release is largely dependent on the length and the composition of the side chain used to link the drug to the polymer backbone [68,78], the nature of the polymer backbone is also known to affect the enzymatic degradation of the pendant chains. Mora and Pato [79] examined the effect of the backbone modification on the enzymatic breakdown of the drug-containing side chains. 6-Aminocaproyl-L-phenylalanyl-4-nitroanilide was bound to poly(*N*-vinylpyrrolidone-*co*-maleic acid) and polyanionic dextran. As the number of carboxyl groups increased, the Michaelis-Menten constant decreased. The interactions can occur between the carboxylates and enzymes. These interactions yield a more stable enzyme-substrate complex. Thus, the constant decreased. An increase in the number of carboxyl groups resulted in a decrease in the catabolic rate constant, i.e., the accessibility of the degradable bond by enzymes is hindered. This may be due to the decreased flexibility of the polymer substrate fixed by its carboxylates at several points on the enzyme surface. This shielding effect of the backbone can be reduced by increasing the length of the spacer [79].

Thus, it appears that the enzymatic hydrolysis of pendant groups and the subsequent release of fragments, such as drugs, can be controlled by varying at least three parameters: the specific sequence of the spacer, spacer length, and the charge on the backbone chain [79–81].

3.4 REFERENCES

1. Baker, R. 1987. *Controlled Release of Biologically Active Agents*. New York, NY: John Wiley & Sons, Chapter 4.
2. Heller, J. 1984. "Bioerodible Systems," Chapter 3 in *Medical Applications of Controlled Release*, R. S. Langer and D. L. Wise, eds., Boca Raton, FL: CRC Press.
3. Sparer, R. V., N. Ehwuribe and A. G. Walton. 1983. "Controlled Release from Glycosaminoglycan Drug Complexes," Chapter 6 in *Controlled Release Delivery Systems*, T. J. Roseman and S. Z. Mansdorf, eds., New York, NY: Marcel Dekker, Inc.
4. Dickinson, H. R., A. Hiltner, D. F. Gibbons and J. M. Anderson. 1981. "Biodegradation of a Poly(α-amino Acid) Hydrogels. I. *In Vivo*," *J. Biomed. Mater. Res.*, 15:577–589.

5. Dinkinson, H. R. and A. Hiltner. 1981. "Biodegradation of a Poly(α-amino Acid) Hydrogels. I. *In Vitro*," *J. Biomed. Mater. Res.*, 15:591–603.
6. Pytela, J., V. Saudek, J. Drobnik and F. Pypacek. 1989. "Poly(N^5-hydroxyalkylglutamines). IV. Enzymatic Degradation of N^5-(2-Hydroxyethyl)-L-glutamine Homopolymers and Copolymers," *J. Controlled Rel.*, 10:17–25.
7. Sperling, L. H. 1981. *Interpenetrating Polymer Networks and Related Materials*. New York, NY: Plenum Press, Chapter 1.
8. Visscher, K. B., I. Manners and H. R. Allcock. 1990. "Synthesis and Properties of Polyphosphazene Interpenetrating Polymer Networks," *Macromolecules*, 23:4885–4886.
9. Kaur, H. and P. R. Chatterji. 1990. "Interpenetrating Hydrogel Networks. 2. Swelling and Mechanical Properties of the Gelatin-Polyacrylamide Interpenetrating Networks," *Macromolecules*, 23:4868–4871.
10. Chatterji, P. R. 1991. "Cross-Link Dimensions in Gelatin-poly(acrylamide) Interpenetrating Hydrogel Networks," *Macromolecules*, 24:4214–4215.
11. Jeyanthi, R. and K. P. Rao. 1990. "Collagen-poly(HEMA) Hydrogels for the Controlled Release of Anticancer Drugs–Preparation and Characterization," *J. Bioact. Compat. Polymers*, 5:194–211.
12. Heller, J., R. W. Baker, R. F. Helwing and M.E. Tuttle. 1983. "Controlled Release of Water-Soluble Macromolecules from Bioerodible Hydrogels," *Biomaterials*, 4:262–266.
13. Baker, R. W., M. E. Tuttle and R. Helwing. 1984. "Novel Erodible Polymers for the Delivery of Macromolecules," *Pharm. Technol.*, 8:26–30.
14. Escbach, F. O. and S. J. Huang. 1991. "Hydrophobic-Hydrophilic IPN and SIPN," *Proc. ACS Div. Polym. Mat.: Sci. & Eng.*, 65:9–10.
15. Davis, P. A., L. Nicolais, L. Ambrosio and S. J. Huang. 1988. "Poly(2-hydroxyethyl Methacrylate)/Poly(caprolactone) Semi-Interpenetrating Polymer Networks," *J. Bioact. Compat. Polymers*, 3:205–218.
16. St. Pierre, T. and E. Chiellini. 1986. "Biodegradability of Synthetic Polymers Used for Medical and Pharmaceutical Applications: Part 2–Backbone Hydrolysis," *J. Bioact. Compat. Polymers*, 2:4–30.
17. Cohn, D. and H. Younes. 1988. "Biodegradable PEO/PLA Block Copolymers," *J. Biomed. Mater. Res.*, 22:993–1009.
18. Zhu, K. J., S. Bihai and Y. Shilin. 1989. " 'Super Microcapsules' (SMC). I. Preparation and Characterization of Star Polyethylene Oxide (PEO)–Polylactide (PLA) Copolymers," *J. Polymer Sci.: Part A: Polymer Chem.*, 27:2151–2159.
19. Kobayashi, H., S.-H. Hyon and Y. Ikada. 1991. "Water-Curable and Biodegradable Prepolymers," *J. Biomed. Mater. Res.*, 25:1481–1494.
20. Gilding, D. K. 1981. "Biodegradable Polymers," Chapter 9 in *Biocompatibility of Clinical Implant Materials, Vol. II,* D. F. Williams, ed., Boca Raton, FL: CRC Press.
21. Reed, A. M., D. K. Gilding and J. Wilson. 1977. "Biodegradable Elastomeric Biomaterials–Polyethylene Oxide/Polyethylene Terephthalate Copolymers," *Trans. Am. Soc. Artif. Intern. Organs*, 23:109–114.
22. Reed, A. M. and D. K. Gilding. 1981. "Biodegradable Polymers for Use in Surgery–Poly(ethylene Oxide)/Poly(ethylene Terephthalate) (PEO/PET) Copolymers: 2. *In vitro* Degradation," *Polymer*, 22:499–504.
23. Sidman, K. R., W. D. Steber, A. D. Schwope, G. R. Schnaper and R. Gayle. 1983.

"Controlled Release of Macromolecules and Pharmaceuticals from Synthetic Polypeptides Based on Glutamic Acid," *Biopolymers*, 22:547–556.

24. Marck, K. W., C. H. Wildevuur, W. L. Sederel, A. Bantjes and J. Feijen. 1977. "Biodegradability and Tissue Reaction of Random Copolymers of L-Leucine, L-Aspartic Acid, and L-Aspartic Acid Esters," *J. Biomed. Mater. Res.*, 11:405–422.
25. Johns, D. B., R. W. Lenz and M. Vert. 1986. "Poly(malic Acid). Part I. Preparation and Polymerization of Benzyl Malolactonate," *J. Bioact. Comp. Polym.*, 1:47–60.
26. Asano, M., M. Yoshida, I. Kaetsu, K. Nakai, H. Yamanaka, H. Yuasa, K. Shida, K. Suzuki and M. Oya. 1983. "Biodegradable Random Copolypeptides of β-Benzyl L-Aspartate and γ-Methyl L-Glutamate for the Controlled Release of Testosterone," *Makromol. Chem.*, 184:1761–1170.
27. Hayashi, T. and Y. Ikada. 1990. "Enzymatic Hydrolysis of Copoly-(*N*-hydroxyalkyl L-Glutamine/γ-Methyl L-Glutamate) Fibres," *Biomaterials*, 11:409–413.
28. Olabisi, O., L. M. Robeson, and M. T. Shaw. 1979. *Polymer-Polymer Miscibility*. New York, NY: Academic Press, Chapter 4.
29. Cross, R. A. 1976. "Polyelectrolyte Complexes for Medical Applications," in *Polyelectrolytes*, K. C. Frisch, ed., Westport, CT: Technomic Publishing Co., Inc., pp. 134–143.
30. Booij, H. L. and H. G. Bungengerg de Jong. 1956. *Biocolloids and Their Interactions*. Vienna, Austria: Spring-Verlag, Chapters 2 and 4.
31. Albertsson, P.-Å. 1986. *Partition of Cell Particles and Macromolecules, 3rd Ed.* New York, NY: John Wiley & Sons, p. 12.
32. Lysaght, M. J. 1976. "Polyelectrolyte Complexes," in *Polyelectrolytes*, K. C. Frisch, ed., Westport, CT: Technomic Publishing Co., Inc., pp. 34–42.
33. Margolin, A. L., S. F. Sherstyuk, V. A. Izumrudov, A. B. Zekin and V. A. Kabanov. 1985. "Enzymes in Polyelectrolyte Complexes. The Effect of Phase Transition on Thermal Stability," *Eur. J. Biochem.*, 146:625–632.
34. Phillip, B., J. Kötz, K.-J. Lindow and H. Dautzenberg. 1991. "Polyanion-Polycation Complexes and Their Areas of Application," *Polymer News*, 16:106–110.
35. Brøndsted, H. and J. Kopeček. 1992. "pH-Sensitive Hydrogels," in *Polyelectrolyte Gels. Properties, Preparations, and Applications, ACS Symp. Ser. 480,* R. S. Harland and R. K. Prud'homme, eds., Washington, DC: American Chemical Society, pp. 285–304.
36. Bakan, J. A. 1980. "Microencapsulation Using Coacervation/Phase Separation Techniques," Chapter 4 in *Controlled Release Technologies: Methods, Theory, and Applications, Vol. II,* A. F. Kydonieus, ed., Boca Raton, FL: CRC Press.
37. Nixon, J. R., ed. 1976. *Microencapsulation*. New York, NY: Marcel Dekker, Inc.
38. Helliwell, M., G. P. Martin and C. Marriott. 1989. "The Release of Arachis Oil from Gelatin-Acacia Microspheres," *J. Pharm. Pharmacol.*, 41:117P.
39. Daly, M. M. and D. Knorr. 1988. "Chitosan-Alginate Complex Coacervate Capsules: Effects of Calcium Chloride, Plasticizers, and Polyelectrolytes on Mechanical Stability," *Biotechnol. Prog.*, 4:76–81.
40. Chavasit, V. and J. A. Torres. 1990. "Chitosan-Poly(acrylic Acid): Mechanisms of Complex Formation and Potential Industrial Applications," *Biotechnol. Prog.*, 6:2–6.
41. Nagasawa, M., T. Murase and K. Kondo. 1965. "Potentiometric Titration of Stereoregular Polyelectrolytes," *J. Phys. Chem.*, 69:4005–4012.
42. Pitt, C. G., Y. Cha, S. S. Shah and K. J. Zhu. 1992. "Blends of PVA and PGLA:

Control of the Permeability and Degradability of Hydrogels by Blending," *J. Controlled Rel.*, 19:189–200.

43. Heller, J. 1980. "Controlled Release of Biologically Active Compounds from Bioerodible Polymers," *Biomaterials*, 1:51–57.
44. Davis, B. K. 1972. "Control of Diabetes with Polyacrylamide Implants Containing Insulin," *Experientia*, 28:348.
45. Torchilin, V. P., E. G. Tischenko, V. N. Smirnov and E. I. Chazov. 1977. "Immobilization of Enzymes on Slowly Soluble Carriers," *J. Biomed. Mater. Res.*, 11:223–235.
46. Bernfield, P. and J. Wan. 1963. "Antigens and Enzymes Made Insoluble by Entrapping Them into Lattices of Synthetic Polymers," *Science*, 142:678.
47. Brown, J. P., G. V. McGarraugh, T. M. Parkinson, R. E. Wignard, Jr. and A. B. Oderdonk. 1983. "A Polymeric Drug for Treatment of Inflammatory Bowel Diseases," *J. Med. Chem.*, 26:1300–1307.
48. Saffran, M., G. S. Kumar, C. Savarian, J. C. Burnham, F. Williams and D. C. Necker. 1986. "A New Approach to the Oral Administration of Insulin and Other Peptide Drugs," *Science*, 233:1081–1084.
49. Saffran, M. 1982. "Oral Administration of Peptides," *Endocrinologia Experimentalis*, 16:327–333.
50. Brøndsted, H. and J. Kopeček. 1991. "Hydrogels for Site-Specific Oral Drug Delivery: Synthesis and Characterization," *Biomaterials*, 12:584–592.
51. Kopeček, J., P. Kopeckova, H. Brøndsted, R. Rathi, B. Řihová, P.-Y. Yeh and K. Ihesue. 1992. "Polymers for Colon-Specific Drug Delivery," *J. Controlled Rel.*, 19:121–130.
52. Gruber, H. and G. Greber. 1990. "Reactive Sucrose Derivatives," *Zuckerind.*, 115:476–482.
53. Gruber, H. and G. Greber. 1991. "Reactive Sucrose Derivatives," in *Carbohydrates as Organic Raw Materials*, F. W. Lichtenthaler, ed., New York, NY: VCH, pp. 95–125.
54. Mattson, F. H. and R. A. Volpenhein. 1972. "Hydrolysis of Fully Esterified Alcohols Containing from One to Eight Hydroxyl Groups by the Lipolytic Enzymes of Rat Pancreatic Juice," *J. Lipid Res.*, 13:325–328.
55. Kopeček, J. 1977. "Reactive Copolymers of *N*-(2-Hydroxypropyl)methacryl-amide with *N*-Methacrylolyated Derivatives of L-Leucine and L-Phenylalanine, 1," *Makromol. Chem.*, 178:2169–2183.
56. Rejmanová, P., B. Obereigner and J. Kopeček. 1981. "Polymers Containing Enzymatically Degradable Bonds, 2. Poly[*N*-(2-hydroxypropyl)methacryl-amide] Chains Connected by Oligopeptide Sequences Cleavable by Chymotrypsin," *Makromol. Chem.*, 182:1899–1915.
57. Ulbrich, K., J. Strohalm and J. Kopeček. 1982. "Polymers Containing Enzymatically Degradable Bonds. VI. Hydrophilic Gels Cleavable by Chymotrypsin," *Biomaterials*, 3:150–154.
58. Ulbrich, K., Zacharieva, B. Obereigner and J. Kopeček. 1980. "Polymers Containing Enzymatically Degradable Bonds. V. Hydrophilic Polymers Degradable by Papain," *Biomaterials*, 1:199–204.
59. Ulbrich, K., J. Strohalm and J. Kopeček. 1981. "Polymers Containing Enzymatically Degradable Bonds. III. Poly[*N*-(2-hydroxypropyl)methacryl-amide] Chains Connected by Oligopeptide Sequences Cleavable by Trypsin," *Makromol. Chem.*, 182:1917–1928.

60. Cartlidge, S. A., R. Duncan, J. B. Lloyd, P. Rejmanová and J. Kopeček. 1986. "Soluble, Crosslinked *N*-(2-Hydroxypropyl)methacrylamide Copolymers as Potential Drug Carriers. 1/ Pinocytosis by Rat Visceral Yolk Sacs and Rat Intestine Cultured *in Vitro*. Effect of Molecular Weight on Uptake and Intracellular Degradation," *J. Controlled Rel.*, 3:55–66.

61. Kopeček, J. and P. Rejmanová. 1979. "Reactive Copolymers of *N*-(2-Hydroxypropyl)methacrylamide with *N*-Methacryloylated Derivatives of L-Leucine and L-Phenylalanine. II. Reaction with the Polymeric Amine and Stability of Cross-Links Towards Chymotrypsin *in Vitro*," *J. Polymer Sci.: Polymer Symp.*, 66:15–32.

62. Park, K. 1988. "Enzyme-Digestible Swelling Hydrogels as Platforms for Long-Term Oral Drug Delivery: Synthesis and Characterization," *Biomaterials*, 9:435–441.

63. Snyder, S. L. and P. Z. Sobocinski. 1975. "An Improved 2,4,6-Trinitrobenzenesulfonic Acid Method for the Determination of Amines," *Anal. Biochem.*, 64:284–288.

64. Dayhoff, M. O. 1978. *Atlas of Protein Sequence and Structure, Vol. 5, Supplement 3.* Washington, DC: The National Biomedical Research Foundation, p. 306.

65. Buyanov, A. L., L. G. Revel'skaya and G. A. Petropavlovskii. 1992. "Formation and Swelling Behavior of Polyacrylate Superabsorbent Hydrogels Crosslinked by Allyl Ethers of Polysaccharides," *Proc. ACS Div. Polym. Mat. Sci. Eng.*, 66:87–88.

66. Buyanov, A. L., L. G. Revel'skaya and G. A. Petropavlovskii. 1989. "Mechanism of Formation and Structural Features of Highly Swollen Acrylate Hydrogels Cross-Linked with Cellulose Allyl Ethers," *J. Appl. Chem. – USSR*, 62:1723–1728.

67. Shalaby, W. S. W., W. E. Blevins and K. Park. 1992. "*In vitro* and *in vivo* Studies of Enzyme-Digestible Hydrogels for Oral Drug Delivery," *J. Controlled Rel.*, 19: 131–144.

68. Stjärnkvist, P., L. Degling and I. Sjöholm. 1991. "Biodegradable Microspheres. XIII: Immune Response to the DNP Hapten Conjugated to Polyacryl Starch Microparticles," *J. Pharm. Sci.*, 80:436–440.

69. Laakso, T., P. Stjärnikvis and I. Sjöholm. 1987. "Biodegradable Microspheres VI: Lysosomal Release of Covalently Bound Antiparasitic Drugs from Starch Microparticles," *J. Pharm. Sci.*, 76:134–140.

70. Peterson, R. V. 1985. "Biodegradable Drug Delivery Systems Based on Polypeptides," in *Bioactive Polymeric Systems. An Overview*, C. G. Gebelein and C. E. Carraher, Jr., eds., New York, NY: Plenum Press, pp. 151–177.

71. Negishi, N., D. B. Bennet, C. S. Cho, S. Y. Jeong, W. A. R. van Heeswijk, J. Feijen and S. W. Kim. 1987. "Coupling of Naltrexone to Poly (α-amino Acids)," *Pharm. Res.*, 4:305–310.

72. Shen, W.-C., H. J. P. Lyser and L. LaManna. 1985. "Disulfide Spacer between Methotrexate and Poly(D-lysine). A Probe for Exploring the Reductive Process in Endocytosis," *J. Biol. Chem.*, 260:10905–10908.

73. Schacht, E., F. Vandoorná, J. Vermeersch and R. Duncan. 1987. "Polysacharides as Drug Carriers. Activation Procedures and Biodegradation Studies," Chapter 14 in *Controlled-Release Technology. Pharmaceutical Applications*, P. I. Lee and W. R. Good, eds., Washington, DC: American Chemical Society.

74. Kopeček, J., P. Rejmanová and V. Chytrý. 1981. "Polymers Containing Enzymatically Degradable Bonds, 1. Chymotrypsin Catalyzed Hydrolysis of *p*-Nitroanilides of Phenylalanine and Tyrosine Attached to Side-Chains of Copolymers of *N*-(2-Hydroxypropyl)methacrylamide," *Makromol. Chem.*, 182:799–809.

75. Duncan, R., J. B. Lloyd and J. Kopeček. "Degradation of Side Chain of *N*-(2-

Hydroxypropyl)methacrylamide Copolymers by Lysosomal Enzymes," *Biochem. Biophys. Res. Commun.*, 94:284–290.

76. Rejmanová, P., J. Kopeček, R. Duncan and J. B. Lloyd. 1985. "Stability in Rat Plasma and Serum of Lysosomally Degradable Oligopeptide Sequences in *N*-(2-Hydroxypropyl)methacrylamide Copolymers," *Biomaterials*, 6:45–48.
77. Subr, V., J. Kopeček, J. Pohl, M. Baudys and V. Kodtka. 1988. "Cleavage of Oligopeptide Side-Chains in *N*-(2-Hydroxypropyl)methacrylamide Copolymers by Mixtures of Lysosomal Enzymes," *J. Controlled Rel.*, 8:133–140.
78. Duncan, R., I. C. Hume, H. J. Yardley, P. A. Flanagan, J. Strohalm, K. Ulbrich and V. Subr. 1991. "Macromolecular Prodrugs for Use in Targeted Cancer Chemotherapy: Melphalan Covalently Coupled to *N*-(2-Hydroxypropyl)methacrylamide Copolymers," *J. Controlled Rel.*, 16:121–136.
79. Mora, M. and J. Pato. 1992. "Effect of Negative Electric Charges of Polymers on the Chymotrypsin Catalyzed Hydrolysis of the Side Chains," *J. Controlled Rel.*, 18:153–158.
80. Kopeček, J. 1984. "Controlled Biodegradability of Polymers – A Key to Drug Delivery Systems," *Biomaterials*, 5:19–25.
81. Bennett, D. B., N. W. Adams, X. Li, J. Feijen and S. W. Kim. 1988. "Drug-Coupled Poly(amino Acids) as Polymeric Drugs," *J. Bioact. Compat. Polymers*, 3:44–52.

CHAPTER 4

Chemical Gels

Chemical gels are those which have covalently crosslinked networks. Thus, chemical gels will not dissolve in water or other organic solvents unless covalent crosslinks are cleaved. At least two different approaches can be used to form chemical gels. First, chemical gels can be made by polymerizing water-soluble monomers in the presence of bi- or multifunctional crosslinking agents. Second, chemical gels can be prepared by crosslinking water-soluble polymers using typical organic chemical reactions which involve functional groups of the polymers.

4.1 POLYMERIZATION IN THE PRESENCE OF CROSSLINKING AGENTS

Polymerization of water-soluble monomers in the presence of crosslinking agents leads to the formation of chemical gels. Examples of water-soluble monomers are acrylic acid, methacrylic acid, acrylamide, *N*-alkylacrylamide, methacrylamide, vinylpyrrolidone, methyl methacrylate, hydroxyethyl methacrylate, and vinylpyridine. The size of the crosslinking agent can vary widely. Not only low molecular weight crosslinking agents such as *N,N'*-methylenebisacrylamide but also macromolecules such as proteins can be used as crosslinking agents. Basically, any molecule that contains at least two C=C bonds should be able to function as a crosslinking agent in the copolymerization with vinyl monomers. Indeed, even macromolecules such as albumin or dextran can readily participate in the copolymerization if they are functionalized (i.e., modified with double bonds) as described in Chapter 3, Section 3.2.3.

4.1.1 Introduction of Double Bonds to Macromolecules

Macromolecules can be easily functionalized by various reagents. If a macromolecule contains only one polymerizable group, then it is called a macromolecular monomer or macromonomer. Linear macromolecules which have two reactive functional groups at both ends are called telechelic polymers. Thus, macromonomers are basically "monotelechelics" [1]. Macromonomers can be used for the synthesis of custom-made, comb-type graft copolymers. Polymers with bioactive macromolecules attached to the backbone chain can be easily prepared using macromonomers [2,3].

Macromolecules with more than one polymerizable group can function as crosslinking agents in copolymerization with monomers. Such functionalized macromolecules can also react with each other in the absence of monomers to form a three-dimensional network. The polymerizable group may be an unsaturated bond such as a vinylic or acrylic group, a heterocycle for ring-opening polymerization, or a dicarboxyl or dihydroxyl group for step-growth polymerization [1]. Here we will focus on the introduction of unsaturated bonds to macromolecules.

Water-soluble macromolecules frequently contain functional groups such as hydroxyl ($-OH$), amino ($-NH_2$), carboxyl ($-COOH$), and sulfate ($-SO_3H$) moieties which can be easily modified by various chemical reagents as shown in Figure 4.1. Hydroxyl groups of macromolecules such as dextran, starch, agarose, or poly(ethylene oxide) have been converted to vinyl groups by reacting with acid chlorides (e.g., acryloyl chloride, allyl chloride, methacryloyl chloride, and *p*-vinylbenzyl chloride) or epoxides in alkaline conditions [4–9]. Commonly used epoxides are glycidyl acrylate, glycidyl methacrylate, butene-2,3-oxide, allyl glycidyl ether, and 1,2-epoxy-5-hexene. Vinyl groups can also be introduced to macromolecules by modifying amine groups using acid chlorides, isocyanates, epoxides, or aldehydes [3,10–14].

Carboxyl groups of macromolecules can be converted to vinyl derivatives by treating the acid with alcohols containing a vinyl group in the presence of water-soluble carbodiimide. Carbodiimides react very well with the carboxylate ion to form O-acryl isourea intermediates which can then react with hydroxyl-, amino-, or thio-containing compounds via nucleophilic substitution [15]. In addition, vinyl ester can be prepared by reacting the acid with vinyl acetate catalyzed by mercuric acetate and sulfuric acid. The acid hydrogen is replaced by the vinyl group in the acetate [16]. In a two-step process, the carboxylic acid can be first converted to the acid chloride by thionyl chloride and then the highly reactive acid chloride reacts with an alcohol containing a vinyl group [16].

The sulfate groups of macromolecules such as heparin can be modified to amine groups which are then converted to vinyl groups [17]. The reaction of the heparin macromonomer practically fails to proceed if the ionic strength of the reaction system is 0.2 or lower. At an ionic strength of 0.33 the negative charges of heparin are screened and the heparin macromonomer enters readily into a reaction of radical polymerization with the hydrophilic monomer.

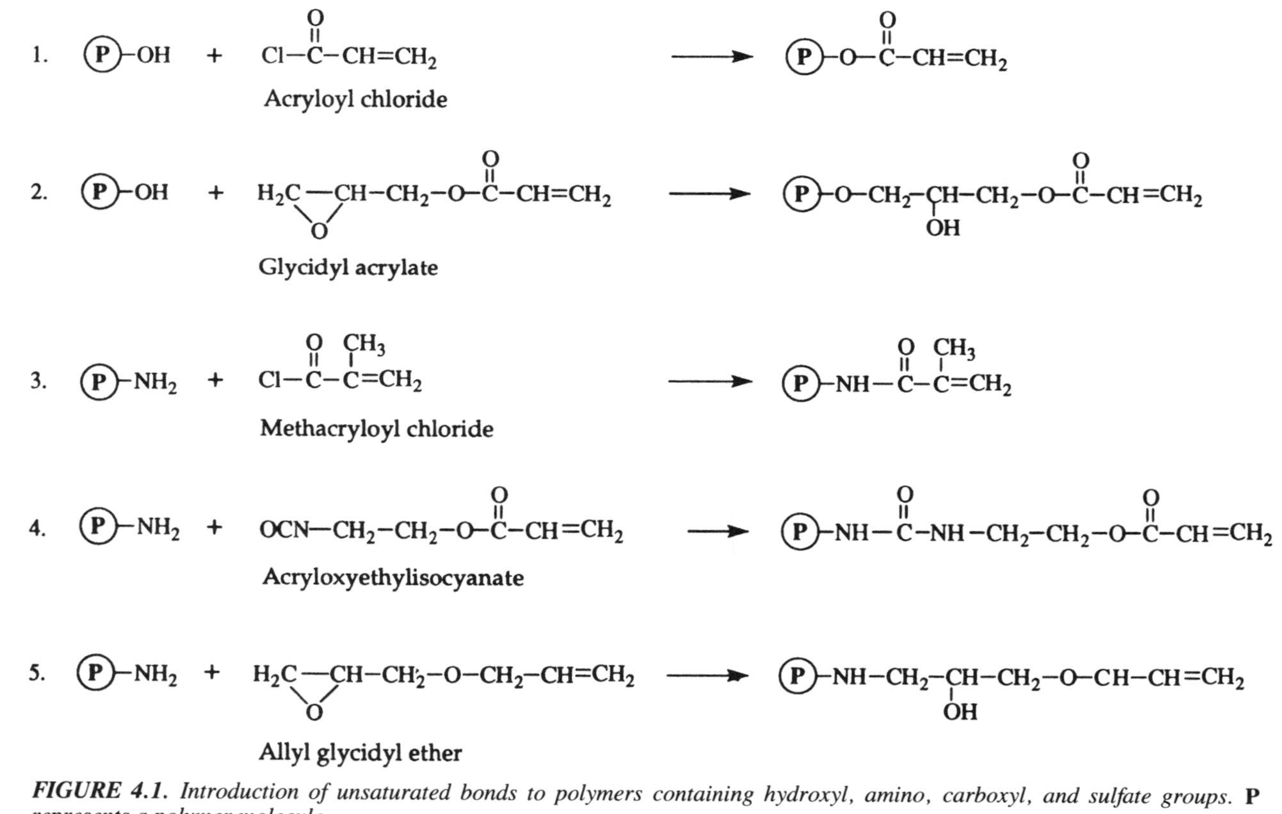

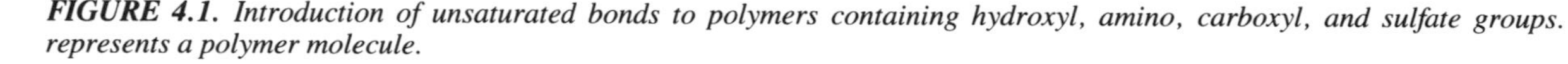

FIGURE 4.1. *Introduction of unsaturated bonds to polymers containing hydroxyl, amino, carboxyl, and sulfate groups.* **P** *represents a polymer molecule.*

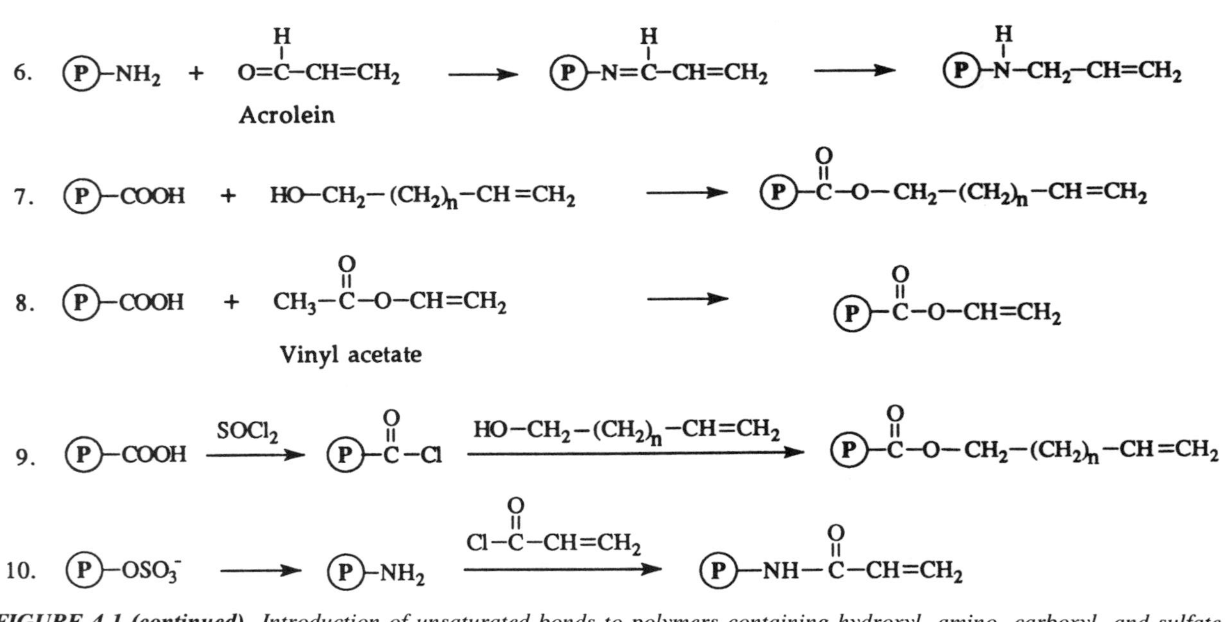

*FIGURE 4.1 **(continued)**. Introduction of unsaturated bonds to polymers containing hydroxyl, amino, carboxyl, and sulfate groups.* **P** *represents a polymer molecule.*

4.1.2 Determination of Double Bonds on the Macromonomers

4.1.2.1 MEASUREMENT OF THE NUMBER OF DOUBLE BONDS

4.1.2.1.1 Halogenation of Double Bonds

The content of double bonds on the modified macromolecules can be measured by a method based on the halogenation of the double bonds. Alkenes are known to be readily converted by bromine or chlorine into saturated compounds that contain two atoms of halogen attached to adjacent carbons [18]. Thus, the number of double bonds in the modified macromolecules can be determined by titration with an aqueous solution of bromine in the presence of potassium bromide [2,3,10,17]. The reaction is carried out simply by mixing together the two reactants at room temperature or below. A solution of bromine is red while the dihalide is colorless. The color reduction can be conveniently measured at 480–500 nm [5]. Potassium bromide is usually added to the aqueous solution to reduce the loss of bromine by volatilization. Potassium bromide forms the nonvolatile tribromide ion [19].

Lepistö et al. [20] also used the bromine method following Hoppe et al. [21] to determine the double bonds in the acryloylated polysaccharides. Typically, 1 ml of freshly prepared bromine-water (0.2 percent v/v) was added to 1 ml of the acryloylated polysaccharide sample in buffer. The mixture was protected from light and after 30 minutes the absorption decrease was measured at 500 nm. The standard curve was made using acrylamide. Hoppe used the bromine method to determine the residual acrylamide in polyacrylamide. Bromine in an 80:20 methanol-water solution was mixed with a solution of the monomer containing polyacrylamide.

4.1.2.1.2 Spectroscopic Methods

The extent of the introduction of vinyl groups can be quantitated by spectroscopic analysis such as infra red (IR) or nuclear magnetic resonance (NMR) spectroscopy. The presence of vinyl groups can be determined by measuring the C=C stretching bands at 1636–1648 cm^{-1} from IR the spectra of samples [22]. Artursson et al. [6,23] analyzed the extent of vinyl group introduction to starch and dextran by proton NMR [20]. The region containing the acrylic and anomeric proton resonances (δ 5.8–6.7 ppm) was integrated and compared with signals from the internal standards of the anomeric proton (δ 4.8–5.8 ppm). Heller et al. [7] used ^{13}C NMR to quantitate the extent of reaction between soluble starch and glycidyl methacrylate. The ratio of oxygen bearing carbons to anomeric carbons is directly related to the degree of substitution. Thus, the extent of reaction can be determined by measuring the ratio of oxygen-bearing carbons to anomeric carbons. The ^{13}C NMR also provided information on the relative reactivities of the various hydroxyl groups on the starch backbone. The hydroxyl at C-6 is more reactive than either of the secondary hydroxyls at C-2 and C-3.

4.1.2.2 MEASUREMENT OF REMAINING FUNCTIONAL GROUPS

After the introduction of double bonds to the selected functional groups of macromolecules, the remaining functional groups can be measured to calculate the extent of modification. For example, the number of amine groups on albumin was measured after interaction with glycidyl acrylate [24]. The amine groups were titrated with 2,4,6-trinitrobenzenesulfonic acid [25]. Unmodified albumin was used as a control and the degree of modification was expressed as the percentage of the total amine groups available for titration. Determination of glycidyl groups remaining after the reaction may be an alternative [26].

4.2 CROSSLINKING OF WATER-SOLUBLE POLYMERS

Crosslinking of water-soluble polymers by the addition of bifunctional or multifunctional reagents results in chemical gels. A crosslinking agent has reactive groups at both ends and joins two molecular components by covalent bonding. If the macromolecules have functional moieties only at their end groups, then bifunctional crosslinking agents simply link macromolecules to result in block copolymers while multifunctional reagents achieve gel formation. If macromolecules possess functional groups along the backbone, however, gel can be formed by bifunctional reagents.

Crosslinking is another form of chemical modification. Polymers undergo the same reactions as their low molecular weight homologs, although the reaction rates and maximum conversions in the reactions of polymer functional groups may differ significantly from those of the corresponding low molecular weight homologs [27]. The efficiency of crosslinking depends on the specificity toward particular groups to be functionalized. Thus, the choice of crosslinking agents is critical to successful crosslinking. While it would be ideal if the reagents possessed high selectivity for a certain functional group, few, if any, reagents are absolutely group-specific. The specificity can be influenced by the experimental conditions, mainly by pH. The lack of specificity may be important in the modification of macromolecules which possess more than one type of functional group as in proteins. The existing functional groups of a polymer can be modified by various chemical reactions to introduce new, specific, or more reactive functional groups which may improve the crosslinking reaction.

Bifunctional crosslinking agents are divided into three classes: homobifunctional, heterobifunctional, and zero length [28]. Homobifunctional crosslinking agents contain two identical functional groups that react with the same functional groups of the polymer. Heterobifunctional agents contain two dissimilar functional groups of different specificity and hence react with different functional groups of the polymer. Zero-length crosslinking agents link polymers without the addition of extrinsic compounds. Many zero-length crosslinking agents condense carboxyl groups with primary amino, hydroxyl, carboxylic, and thio groups to form amide, ester, and thioester bonds. Thus, zero-length crosslinking agents are useful in crosslinking polymers with various functional groups such as proteins.

The use of chemical crosslinking agents is important not only for connecting two macromolecules and forming hydrogels but also for immobilizing bioactive molecules, such as drugs and enzymes, to hydrogel matrices or water-soluble polymers [29−32]. Chemical modification of polymer chains and immobilization of enzymes are widely used in the development of environment-sensitive drug delivery systems [33].

4.2.1 Crosslinking with Functional Groups

4.2.1.1 CROSSLINKING WITH HYDROXYL GROUPS

A variety of crosslinking agents have been used to crosslink hydroxyl groups of polymer molecules, especially polysaccharides. Examples of crosslinking agents which have been used for the crosslinking of hydroxyl containing polymers are shown in Figure 4.2. Bis-epoxides (bisoxiranes) are commonly used in the crosslinking of polysaccharides containing hydroxyl groups. Epoxides undergo ring opening reactions with nucleophiles including hydroxyl groups at alkaline pH (e.g., pH 11) [34]. Oxirane-coupled ligands are very stable. Divinyl sulfone reacts with hydroxyl groups at alkaline pH values (e.g., pH 11). Vinylsulfonyl groups also react with thiols and amino groups at lower pH than for oxyrane coupling (pH 9−13) and reaction proceeds at a higher rate [15].

Carbonylating agents such as *N,N'*-carbonyldiimidazole are used to crosslink hydroxyl groups of polysaccharides. Imidazole can also couple with amines to form stable carbamate (urethane) derivatives [15]. Epichlorohydrin is commonly used to crosslink hydroxyl group-containing polysaccharides such as dextran, chitin, alginate, or hydroxyethylcellulose [35−38]. Cyanuric chloride (2,4,6-trichloro-*s*-triazine) is also widely used for modifying hydroxyl groups. The chlorine atoms of the *s*-triazine series are very reactive toward nucleophiles including hydroxyl groups of carbohydrates. Because of this reactivity, they are rapidly hydrolyzed in aqueous solutions resulting in poor crosslinking [28]. Terephthaloyl chloride is used to crosslink the surface layer of starch microcapsules

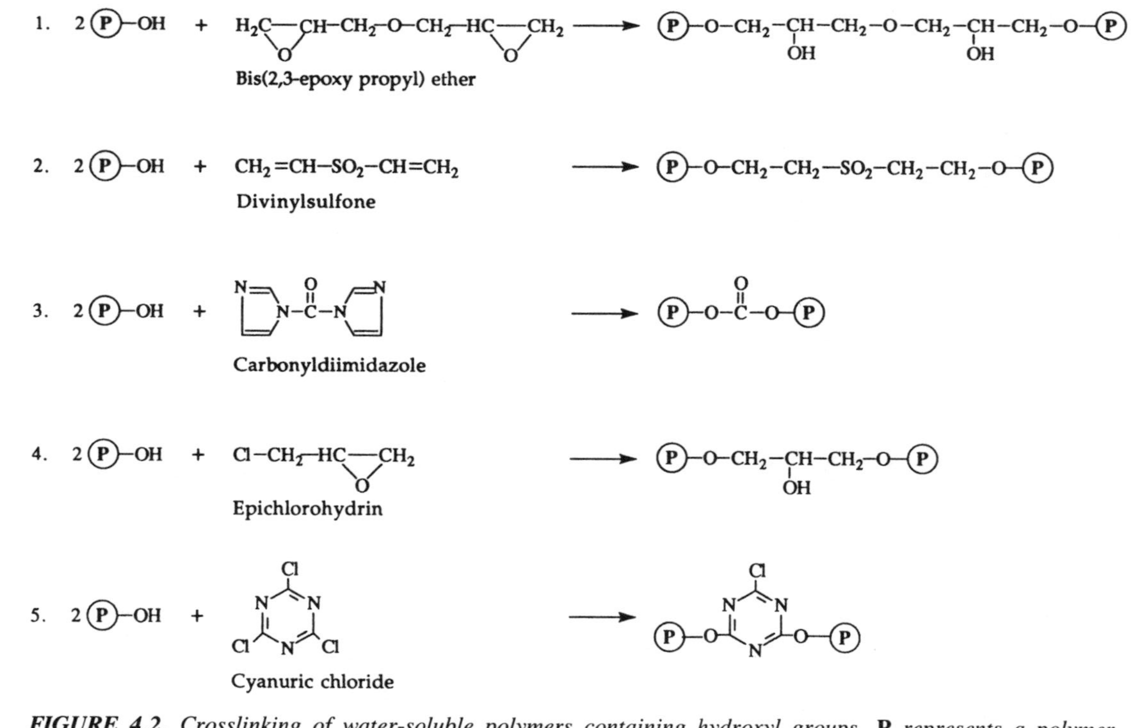

FIGURE 4.2. *Crosslinking of water-soluble polymers containing hydroxyl groups.* **P** *represents a polymer molecule.*

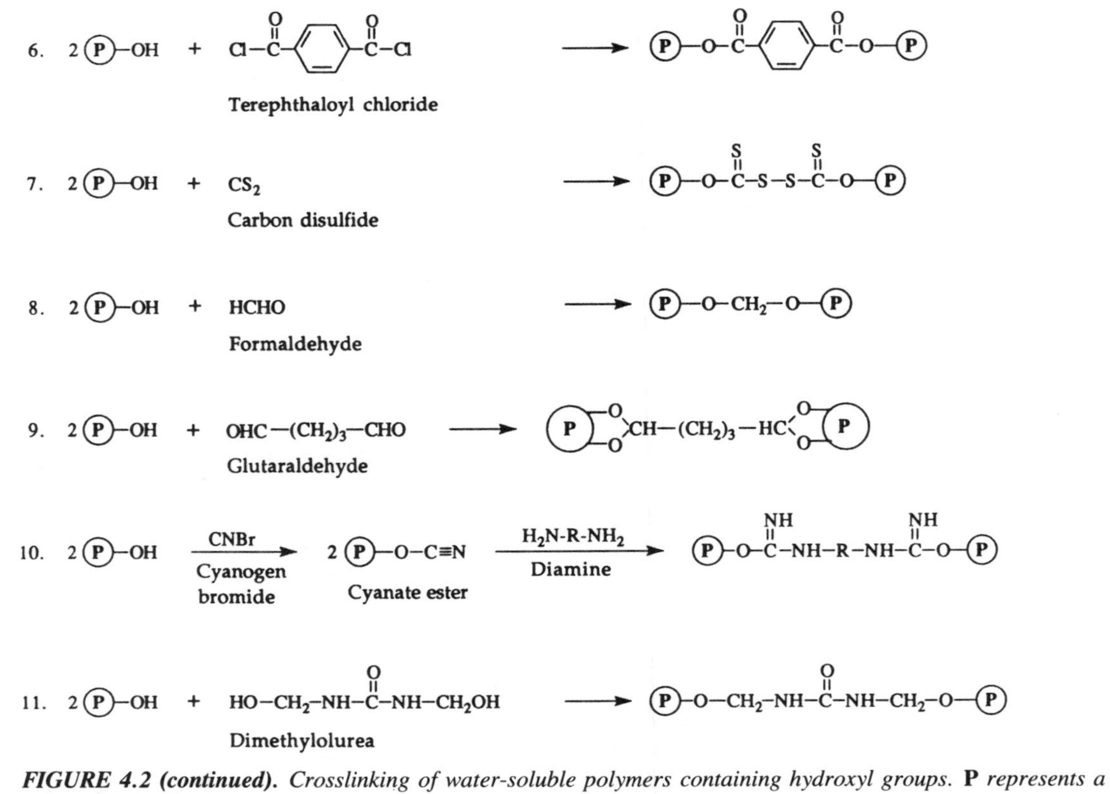

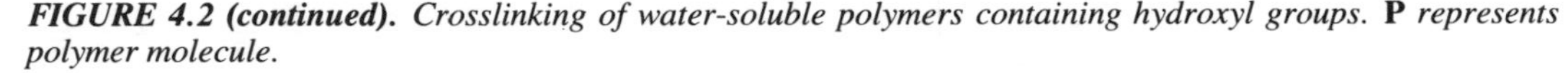

FIGURE 4.2 (continued). *Crosslinking of water-soluble polymers containing hydroxyl groups.* **P** *represents a polymer molecule.*

dispersed in organic solvent [39]. Carbon disulfide (CS_2) is used to derivatize polysaccharides such as starch. CS_2-derivatized starch can be crosslinked to yield starch xanthate. Starch xanthate is oxidatively crosslinked in an aqueous system in the presence of biologically active material [40]. Formaldehyde is not bifunctional, and yet is able to crosslink. Glutaraldehyde is also used to crosslink hydroxyl groups of polymers such as PVA [41,42]. Hydroxyl groups of polysaccharides can be activated using cyanogen bromide to form cyanate ester which is very reactive with primary amine groups. Thus, diamines can crosslink polysaccharides by reacting with the cyanate esters at alkaline pH [15]. Many of the crosslinking agents shown in Figure 4.2 also react with amine groups. For example, chlorine atoms in triazine readily react with primary amine groups. Thus, diamines can be used to crosslink triazine activated polysaccharides.

Urea-formaldehyde is commonly used to crosslink hydroxyl groups of polysaccharides such as starch. Urea reacts with formaldehyde to form dimethylolurea which then crosslinks with starch [43]. The NMR study showed that chemical crosslinking occurred through the primary hydroxyl groups located at the C-6 position of starches [43]. Dimethylolurea may undergo self-condensation to form the urea formaldehyde prepolymer (polynoxylin). The prepolymer can be prepared by refluxing a neutralized formalin solution with urea for 15–20 minutes. This prepolymer was slowly added to the starch paste which was gelatinized by heating to 80°C. Dilute formic acid was added dropwise until the pH was adjusted to 3.5 [12,44].

Once polymers are crosslinked, the remaining hydroxyl groups can be used to couple low molecular weight drugs which otherwise will be released from the hydrogel quite rapidly. Laakso et al. coupled amino group-containing drugs (primaquine and trimethoprim) to starch microparticles after activating hydroxyl groups with carbonyldiimidazole [45].

Hydroxyl groups can be converted to other more reactive groups. Reaction with organic sulfonyl chlorides, such as 4-toluene sulfonyl chloride (tosyl chloride) and 2,2,2-trifluoroethane sulfonyl chloride (tresyl chloride), converts hydroxyl groups to sulfonates. The introduction of a tosyl reagent to the polymers can be followed conveniently by UV spectroscopy. The reactivity of tosyl chloride to hydroxyl groups at pH 9–10.5 is slightly greater than that of epoxy groups. Tresyl chloride forms sulfonates much more readily than tosyl chloride [15].

4.2.1.2 CROSSLINKING WITH AMINE GROUPS

Many crosslinking agents shown in Figure 4.2 are not absolutely group specific and have crossreactivity. For example, *N,N'*-carbonyldiimidazole, bis-oxyrane, divinylsulfone, triazine, carbon disulfide, and urea-formaldehyde prepolymers readily react with amine groups [29,34,46]. Thus, these agents are useful in the crosslinking of polymer molecules containing hydroxyl or amino groups, or both.

Examples of crosslinking agents directed to amine groups are shown in Figure 4.3. The imidates and *N*-succinimidyl derivatives are considered the most selective for amino groups, followed by aryl halides [27]. Imidoesters readily react with amino groups with a high degree of specificity under mild conditions (pH 7 – 10) to form amidine derivatives [47]. Bis-succinimidyl derivatives readily react with amino groups at pH values of 6 to 9 [48]. Aryl halides also react preferentially with amine groups. Albumin was crosslinked with bis(3-nitro-4-fluorophenyl)sulfone, or (4,4′-difluoro-3,3′-dinitrodiphenylsulfone) [49]. Bifunctional aryl halides also react with tyrosine phenolic, thio, and imidazole groups [27].

The diisocyanates and diisothiocyanates generally react with amino groups to form stable urea and thiourea derivatives, respectively. Bis-nitrophenyl esters react with amino groups most rapidly, although their specificity is not very high [50]. Since the liberated nitrophenol has a distinct yellow color above pH 7, absorbance at 410 nm may be used to quantitate the reaction [27]. The same reaction was used to crosslink polymers containing *p*-nitrophenoxy groups with diamines as described in Chapter 3, Sections 3.2.2 and 3.3. The most important competitive reaction in an aqueous solution is hydrolysis of the esters. Bifunctional acylazides such as tartryl diazide also readily react with amine groups to produce amide bonds. Bis-epoxides were used as a crosslinking agent that reacts with primary amines [51].

Water-insoluble crosslinking agents are useful in the preparation of microparticles which are crosslinked only at the outer wall of the particles. Water-insoluble acyldichlorides, such as sebacoyl chloride, succinyl-dichloride, and terephthaloylchloride, have been used to prepare microcapsules crosslinked only at the outlayer. Interfacial polymerization and the emulsion-reticulation procedure were used to prepare interfacially prepared polyamide and crosslinked proteins, respectively [52,53].

Formaldehyde is not a dialdehyde but is capable of crosslinking amine functional groups [27]. Gelatin beads were prepared by crosslinking with formaldehyde [54]. Formaldehyde was used to prepare crosslinked gelatin beads and gelatin sponges or films which are used in arresting surgical hemorrhage [55]. Tissue-simulating gelatin gel was prepared by crosslinking gelatin with formaldehyde in the presence of ethylene glycol [56].

Of the many dialdehydes probably the most extensively used crosslinking agent is glutaraldehyde, which forms linkages resistant to extremes of pH and temperature [15]. The glutaraldehyde reaction is not a straightforward reaction of forming Schiff' s bases between aldehyde and amine groups. The stability of the glutaraldehyde-ligand linkages is greater than that of Schiff

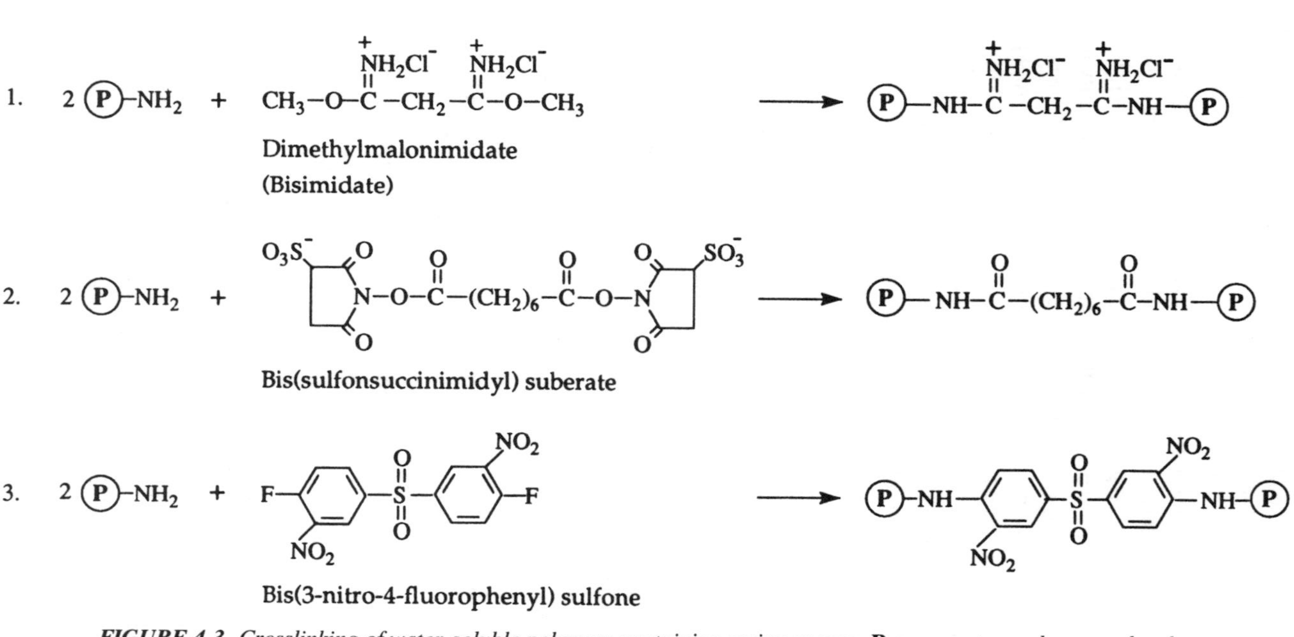

FIGURE 4.3. *Crosslinking of water-soluble polymers containing amine groups.* **P** *represents a polymer molecule.*

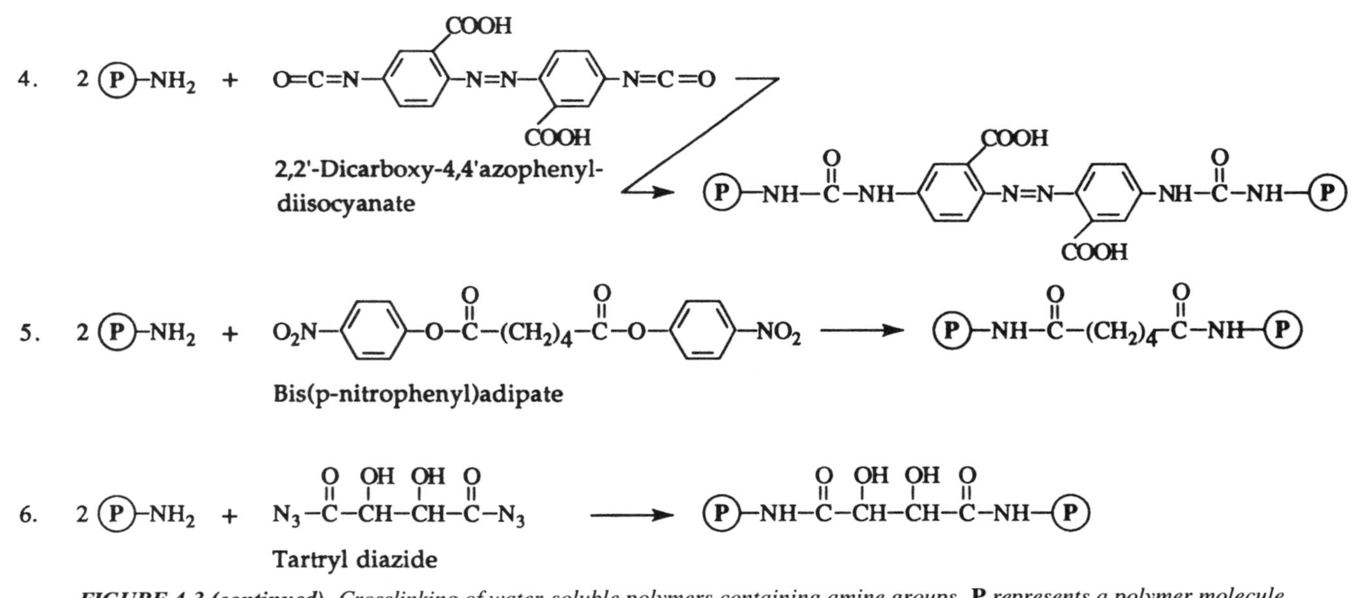

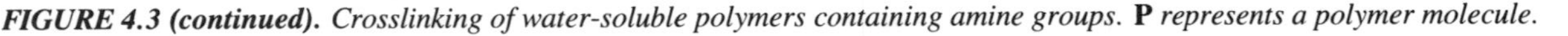

FIGURE 4.3 (continued). *Crosslinking of water-soluble polymers containing amine groups.* **P** *represents a polymer molecule.*

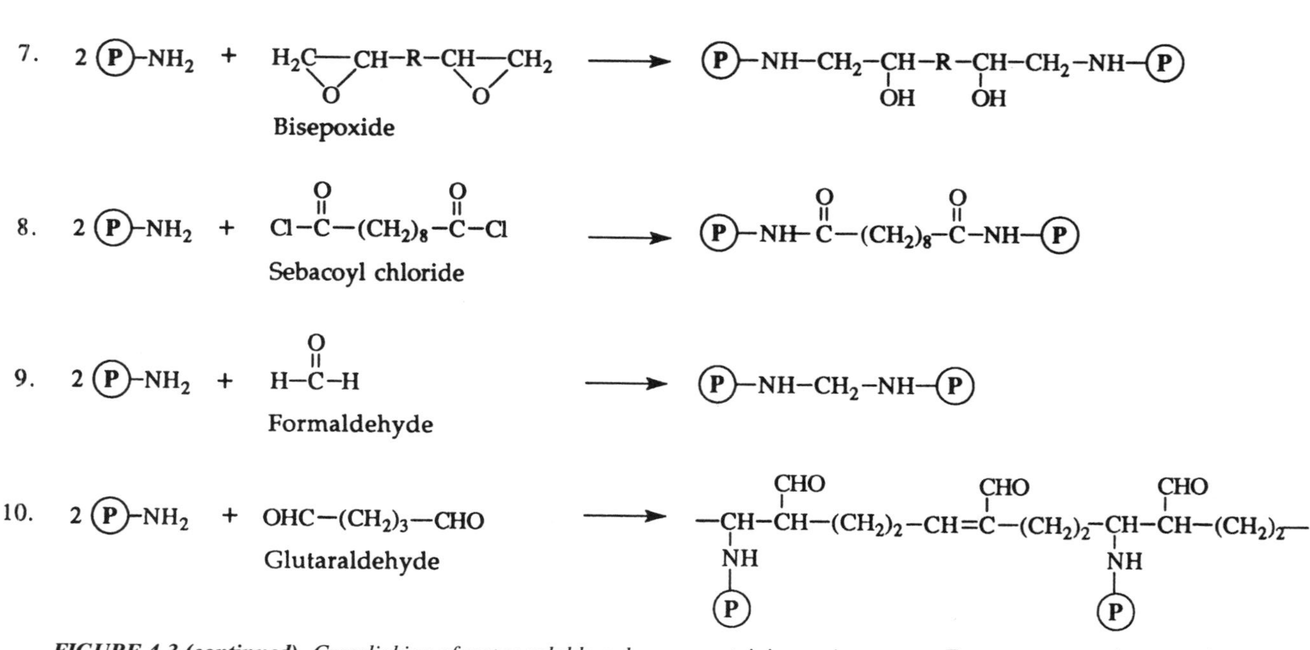

FIGURE 4.3 (continued). *Crosslinking of water-soluble polymers containing amine groups.* **P** *represents a polymer molecule.*

base linkages [15]. Most often an irreversible product is formed by glutaraldehyde. Glutaraldehyde is thought to be in equilibrium with α,β-unsaturated oligomers which react with amino groups [57].

A glutaraldehyde-crosslinked gelatin layer was used to enhance the biocompatibility of biomaterials such as a total artificial heart [58]. Gelatin microspheres crosslinked with glutaraldehyde were used in the delivery of low molecular weight drugs such as phenytoin [59], anticancer agents [60], and high molecular weight drugs such as interferon [61]. Hydrogels of partially-deacetylated chitin and chitosan (fully-deacetylated chitin) can be prepared by crosslinking with glutaraldehyde [62]. The chitin (or chitosan) hydrogel can be degraded by lysozyme. Bioactive compounds can be immobilized to chitosan gel microspheres by glutaraldehyde crosslinking. Albumin hydrogel containing urease was prepared by crosslinking with glutaraldehyde for enzyme-mediated drug delivery [63]. Albumin beads which were prepared using glutaraldehyde were also used for controlled delivery of a variety of drugs [64,65]. If crosslinked albumin is to be degraded by enzymes, the concentration of glutaraldehyde needs to be lower than 1 percent [66]. Albumin was also crosslinked to other proteins such as calmodulin using glutaraldehyde to form a Ca^{++}-responsible, extensible monolayer membrane [67]. Sometimes more than one crosslinking agent is used to enhance the intermolecular crosslinking. Free amino groups of collagenous materials were initially reacted with glutaraldehyde and then bridged by a diamino compound such as lysine. This approach produced more extensive intermolecular crosslinking than using glutaraldehyde alone [68].

4.2.1.3 CROSSLINKING WITH CARBOXYL GROUPS

There are relatively few crosslinking agents directed toward carboxyl groups. The epoxy group reacts, not only with amine groups, but also with carboxyl groups [69]. Polyepoxides were used to crosslink carboxyl groups of collagen, and bioprostheses such as porcine aortic leaflets or canine carotid arteries [69,70]. The treatment with polyepoxy compounds resulted in pronounced improvement in the biocompatibility of the bioprostheses.

In the presence of diamines, polymers containing carboxyl groups can be crosslinked by forming amide bonds. Amide bonds between carboxyl and amino groups can be formed by many reagents, such as carbodiimide, Woodward's reagent K, ethylchloroformate, carbonyldiimidazole, or *N*-carbalkoxydihydroquinoline [27]. These crosslinking agents induce amide bonds by removing atoms from the carboxyl and amine groups. Since no extrinsic spacer is added by these crosslinking agents, they are known as zero-length crosslinking agents.

Chondroitin sulfate was crosslinked with diaminododecane in the presence of dicyclohexylcarbodiimide [71]. Chondroitin sulfate possesses both sulfate and carboxyl groups and is readily soluble in water. The crosslinked chondroitin sulfate was used for colonic drug delivery, since it is degraded by the bacteroids of the large intestine.

4.2.1.4 CROSSLINKING WITH VINYL GROUPS

The macromolecules modified with vinyl groups (functionalized macromolecules) can be polymerized to produce hydrogels upon the initiation of radical formation by chemical initiation or γ-irradiation. Proteins such as albumin or gelatin and polysaccharides such as starch or dextran have been crosslinked in this way [6,7,23,72]. The crosslinking of functionalized macromolecules is schematically described in Figure 4.4.

4.2.1.4.1 Crosslinking by Free Radical Copolymerization

In addition to direct interaction between radicals on macromolecules, the functionalized macromolecules can be crosslinked by free radical copolymerization with a crosslinking agent containing vinyl groups. Glycidyl acrylate-derivatized hydroxyethyl starch was crosslinked with *N,N'*-methylenebisacrylamide by free radical formation [6].

In addition to bifunctional crosslinking agents, vinyl monomers can be copolymerized with macromolecules containing double bonds. As described in Chapter 3, Section 3.1.1.2, monomers such as vinylpyrrolidone can be polymerized to crosslink polyesters. If the concentration of vinyl monomers is larger than that of the functionalized macromolecule, then the latter will crosslink the vinyl polymer chains. Whether functional-

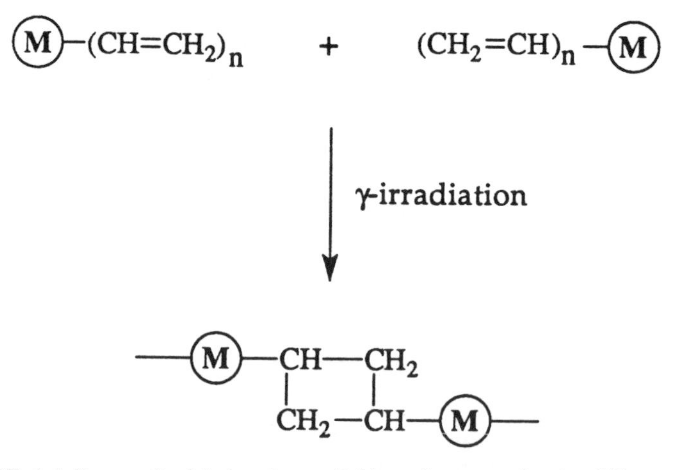

FIGURE 4.4. Free radical induced crosslinking of macromolecules (M) *containing unsaturated bonds.*

ized macromolecules become crosslinked or crosslink other polymer chains depends on the relative concentration of vinyl monomers. The choice of vinyl monomers is important, since it affects the overall hydrophilicity of the hydrogel.

4.2.1.5 THERMAL CROSSLINKING

Thermal denaturation has been used to crosslink proteins. Albumin, gelatin, and collagen have been crosslinked by thermal treatment [55,73–78]. Proteins become insoluble by formation of interchain amide links [73,74]. Proteins are usually heated at a high temperature (around 140°C) for a day and dehydrated at greater than 100°C for a few days.

The heat crosslinking method was used to prepare albumin microspheres [79]. The extent of crosslinking depends on the temperature and the time. Since the extent of crosslinking affects the *in vivo* biodegradation of albumin microspheres, it is possible to tailor the degradation characteristics of heat crosslinked albumin microspheres to meet specific drug delivery needs [79]. For example, albumin microspheres crosslinked at 135°C for 40 minutes degraded in two days while microspheres heated at 180°C for 18 hours remained intact for six months after intravenous injection. Heat denatured albumin microspheres were used to deliver various drugs including anticancer agents such as 5-fluorouracil or mitomycin C [79,80]. Heat denatured gelatin becomes water-insoluble, though it still swells to a certain extent [81].

Sodium carboxymethylcellulose, having both hydroxyl groups and carboxyl groups, was also crosslinked by thermal treatment ranging from about eight hours at 130°C to a few minutes at 210°C [82,83]. Direct esterification occurred between carboxylic acid groups and hydroxyl groups on the polymer chain. A portion of the carboxyl groups must be in the free acid form and the remaining portion must be in the salt form. The formed gel particles dissolved to give a smooth viscous solution at pH 10.5. There was a correlation between the degree of neutralization of the carboxyls and the extent of crosslinking [82,84].

4.2.1.6 CROSSLINKING BY ULTRASOUND

Ultrasound was used to prepare protein microspheres [85]. Ultrasound first disperses small droplets of oil or other organic liquids containing protein in water, similar to the emulsification process used to make mayonnaise. Ultrasound produces cavitation which can lead to homolysis of water molecules and produce radicals, HO· and H· [86]. In the presence of oxygen, the hydroxyl radicals undergo various reactions including the formation of superoxide ($HO_2\cdot$), a highly reactive species. The superoxide can crosslink protein molecules in the dispersed microspheres and form permanent structures. Crosslinking occurs when cysteine residues on the protein are oxidized to disulfide groups by superoxide [85].

4.2.2 Nonspecific Crosslinking

Irradiation of relatively inert molecules can generate two types of reactive intermediates: free radicals and carbenes or nitrenes [87].

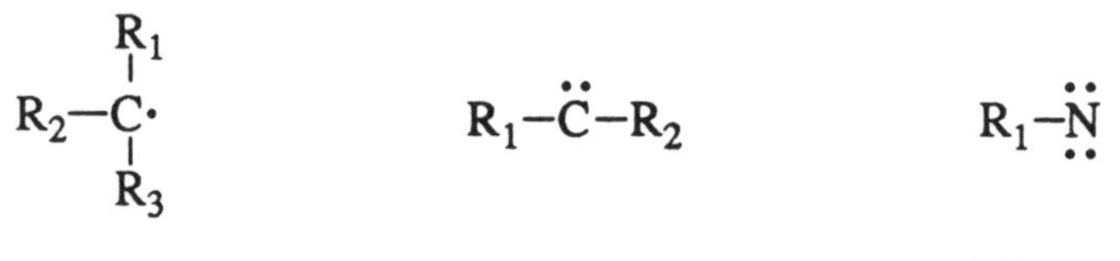

Free radical Carbene Nitrene

Homolytic cleavages of a single bond and a double bond produce free radicals and carbene or nitrene, respectively. Chemically stable polymers can be activated to form radicals by ionizing irradiation, such as gamma ray, or by small radicals. The radicals on the polymer interact to form covalent bonds. Carbenes and nitrenes are formed from photoactivatable groups. They are capable of attacking various chemical bonds including carbon-hydrogen bonds upon irradiation. Thus, the crosslinking by carbenes or nitrenes is nonspecific and nondiscriminatory.

4.2.2.1 CROSSLINKING BY FREE RADICALS

4.2.2.1.1 Ionizing Radiation

High energy radiation is radiation which can cause ionization in chemical compounds. It is the ionization that distinguishes the high energy radiation from photolysis. The ionization requires a minimum of about 10 eV, corresponding to 1,000 kJ/mol and a wavelength of 100 nm. The energy in the radiation is enormously in excess compared to the strength of chemical bonds which is in the range of 200–600 kJ/mol. Radiation, however, does not cause random cleavage of chemical bonds in molecules. Instead, radiation causes highly specific bond cleavage in molecules. Some chemical groups in the molecule, and certain bonds within these groups, have a particularly high probability of being broken [88]. The bond cleavage results in the formation of radicals. The radicals undergo various reactions which may crosslink or degrade polymers depending on their chemical structure. In fact, both crosslinking and degradation occur at the same time and the net effect depends on the process which predominates under irradiation conditions.

Table 4.1 shows two different types of polymers. The polymer tends to crosslink if each carbon atom of the main chain carries at least one hydrogen atom, and tends to degrade if a tetrasubstituted carbon atom is present [89]. The ultimate effect of crosslinking is the formation of a three-dimensional network. Thus, solutions of polymers could form hydrogels under irradiation if the polymer concentration is above a critical value. For acrylic acid to form gels by irradiation, the polymer should be predominantly in its free acid form, i.e., not charged [90]. Figure 4.5 shows the crosslinking of polymers by γ-irradiation. Radiation-induced degradation of polymers is due to random chain scissions. If oxygen is present during irradiation, additional reactions may take place leading to chain scissions.

Polypeptides can also be crosslinked to form hydrogels by high energy radiation. Recently, elastomeric polypeptide biomaterials were synthesized from sequential polypeptides such as poly(Val-Pro-Gly-Val-Gly), poly(Val-Pro-Gly-Gly), or poly(Gly-Ser-Asn-His-Gly) [91]. The polypeptides were crosslinked by 20 Mrads of γ-irradiation.

4.2.2.1.2 UV Radiation

In addition to ionizing radiation, ultraviolet (UV) radiation can also be used to form free radicals. UV radiation, however, is more limited in that the depth of penetration of a sample by UV light is considerably less than that by ionizing radiation. Thus, the UV-induced crosslinking reaction of macromolecules occurs only in the thin surface layer. This property, however, may be useful if it is desirable to crosslink only the surface layer of the polymer.

Polymers which are resistant to UV light can be crosslinked if they are irradiated in the presence of a photosensitizer such as benzophenone or benzoin. A photosensitizer, or photoinitiator, in its excited state can abstract hydrogen from a hydrocarbon chain to form free radicals [92]. The free

TABLE 4.1. Crosslinking versus Degradation in Irradiated Polymers [89].

Crosslinking Polymers	Degrading Polymers
Poly(acrylic acid)	Poly(methacrylic acid)
Poly(methyl acrylate)	Poly(methyl methacrylate)
Polyacrylamide	Polymethacrylamide
Polyvinylpyrrolidone	
Poly(vinyl alcohol)	
Polyacrolein	
Polyamides	
Polyesters	

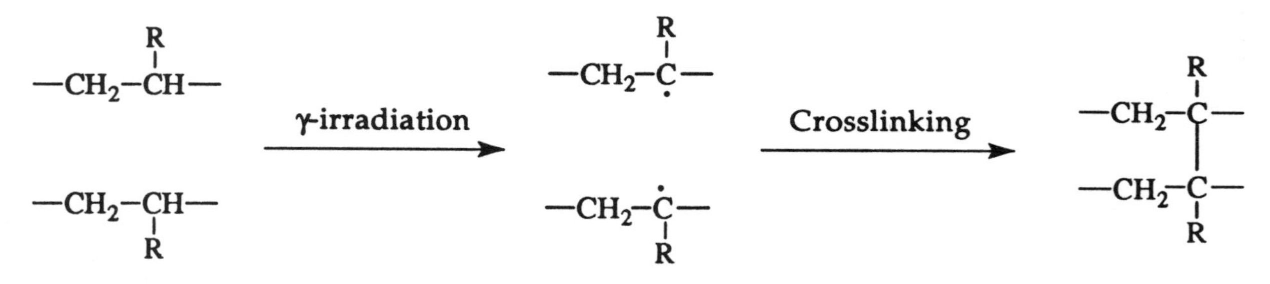

FIGURE 4.5. Crosslinking of polymers by γ-irradiation.

radicals prefer to abstract a hydrogen atom from C–H bonds rather than from O–H bonds. Thus, the presence of solvent water molecules will not hinder the reaction of radicals [87]. The recombination of the formed macroradicals results in crosslinking in much the same way as radiation-induced crosslinking.

UV irradiation of proteins at 254 nm produces radicals in the form of unpaired electrons located predominantly on the nuclei of the aromatic residues such as tyrosine and phenylalanine. Bonding between these two radicals introduces crosslinking. Thus, proteins can be crosslinked by exposure to UV light [93,94].

4.2.2.2 CROSSLINKING BY NITRENES OR CARBENES

Nitrenes are obtained by photolysis of several azido compounds, such as alkyl azides, acyl azides, or aryl azides. Carbenes are formed upon irradiation of diazo compounds such as α-diazoketones, aryldiazomethanes, or aryldiazirines [87].

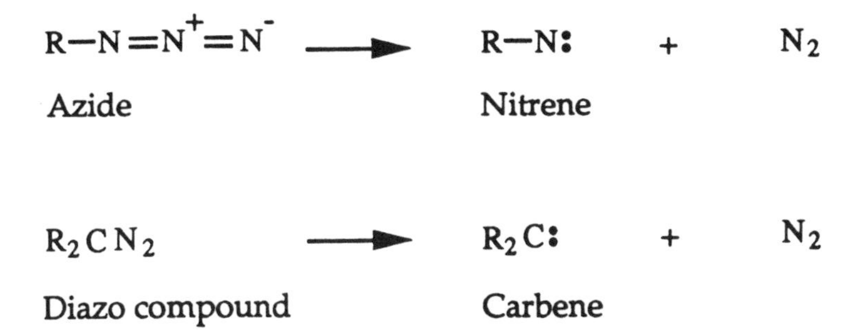

Carbenes and nitrenes have similar reactions, although the reactivity of nitrenes is much lower than that of carbenes. Nitrenes discriminate much more between primary, secondary, and tertiary C–H bonds [87]. They are capable of attacking various chemical bonds including carbon-hydrogen bonds upon irradiation [95]. Among the precursors of carbenes and nitrenes, aryl azides are most widely used since they are chemically stable and much less susceptible to rearrangements to less reactive intermediates [95]. In addition, aryl azides can be photolyzed at longer wavelengths ($\lambda > 300$ nm) [87]. Alkyl azides have absorption maxima in the UV region. Acyl azides, sulfonyl azides, and phosphoryl azides are not used as photoactivatable reagents because of their nucleophilic reactivity in the dark [87].

Photolysis of azides in the 250–380 nm region in dilute solution leads to the elimination of nitrogen and the formation of electron-deficient nitrenes. Nitrenes, which can exist in both the singlet and the triplet states, are extremely reactive and can undergo a multitude of reactions, such as group migration yielding imines, hydrogen abstraction yielding amines, coupling reactions yielding azo-compounds, insertion into double bonds yielding aziridines, and insertion into C–H, N–H, and O–H bonds [96,97]. Due to this nondiscriminatory reactivity, reaction in water is significantly hindered by abundant water molecules [95,98].

Crosslinking can also be achieved by reagents with one photoreactive group and one conventional functional group, or with two photoreactive groups. Photoreactive heterobifunctional crosslinking agents are first linked to polymers in the dark through group directed reaction. The labeled polymer is then irradiated to activate the photoreactive group. Since photoreactive functional groups are inert until they are photolyzed, this approach provides crosslinking of polymers at a particular moment.

Photoreactive crosslinking agents have been used in various applications including ligand-receptor interactions [95], protein-protein interactions [99], polymer grafting to the surface [100], and modification of polymer surfaces [101].

4.2.3 Determination of the Degree of Crosslinking

The crosslinking density is important in relating the molecular structure of a hydrogel to the gel properties [12]. The degree of crosslinking in hydrogels can be determined by measuring either equilibrium volume swelling or equilibrium elastic moduli [102]. The latter values can be obtained from elongation or compression experiments.

4.2.3.1 EQUILIBRIUM SWELLING METHOD

A hydrogel swells in water just like uncrosslinked polymer molecules dissolve in water. A hydrogel reaches a state of equilibrium swelling in which the swelling and elastic retractive forces are in balance [102].

If the crosslinks are introduced in the unswollen polymer (i.e., dry polymer), the theory on the conditions for equilibrium leads to

$$\ln(1-\phi_2)+\phi_2+\chi\phi_2{}^2+v_1(\nu_e/V_o)(\phi_2{}^{1/3}-\phi_2/2)=0 \qquad (4.1)$$

where ϕ_2 is the volume fraction of polymer at equilibrium (i.e., in the fully swollen sample), χ is the Flory-Huggins interaction parameter, v_1 is the molar volume of water, ν_e is the effective number of chains in the network, and V_o is the volume of the unswollen (i.e., dry) polymer network [102]. The ν_e/V_o is the effective network chain concentration or effective chain density (in the unit of mol/cm^3), which is directly proportional to crosslink density. The relationship between the effective number of chains ν_e and the number of crosslinks μ depends on the crosslink functionality Ψ in the following way [103].

$$\mu = (2/\Psi)\nu_e \tag{4.2}$$

Thus, for a tetrafunctional crosslinking ($\Psi = 4$) where four chain segments emanate from the crosslink, the crosslink density μ is simply a half of the effective number of chains in the network ν_e.

In the preparation of hydrogels, crosslinks are frequently introduced while the polymers are in a diluted state, i.e., in the presence of water. For example, polymer chains dissolved in water are crosslinked by added crosslinking agents. Monomers dissolved in water are polymerized in the presence of crosslinking agents. In this case, it is necessary to add to Equation (4.1) a factor $Q_o^{-2/3}$, where Q_o represents a reference degree of swelling [104].

$$\ln(1 - \phi_2) + \phi_2 + \chi\phi_2^{\,2} + v_1(\nu_e/V_o)(\phi_2^{\,1/3}Q_o^{\,-2/3} - \phi_2/2) = 0 \tag{4.3}$$

$1/Q_o$ is equal to ϕ_o, which is the volume fraction of the polymer in the system or the monomer in the polymerization mixture during the crosslinking procedure. Thus, $Q_o^{-2/3}$ can be substituted with $\phi_o^{2/3}$ and we have

$$\ln(1 - \phi_2) + \phi_2 + \chi\phi_2^{\,2} + v_1(\nu_e/V_o)(\phi_2^{\,1/3}\,\phi_o^{\,2/3} - \phi_2/2) = 0 \tag{4.4}$$

The values of the molar volume of water v_1 (in ml/mol) at different temperatures T (in absolute temperature K) can be obtained from the following equation [105].

$$v_1 = 18.05 + 3.6 \times 10^{-3}(T - 298) \tag{4.5}$$

If the degree of swelling is large, so that the volume fraction ϕ_2 is small, then Equation (4.4) can be simplified to [102,106,107]

$$(0.5 - \chi)\phi_2^{\,2} = v_1(\nu_e/V_o)\,\phi_2^{\,1/3}\,\phi_o^{\,2/3} \tag{4.6}$$

The volume fraction of polymer ϕ_2 is inversely related to the volume swelling ratio Q, which is the volume ratio of the swollen to the unswollen hydrogel. If ϕ_2 is replaced by $1/Q$, then Equation (4.6) becomes on rearrangement

$$\nu_e/V_o = (0.5 - \chi)/(v_1\phi_o^{\,2/3}\,Q^{5/3}) \tag{4.7}$$

Thus, the crosslinking density can be determined easily by measuring the volume swelling ratio of hydrogels and from a known value of χ. This allows calculation of the average molecular weight between crosslinks M_c, which is simply the density of the polymer ϱ_2 (g/cm^3) divided by ν_e/V_o

[103,106]. Alternatively, if ν_e/V_o is known from the measurement of the modulus (see below), χ can be calculated from the swelling data [108]. More detailed information on the calculation of crosslink density and M_c can be found in a recent review [109].

The χ values of some water-soluble polymers are listed in Table 4.2. Although the χ value of a polymer is supposed to be independent of the polymer concentration in solution and the temperature, the χ value is known to depend on both parameters [110].

4.2.3.2 ELASTIC MODULI METHOD

Both elastic and shear moduli are used for the determination of the crosslink density. The modulus can be measured on the dried hydrogel if it remains rubbery. Usually, the modulus is measured directly on the equilibrium swollen hydrogels. The modulus equations derived from the theory of rubber elasticity are for dry materials, but they can be used to describe the modulus of the swollen gels [108].

If the polymers are crosslinked in the dry state, the effective network chain concentration ν_e/V_o of crosslinked polymer can be calculated from the elongation experiments using the following equation [41,102,105]

$$\tau = (\nu_e/V_o)RT\phi_2^{\ -1/3}(\alpha - \alpha^{-2}) \tag{4.8}$$

where τ is the applied force per unit area of dry, unstrained hydrogel (i.e., stress), R is the gas constant, T is absolute temperature, and α is the deformation ratio. The deformation ratio α is the ratio of elastically deformed length L to undeformed length L_o of the hydrogel (i.e., L/L_o).

Equation (4.8) was transformed into an equation which is suitable for the measurement of the crosslinking density from a simple compression experiment [41]. The following is a procedure described by Cluff et al. [42]. A

TABLE 4.2. Flory-Huggins Interaction Parameters (χ) of Selected Polymers [111 – 115].

Polymer	χ
Hydroxypropylcellulose	0.482
Poly(acrylic acid)	0.498
Poly(methacrylic acid)	0.499
Polyacrylamide	0.44, 0.48
Polymethacrylamide	0.490
Poly(ethylene oxide)	0.45
Polyvinylpyrrolidone	0.48
Poly[*N*-(2-hydroxypropyl)methacrylamide]	0.48
Poly(vinyl alcohol)	0.8
Poly(hydroxyethyl methacrylate)	0.88

hydrogel in a cylindrical or disc form is allowed to swell to equilibrium in water. A swollen gel is then placed beneath the plate which is connected to a micrometer gauge. The hydrogel and the plate may be immersed in water to prevent evaporation of water during measurement. Weights are applied in increasing amounts and the deflection (difference between undeformed and deformed swollen heights) is recorded for each weight. After each measurement, the weights are completely removed and the hydrogel is allowed to recover to its initial height before the next weight is added. The maximum load is chosen by the amount of compression which usually does not exceed 5 percent of the initial height. The force F corresponding to deflection ΔL is plotted, and the slope of the F-ΔL plot is used to compute the effective network chain concentration ν_e/V_o using the following equation

$$\nu_e/V_o = L_oS/3A_oRT \tag{4.9}$$

where L_o is the height of the undeformed, unswollen hydrogel, S is the slope of the F-ΔL plot, A_o is the cross-sectional area of unswollen gel, R is the gas constant, and T is absolute temperature. Compression modulus is calculated by Sh_s/A_s where h_s is the undeformed swollen height and A_s is the swollen area. This method was used to calculate the effective network chain concentration of poly(vinyl alcohol) gels crosslinked in the dried state [41].

As mentioned in the previous section, most hydrogels are prepared while polymer chains or monomers are dissolved in water. Thus, in the calculation of ν_e/V_o for most hydrogels, Equation (4.8) should contain the term $\phi_o^{2/3}$ as in Equation (4.3) [103,104]. Thus, we have

$$\tau = (\nu_e/V_o)RT\phi_2^{-1/3}\phi_o^{2/3}(\alpha - \alpha^{-2}) \tag{4.10}$$

From a plot of τ vs. $(\alpha - \alpha^{-2})$, which is linear at low deformations (i.e., strains), ν_e/V_o can be calculated. Please note that V_o is the volume of dried hydrogel as defined earlier. Equation (4.10) does not consider volume changes during deformation. If the swelling ratio changes significantly from Q to Q' (or volume fraction changes from ϕ_2 to ϕ) when the freely swollen gel is subjected to a unidirectional strain, the last term in Equation (4.10) has to be changed to $(\alpha - \phi_2/\phi\alpha^2)$ [104]. It should be noted that τ is the applied force per unit area of dry, unstrained hydrogel. If the applied force per unit area of the swollen, unstretched sample, τ_s, is used, then Equation (4.10) is changed to

$$\tau_s = (\nu_e/V_o)RT\phi_2^{1/3}\phi_o^{2/3}(\alpha - \alpha^{-2}) \tag{4.11}$$

Since $\phi_2 = V_o/V$, Equation (4.11) is also equivalent to

$$\tau_s = (\nu_e/V)RT\phi_2^{-2/3}\phi_o^{2/3}(\alpha - \alpha^{-2}) \tag{4.12}$$

where V is the volume of the equilibrium swollen hydrogel. The effective network chain concentration ν_e/V_o is decreased in the swollen gel simply by a factor of ϕ_2 [108]. Equations (4.8), (4.10), and (4.11) can also be used for the compression experiments. In that case, $(\alpha - \alpha^{-2})$ is replaced by $-(\alpha - \alpha^{-2})$ [116].

Since most hydrogels have poor mechanical strength and toughness, it is easier to measure the compression modulus than the extension modulus [41]. Compression measurements, however, may not be as straightforward as the elongation measurements due to sample distortion resulting from friction between the sample and the compression plates [103]. In the case of an unswollen network, compression modulus approximates the low deformation elongation modulus. On the other hand, the compression modulus of highly swollen hydrogels is about the same as the elongation modulus [103].

4.3 REFERENCES

1. Chujo, Y. and Y. Yamashita. 1989. "Macromonomers," Chapter 8 in *Telechelic Polymers: Synthesis and Applications*, E. J. Goethals, ed., Boca Raton, FL: CRC Press.
2. Plate, N. A., V. A. Postnikov, N. Yu. Lukin, M. Yu. Eismont and G. Grudek. 1982. "Acylation of Serum Albumin by Unsaturated Acid Chlorides," *Polymer Sci. U.S.S.R.*, 24:2668–2671.
3. Plate, N. A., A. V. Malykh, L. D. Uzhinova and V. V. Mozhayev. 1989. "Trypsin Macromonomer and Its Copolymerization with Hydrophilic Monomers," *Vysokomol. Soyed.*, A.31:195–197.
4. Laakso, T. and I. Sjöholm. 1987. "Biodegradable Microspheres X: Some Properties of Polyacryl Starch Microparticles Prepared from Acrylic Acid-Esterified Starch," *J. Pharm. Sci.*, 76:935–939.
5. Edman, P., B. Eckman and I. Sjöholm. 1980. "Immobilization of Proteins in Microspheres of Biodegradable Polyacryldextran," *J. Pharm. Sci.*, 69:838–842.
6. Artursson, P., P. Edman, T. Laakso and I. Sjöholm. 1984. "Characterization of Polyacryl Starch Microparticles as Carriers for Proteins and Drugs," *J. Pharm. Sci.*, 73:1507–1513.
7. Heller, J., S. H. Pangburn and K. V. Roskos. 1990. "Development of Enzymatically Degradable Protective Coatings for Use in Triggered Drug Delivery Systems: Derivatized Starch Hydrogels," *Biomaterials*, 11:345–350.
8. Guiseley, K. B. 1987. "Natural and Synthetic Derivatives of Agarose and Their Use in Biomedical Separations," in *Industrial Polysaccharides: Genetic Engineering, Structure/Property Relations and Applications*, M. Yalpani, ed., Amsterdam, The Netherlands: Elsevier Science Publishers, pp. 139–147.
9. Ito, K., K. Tanaka, H. Tanaka, G. Imai, S. Kawaguchi and S. Itsuno. 1991. "Poly(ethylene Oxide) Macromonomers. 7. Micellar Polymerization in Water," *Macromolecules*, 24:2348–2354.
10. Plate, N. A., A. V. Malykh, L. D. Uzhinova and V. V. Mozhayev. 1989. "Trypsin

Macromonomer and Its Copolymerization with Hydrophilic Monomers,'' *Polymer Sci. U.S.S.R.*, 31:216–219.

11. Jaworek, D. 1974. ''New Methods for Covalent Binding of Proteins to Synthetic Polymers,'' in *Insolubilized Enzymes*, M. Salmona, C. Saronio and S. Garattini, eds., New York, NY: Raven Press, pp. 65–76.
12. Finch, C. A. 1983. ''Chemical Modification and Some Cross-Linking Reactions of Water-Soluble Polymers,'' in *Chemistry and Technology of Water-Soluble Polymers*, C. A. Finch, ed., New York, NY: Plenum Press, pp. 81–111.
13. Torchilin, V. P., A. V. Maksimenko, V. N. Smirnov, I. V. Berezin, A. M. Klibanov and K. Martinek. 1979. ''The Principles of Enzyme Stabilization. IV. Modification of 'Key' Functional Groups in the Tertiary Structure of Proteins,'' *Biochim. Biphys. Acta*, 567:1–11.
14. Torchilin, V. P., E. G. Tischenko, V. N. Smirnov and E. I. Chazov. 1977. ''Immobilization of Enzymes on Slowly Soluble Carriers,'' *J. Biomed. Mater. Res.*, 11:223–235.
15. Sturgeon, C. M. 1988. ''The Synthesis of Polysaccharide Derivatives,'' Chapter 13 in *Carbohydrate Chemistry*, J. F. Kennedy, ed., New York, NY: Oxford University Press.
16. Harris, F. W. 1980. ''Polymers Containing Pendent Pesticide Substituents,'' Chapter 3 in *Controlled Release Technologies: Methods, Theory, and Applications, Vol. II*, A. F. Kydonieus, ed., Boca Raton, FL: CRC Press.
17. Plate, N. A., A. V. Malykh, L. D. Uzhinova, V. P. Panov and M. A. Rozenfel'd. 1989. ''Structure of the Heparin Macromonomer and Features of Its Radical Polymerization,'' *Polymer Sci. U.S.S.R.*, 31:220–226.
18. Morrison, R. T. and R. N. Boyd. 1973. *Organic Chemistry, Third Edition*. Boston, MA: Allyn and Bacon, Inc., p. 186.
19. Connors, K. A. 1982. *A Textbook of Pharmaceutical Analysis, Third Edition*. New York, NY: A Wiley-Interscience Publication, pp. 568–573.
20. Lepistö, M., P. Artursson, P. Edman, T. Laakso and I. Sjöholm. 1983. ''Determination of the Degree of Derivatization of Acryloylated Polysaccharides by Fourier Transform Protion NMR Spectroscopy,'' *Anal. Biochem.*, 133:132–135.
21. Hoppe, H., J. Koppe and F. Winkler. 1977. ''Eine Verbesserte Methode zur Bestimmung des Akrylamids im Polyakrylamid,'' *Plaste Kautsch*, 24:105.
22. Schirmer, R. E. 1982. *Modern Methods of Pharmaceutical Analysis, Vol. I*. Boca Raton, FL: CRC Press, p. 137.
23. Artursson, P., P. Edman and I. Sjöholm. 1984. ''Biodegradable Microspheres. I. Duration of Action of Dextranase Entrapped in Polyacrylstarch Microparticles *in Vivo*,'' *J. Pharmacol. Exp. Ther.*, 231:705–712.
24. Park, K. 1988. ''Enzyme-Digestible Swelling Hydrogels as Platforms for Long-Term Oral Drug Delivery: Synthesis and Characterization,'' *Biomaterials*, 9:435–441.
25. Snyder, S. L. and P. Z. Sobocinski. 1975. ''An Improved 2,4,6-Trinitrobenzenesulfonic Acid Method for the Determination of Amines,'' *Anal. Biochem.*, 64:284–288.
26. Kondo, T., A. Ishizu and J. Nakano. 1989. ''Preparation of Glycidyl Celluloses from Completely Allylated Methylcellulose and Tri-O-allylcellulose,'' *J. Appl. Polymer Sci.*, 37:3003–3009.
27. Odian, G. 1991. *Principles of Polymerization, Third Edition*. New York, NY: John Wiley & Sons, Inc., Chapter 9.

28. Wong, S. S. 1991. *Chemistry of Protein Conjugation and Cross-Linking*. Boca Raton, FL: CRC Press, Chapters 4, 5 and 6.

29. White, C. A. and J. F. Kennedy. 1980. "Popular Matrices for Enzyme and Other Immobilization," *Enzyme Microb. Technol.*, 2:82–90.

30. Schacht, E. H., G. E. Desmarets and E. J. Goethals. 1981. "Synthesis and Properties of Urea Formaldehyde Resins Generating 2,6-Dichlorobenzonitrile," in *Controlled Release of Pesticides and Pharmaceuticals*, D. H. Lewis, ed., New York, NY: Plenum Press, pp. 159–170.

31. Maksimenko, A. V., L. A. Nadirashvili, A. D. Romaschenko, G. S. Erkomaishvili, V. V. Abramova and V. P. Torchilin. 1989. "Complex Papaya Proteinases Modified with Soluble Polymer and Its Possible Medical Application," *J. Controlled Rel.*, 10:131–143.

32. Larsen, C. 1989. "Dextran Prodrugs–Structure and Stability in Relation to Therapeutic Activity," *Adv. Drug Delivery Reviews*, 3:103–154.

33. Heller, J. 1988. "Chemically Self-Regulated Drug Delivery Systems," *J. Controlled Rel.*, 8:111–125.

34. Amiya, T. and T. Tanaka. 1987. "Phase Transition in Cross-Linked Gels of Natural Polymers," *Macromolecules*, 20:1162–1164.

35. Rehab, A., A. Akelah, R. Issa, S. D'Antone, R. Solaro and E. Chiellini. 1991. "Controlled Release of Herbicides Supported on Polysaccharide Based Hydrogels," *J. Bioact. Compat. Polymers*, 6:52–63.

36. Rothman, U., K.-E. Arfors, K. F. Aronsen, B. Lindell and G. Nylander. 1976. "Enzymatically Degradable Microspheres for Experimental and Clinical Uses," *Microvasc. Res.*, 11:421–430.

37. Mateescu, M. A. and H. D. Schell. 1983. "A New Amyloclastic Method for the Selective Determination of a-Amylase Using Cross-Linked Amylose as an Insoluble Substrate," *Carbohydrate Res.*, 124:319–323.

38. Lenaerts, V., Y. Dumoulin and M. A. Mateescu. 1991. "Controlled Release of Theophylline from Cross-Linked Amylose Tablets," *J. Controlled Rel.*, 15:39–46.

39. Lévy, M.-C. and M.-C. Andry. 1990. "Microcapsules Prepared through Interfacial Cross-Linking of Starch Derivatives," *Int. J. Pharm.*, 62:27–35.

40. Doane, W. M., B. S. Shasha and C. R. Russell. 1977. "Encapsulation of Pesticides within a Starch Matrix," in *Controlled Release Pesticides, ACS Symp. Ser., Vol. 53*, H. B. Sher, ed., pp. 74–83.

41. Watler, P. K., C. H. Cholakis and M. V. Sefton. 1988. "Water Content and Compression Modulus of Some Heparin-PVA Hydrogels," *Biomaterials*, 9:150–154.

42. Cluff, E. F., E. K. Gladding and R. Pariser. 1960. "A New Method for Measuring the Degree of Crosslinking in Elastomers," *J. Polymer Sci.*, 45:341–345.

43. Shukla, P., S. Sivaram and B. Mohanty. 1992. "Structure of Dynamics of Starch Cross-Linked with Urea-Formaldehyde Polymers by ^{13}C CP/MAS NMR Spectroscopy," *Macromolecules*, 25:2746–2751.

44. Shukla, P. G., N. Rajagopalan, C. Bhaskar and S. Sivaram. 1991. "Crosslinked Starch-Urea Formaldehyde (St-UF) as a Hydrophilic Matrix for Encapsulation: Studies in Swelling and Release of Carbofuran," *J. Controlled Release*, 15:153–166.

45. Laakso, T., P. Stjärnikvis and I. Sjöholm. 1987. "Biodegradable Microspheres VI: Lysosomal Release of Covalently Bound Antiparasitic Drugs from Starch Microparticles," *J. Pharm. Sci.*, 76:134–140.

46. Patwardhan, S. A. and K. G. Das. 1983. "Chemical Methods of Controlled Release," Chapter 3 in *Controlled-Release Technology. Bioengineering Aspects*, K. G. Das, ed., New York, NY: John Wiley & Sons.

47. Hunter, M. J. and M. L Ludwig. 1972. "Amidination," *Methods Enzymol.*, 25:585–596.

48. Cuatrecasas, P. and I. Parikh. 1972. "Adsorbents for Affinity Chromatography. Use of *N*-Hydroxysuccinimide Esters of Agarose," *Biochem.*, 11:2291–2299.

49. Wold, F. 1961. "Reaction of Bovine Serum Albumin with the Bifunctional Reagent *p,p'*-Difluoro-*m,m'*-dinitrodiphenylsulfone," *J. Biol. Chem.*, 236:106–111.

50. Plotz, P. H. 1977. "Bivalent Affinity Labeling Haptens in the Formation of Model Immune Complexes," *Methods Enzymol.*, 46:505–508.

51. Pishko, M. V., A. C. Michael and A. Heller. 1991. "Amperometric Glucose Microelectrodes Prepared through Immobilization of Glucose Oxidase in Redox Hydrogels," *Anal. Chem.*, 63:2268–2272.

52. Lévy, M.-C., P. Rambourg, J. Lévy and G. Porton. 1982. "Microencapsulation IV: Cross-Linked Hemoglobin Microcapsules," *J. Pharm. Sci.*, 71:759–762.

53. Rambourg, P., J. Lévy and M.-C. Lévy. 1982. "Microencapsulation III: Preparation of Invertase Microcapsules," *J. Pharm. Sci.*, 71:753–758.

54. Chiao, C. S. L. and J. C. Price. 1989. "Modification of Gelatin Beadlets for Zero-Order Sustained Release," *Pharm. Res.*, 6:517–520.

55. Heller, J. 1987. "Bioerodible Hydrogels," Chapter 7 in *Hydrogels in Medicine and Pharmacy, Vol. III. Properties and Applications*, N. A Peppas, ed., Boca Raton, FL: CRC Press.

56. Companion, J. A. 1992. "Tissue-Simulating Gel for Medical Research," *NASA Tech Briefs*, 16:80.

57. Hardy, P. M., A. C. Nicholls and H. N. Rydon. 1976. "The Nature of the Cross-linking of Proteins by Glutaraldehyde. Part I. Interaction of Glutaraldehyde with the Amino Groups of 6-Aminohexanoic Acid and of a-*N*-Acetyl-lysine," *J. Chem. Soc. Perkin Trans.*, 1:958–962.

58. Emoto, H., H. Kambic, H. Harasaki and Y. Nose. 1990. "*In vitro* Analysis of Plasma Protein Diffusion in Crosslinked Gelatin Coatings Used for Blood Pumps," in *Progress in Biomedical Polymers*, C. G. Gebelein and R. L. Dunn, eds., New York, NY: Plenum Press, pp. 229–238.

59. Raymond, G., M. Degennard and R. Mikeal. 1990. "Preparation of Gelatin:Phenytoin Sodium Microspheres: An *in vitro* and *in vivo* Evaluation," *Drug Dev. Ind. Pharm.*, 16:1025–1051.

60. Yan, C., X. Li, X. Chen, D. Wang, D. Zhong, T. Tan and H. Kitano. 1991. "Anticancer Gelatin Microspheres with Multiple Functions," *Biomaterials*, 12:640–644.

61. Tabata, Y. and Y. Ikada. "Synthesis of Gelatin Microspheres Containing Interferon," *Pharm. Res.*, 6:422–427.

62. Pangburn, S. H., P. V. Trescony and J. Heller. 1982. "Lysozyme Degradation of Partially Deacetylated Chitin, Its Films and Hydrogels," *Biomaterials*, 3:105–108.

63. Heller, J. and P. V. Trescony. 1979. "Controlled Drug Release by Polymer Dissolution II: Enzyme-Mediated Delivery Device," *J. Pharm. Sci.*, 68:919–921.

64. Morimoto, Y. and S. Fujimoto. 1985. "Albumin Microspheres as Drug Carriers," *Critical Reviews in Therapeutic Drug Carrier Systems*, 2:19–63.

65. Sheu, M.-T. and T. D. Sokoloski. 1991. "Entrapment of Bioactive Compounds within Native Albumin Beads: IV. Characterization of Drug Release from Polydisperse Systems," *Int. J. Pharm.*, 71:7–18.
66. Lee, T. K., T. D. Sokoloski and G. P. Royer. 1981. "Serum Albumin Beads: An Injectible, Biodegradable System for the Sustained Release of Drugs," *Science*, 213:233–235.
67. Miwa, T., E. Kobatake, Y. Ikariyama and M. Aizawa. 1991. "Ca^{2+}-Responsive Extensible Monolayer Membrane of Calmodulin-Albumin Conjugates," *Bioconjugates Chem.*, 2:270–274.
68. Simionescu, A., D. Simionescu and R. Deac. 1991. "Lysine-Enhanced Glutaraldehyde Crosslinking of Collagenous Biomaterials," *J. Biomed. Mater. Res.*, 25:1495–1505.
69. Imamura, E., O. Sawatani, H. Koyangi, Y. Noishiki and T. Miyata. 1989. "Epoxy Compounds as a New Crosslinking Agent for Porcine Aortic Leaflets: Subcutaneous Implant Studies in Rats," *J. Cardiac Surg.*, 4:50–57.
70. Nojiri, C., Y. Noishiki and H. Koyanagi. 1987. "Aorta-Coronary Bypass Grafting with Heparinized Vascular Grafts in Dogs," *J. Thorac. Cardiovasc. Surg.*, 93:867–877.
71. Rubinstein, A., D. Nakar and A. Sintov. 1992. "Colonic Drug Delivery: Enhanced Release of Indomethacin from Cross-Linked Chondroitin Matrix in Rat Cecal Content," *Pharm. Res.*, 9:276–278.
72. Kamath, K. R. and K. Park. In press. "Use of Gamma-Irradiation for the Preparation of Hydrogels from Natural Polymers," *Proceed. Intern. Symp. Control. Rel. Bioact. Mater.*
73. Yannas, I. V. and A. V. Tobolsky. 1967. "Cross-Linking of Gelatine by Dehydration," *Nature*, 215:509–510.
74. Yannas, J. B. and A. V. Tobolsky. 1966. "Transitions in Gelatin-Nonaqueous-Diluent Systems," *J. Macromol. Chem.*, 1:723–737.
75. Wetzel, R., M. Becker, J. Behlke, H. Billwitz, S. Bohm, B. Ebert, H. Hamann, J. Krumbiegel and G. Lassmann. 1980. "Temperature Behavior of Human Serum Albumin," *Eur. J. Biochem.*, 104:469–478.
76. Senyei, A. E., S. D. Reich, C. Gonczy and K. J. Widder. 1981. "*In vivo* Kinetics of Magnetically Targeted Low-Dose Doxorubicin," *J. Pharm. Sci.*, 70:389–391.
77. Burgess, D. J. and J. E. Carless. 1987. "Manufacture of Gelatin/Gelatin Coacervate Microcapsules," *Int. J. Pharm.*, 27:61–70.
78. Burgess, D. J., S. S. Davis and E. Tomlinson. 1987. "Potential Use of Albumin Microspheres as a Drug Delivery System. I. Preparation and *in vitro* Release of Steroids," *Int. J. Pharm.*, 39:129–136.
79. Yapel, A. F., Jr. 1985. "Albumin Microspheres: Heat and Chemical Stabilization," *Meth. Enzymology*, 112:3–18.
80. Fujimoto, S., M. Miyazaki, F. Endoh, O. Takahashi, K. Okui and Y. Morimoto. 1985. "Biodegradable Mitomycin C Microspheres Given Intraarterially for Inoperable Hepatic Cancer," *Cancer*, 56:2404–2410.
81. Alexander, J. 1937. *Colloid Chemistry. Principles and Applications, 4th Edition*. New York, NY: D. Van Nostrand Company, p. 147.
82. Reid, A. R. U.S. patent 3,379,720, April 23, 1968.
83. Meltzer, Y. L. 1976. *Water-Soluble Resins and Polymers.* Park Ridge, NJ: Noyes Data Corp., pp. 78–80.

84. Gross, J. R. 1990. "The Evolution of Absorbent Materials," in *Absorbent Polymer Technology*, L. Brannon-Peppas and R. S. Harland, eds., New York, NY: Elsevier, pp. 3–22.

85. Suslick, K. S. and M. W. Grinstaff. 1990. "Protein Microencapsulation of Nonaqueous Liquid," *J. Am. Chem. Soc.*, 112:7807–7809.

86. Davidson, R. S. 1990. "Ultrasonically Assisted Organic Synthesis," Chapter 3 in *Chemistry with Ultrasound*, T. J. Mason, ed., New York, NY: Elsevier Applied Science.

87. Schäfer, H.-J. 1987. "Photoaffinity Labeling and Photoaffinity Crosslinking of Enzymes," in *Chemical Modification of Enzymes: Active Site Studies*, J. Eyzaguirre, ed., New York, NY: Ellis Horwood Ltd., pp. 45–62.

88. O'Donnell, J. H. 1987. "Radiation Effects on Polymers," in *Polymer Science in the Next Decades. Trends, Opportunities, Promises*, O. Vogl and E. H. Immergut, eds., New York, NY: John Wiley & Sons, pp. 203–216.

89. Chapiro, A. 1962. *Radiation Chemistry of Polymeric Systems*. New York, NY: John Wiley & Sons, pp. 352–364.

90. Alexander, P. and A. Charlesby. 1957. "Effects of X-rays and G-rays on Synthetic Polymers in Aqueous Solutions," *J. Polym. Sci.*, 23:355–375.

91. Urry, D. W. 1988. "Entropic Elastic Processes in Protein Mechanisms. II. Simple (Passive) and Coupled (Active) Development of Elastic Forces," *J. Protein Chem.*, 7:81–114.

92. Lazár, M., T. Bleha and J. Rychly. 1989. *Chemical Reactions of Natural and Synthetic Polymers*. West Sussex, England: Ellis Horwood Ltd., Chapter V.

93. Forbes, W. F. and P. D. Sullivan. 1966. "The Effect of Radiation on Collagen. I. Electron-Spin Resonance Spectra of 2537-Å-Irradiated Collagen," *Biochim. Biophys. Acta*, 120:222–228.

94. Weadock, K., R. M. Olson and F. H. Silver. 1984. "Evaluation of Collagen Crosslinking Techniques," *Biomat., Med. Dev., Art. Org.*, 11:293–318.

95. Bayley, H. L. 1983. *Photogenerated Reagents in Biochemistry and Molecular Biology*. Amsterdam, The Netherlands: Elsevier, Chapters 1, 2 and 5.

96. Harmer, M. A. 1991. "Photomodification of Surfaces Using Heterocyclic Azides," *Langmuir*, 7:2010–2012.

97. Scriven, E. F. V. 1984. *Azides and Nitrenes*. New York, NY: Academic Press.

98. Smith, P. A. S. 1984. "Aryl and Heteroaryl Azides and Nitrenes," Chapter 3 in *Azides and Nitrenes: Reactivity and Utility*, E. F. V. Scriven, ed., Orlando, FL: Academic Press.

99. Park, K., S. J. Gerndt and H. Park. 1988. "Patchwise Adsorption of Fibrinogen on Glass Surfaces and Its Implication in Platelet Adhesion," *J. Colloid Interf. Sci.*, 125:702–711.

100. Aiba, S., N. Minoura, K. Taguchi and Y. Fujiwara. 1987. "Covalent Immobilization of Chitosan Derivatives onto Polymeric Film Surfaces with the Use of a Photosensitive Hetero-Bifunctional Crosslinking Reagent," *Biomaterials*, 8:481–488.

101. Tseng, Y.-C. and K. Park. 1992. "Synthesis of Photoreactive Poly(ethylene Glycol) and Its Application to the Prevention of Surface-Induced Platelet Activation," *J. Biomed. Mater. Res.*, 26:373–391.

102. Flory, P. J. 1953. *Principles of Polymer Chemistry*. Ithaca, NY: Cornell University Press, Chapters 11 and 13.

103. Mark, J. E. 1982. "Experimental Determination of Crosslink Densities," *Rubber Chem. Technol.*, 55:762–768.

104. Dusek, K. and W. Prins. 1969. "Structure and Elasticity of Non-Crystalline Polymer Networks," *Adv. Polymer Sci.*, 6:1–102.

105. Davis, T. P., M. B. Huglin and D. C. F. Yip. 1988. "Properties of Poly(*N*-vinyl-2-pyrrolidone) Hydrogels Crosslinked with Ethyleneglycol Dimethacrylate," *Polymer*, 29:701–706.

106. Guillet, J. 1985. *Polymer Photophysics and Photochemistry*. New York, NY: Cambridge University Press, Chapter 4.

107. Treloar, L. R. G. 1975. *The Physics of Rubber Elasticity, 3rd Edition*. New York, NY: Oxford University Press, pp. 140–147.

108. Munk, P. 1989. *Introduction to Macromolecular Science*. New York, NY: John Wiley & Sons, pp. 423–431.

109. Peppas, N. A. and B. D. Barr-Howell. 1986. "Characterization of the Cross-Linked Structure of Hydrogels," Chapter 2 in *Hydrogels in Medicine and Pharmacy, Vol. I*, N. A. Peppas, ed., Boca Raton, FL: CRC Press.

110. Piirma, I. 1992. *Polymeric Surfactants*. New York, NY: Marcel Dekker, Inc., Chapter 4.

111. Franks, F. 1983. "Water Solubility and Sensitivity–Hydration Effect," in *Chemistry and Technology of Water-Soluble Polymers*, C. A. Finch, ed., New York, NY: Plenum Press, pp. 157–178.

112. Brandrup, J. and E. H. Immergut, eds. 1975. *Polymer Handbook, Second Edition*. New York, NY: John Wiley & Sons, p. IV-133.

113. Hughlin, M. B. and J. M. Rego. 1991. "Influence of a Salt on Some Properties of Hydrophilic Methacrylate Hydrogels," *Macromolecules*, 24:2556–2563.

114. Chatterji, P. R. 1991. "Cross-Link Dimensions in Gelatin-Poly(acrylamide) Interpenetrating Hydrogel Networks," *Macromolecules*, 24:4214–4215.

115. Subr, V., R. Duncan and J. Kopeček. 1990. "Release of Macromolecules and Daunomycin from Hydrophilic Gels Containing Enzymatically Degradable Bonds," *J. Biomater. Sci. Polymer Edn.*, 1:261–278.

116. Brøndsted, H. and J. Kopeček. 1991. "Hydrogels for Site-Specific Oral Drug Delivery: Synthesis and Characterization," *Biomaterials*, 12:584–592.

CHAPTER 5

Physical Gels

Physical gels (also called physical networks, association networks, or pseudo gels) are the continuous, disordered, three-dimensional networks formed by associative forces capable of forming noncovalent crosslinks. The point covalent crosslinks often found in synthetic polymer networks are replaced by weaker and potentially more reversible forms of chain-chain interactions. These interactions include hydrogen bonding, ionic association, hydrophobic interaction, stereocomplex formation, crosslinking by the crystalline segments, and solvent complexation [1]. Since physical gels are not covalently crosslinked, they behave as networks only at short time periods. They will eventually relax and show the viscoelastic liquid-like behavior at longer times [2,3]. Thus, physical gels can be viewed as liquids with long relaxation times. According to Ross-Murphy [4], the term "physical gel" was introduced by de Gennes who used the term to describe polymers with thermoreversibility [5]. The term physical gel is generally used to describe any noncovalently crosslinked system.

In physical gels, a substantial fraction of a polymer chain is involved in the formation of stable contacts between polymer chains. Association of certain linear segments of long polymer molecules form extended "junction zones." This distinguishes them from chemical gels which have well-defined point crosslinks as shown in Figure 5.1 [6]. The junction zones are expected to maintain ordered structures [7,8]. The junction zones hold together the amorphous regions of the polymer chains which are in a random coil conformation [9]. The formation of junction zones among polymer chains is usually induced by a modification of the thermodynamic parameters of the medium, i.e., changes in temperature, pH, salt type, ionic strength, or addition of a nonsolvent [2,10]. The junction zones become more stable as the chain segments involved in the junction zones reach 15–25 residues long or longer, due to the cooperative interaction between the polymer molecules [11]. Differences between physical gels are determined by the nature and number of junction zones [9].

Unlike polysaccharides and fibrous proteins such as gelatin, globular proteins usually have poor heat-gelling properties and require substantial

(a)

(b)

FIGURE 5.1. *Schematic description of a chemical gel with point crosslinks (a) and physical gels with multiple junction zones (b) (from Reference [6]).*

protein concentration to form three-dimensional networks [12]. In general, the concentration of globular proteins necessary for gelation is an order of magnitude higher than that of fibrous proteins and polysaccharides [13]. Gelation of globular proteins typically requires denaturation of the native protein structure either by heating or by chemical agents [13]. Heat denaturation, at a pH which increases the net charge on the protein, maximizes protein unfolding. The unfolded protein may form a large number of intermolecular contacts, e.g., through exposed hydrophobic patches, and physical entanglements to form gels [14].

For synthetic block copolymers, gelation is initiated by a transition of one of the blocks to domains, or junctions, where polymer chains cannot move along each other freely. Such a transition may be caused by thermally induced glass transition, phase separation, partial crystallization, or intermolecular ionic bonding of the chains of one of the blocks [15].

In most physical gels, the nature of the reversible interactions responsible for the self-supporting gel is not clearly understood [14]. The conformation and state of aggregation of the polymer at the junction zones are also not clearly known [16]. For this reason, the relationship between the physical network structure and functional properties is not clearly established [17]. The pore size within the gel is not well characterized. These aspects are important in the application of physical gels in the controlled release drug delivery area. The structure of the physical network and the size of the pores will undoubtedly determine the extent and kinetics of drug release.

5.1 THERMOREVERSIBLE GELLING SYSTEM

Many polymers have a tendency to form aggregates in dilute solutions or gels in moderately concentrated solutions upon temperature change. As the temperature changes, the polymer molecules freely moving in a random spatial distribution in solution may be held together by noncovalent crosslinks to form a gel. Those physical gels which are induced by temperature change are called thermoreversible gels [18]. According to Flory's gel theory, transition from a viscous liquid of individual polymer chains to an elastic gel occurs rather abruptly at the gel point [19]. The gel point is defined as the instant at which the largest cluster extends throughout the entire sample. The polymer at the gel point is called a "critical gel" [20]. The critical gel is not an equilibrium structure. The network continues to grow and this continuing gelation process is known as physical ageing [7]. The gel continues to mature, and additional and stronger bonding points develop through diffusion and reorientation of the macromolecular chains [21].

The sol-gel transformation (also called thermoreversible gelling and remelting, or order-disorder process) can be repeated many times [22]. The higher the concentration of polymer in solution, the greater the number of junctions in the physical network [18]. Thus, the energy necessary for gelling or melting thermoreversible gels increases as the concentration increases. For this reason, gelling or melting temperatures need to be expressed at a given concentration. The energy necessary for melting a gel is usually greater than that available at the gelling temperature. Thus, the gel needs to be heated to a higher temperature for remelting. The difference between the temperatures necessary for gelling and for remelting is known as the hysteresis effect. The hysteresis effect is particularly pronounced with agarose. There is about 50°C difference between gel setting (~30°C) and gel melting (~80°C) temperatures [2].

5.1.1 Gels with Thermomelting Properties

Gelatin is a "classical" thermoreversible gel in that it becomes a gel by forming quasicrystalline junction zones upon cooling and melts without substantial hysteresis [23]. The quasicrystalline junction zones in gelatin gels consist of triple helices formed by interchain hydrogen bonding [24]. A certain amount of helix formation is necessary before the network starts to grow. For example, a 4.7 percent aqueous solution of gelatin with a molecular weight of 100,000 requires about 6.5 percent helix formation to reach a gel point [7,25]. The physical ageing process of gelatin occurs faster as the temperature becomes lower. Most of the jello that we enjoy as a dessert is made of gelatin. Jello can be made by simply dissolving gelatin in hot water and cooling it in a refrigerator. If one does not like the shape of the jello, one can remelt and reshape it quite easily.

Agarose is another example of a thermoreversible gel. Agarose is obtained from agar by removing charged polysaccharides [26]. Agarose becomes a gel by forming double helices which subsequently aggregate to form junction zones (or bundles) and then to form a gel [9,26,27]. Some agarose gels appear whitish due to light scattering by the microcrystalline junction zones [28]. After the gel point, a slow tightening of the agarose helices (i.e., contraction of gels) continues and water molecules are exuded from the gel. The separation of water from the network is known as syneresis [10]. The syneresis in agarose gel, however, is slow enough not to interfere with the use of the gel. The degree of syneresis is inversely proportional to the concentration of agarose. Syneresis occurs when the polymer network is formed so quickly that water, and the polymer molecules dissolved in the water, are embedded in the gel phase [10]. Agarose gel is commonly used in gel electrophoresis for the separation of macromolecules such as DNA or high molecular weight proteins. Agarose is also used in the microencapsulation of islets. *In vitro* study suggested that entrapped islets in agarose microbeads satisfied most of the requirements of a bioartificial pancreas [29].

Gels made of synthetic polymers such as poly(vinyl alcohol) (PVA) also have thermoreversible properties. Concentrated aqueous solutions of PVA can be prepared at temperatures well above about 70°C. When such a solution is cooled to below room temperature, a transparent gel is formed as a consequence of the crystallization of the PVA molecule [21,30]. Uncrosslinked PVA gel can be strengthened by repeated freeze-thaw processes [31].

5.1.1.1 HYDROGEN BONDING

It is well known that hydrogen bonding is responsible for maintaining the structures of proteins and the stable helical structures of negatively charged DNA molecules [32]. Hydrogen bonding is also responsible for the stabilization of many thermomelting gels. For example, agarose gel is stabilized primarily by hydrogen bonding at temperatures below 40°C and melts at higher temperatures [26]. Hydrogen bonding occurs primarily at low temperatures [33,34] and is disrupted by heating [35]. Hydrogen bonding results in a lowering of the total energy of the system as well as a lowering of the total entropy. When the system is at low temperatures, the lowered energy brought about by hydrogen bonding has a large effect on the free energy of the system and the lowered entropy has only a small effect [34].

Hydrogen bonding, which is predominantly an electrostatic interaction, exists between electronegative atoms (e.g., O, N, F, and Cl) and H atoms which are covalently bound to similar electronegative atoms [36]. The strength of hydrogen bonding (<10 kcal/mol) is far weaker than covalent bond energy (>100 kcal/mol) but still much stronger than the van der Waals interaction energies (~ 1 kcal/mol) [36]. The formation of multiple hydrogen bonds between two water-soluble macromolecules may result in strong intermolecular interaction. The driving force for the hydrogen bond formation between macromolecules in water is the cooperative interaction, i.e., the formation of multiple, simultaneous hydrogen bonds, between macromolecules. The cooperative interaction requires that the chain length of the interacting polymers be above a certain minimum value [37]. Cooperativity and pH-dependency are two main characteristics of complex formation by hydrogen bonding [38].

5.1.2 Gels with Thermogelation Properties

Gel is formed by lowering the temperature in most cases, but not always. Some polymers form a gel upon heating above a certain temperature. Polymers, such as methylcellulose, hydroxypropylmethylcellulose, or certain PEO/PPO/PEO triblock copolymers, dissolve only in cold water forming a viscous solution [39]. On raising the temperature these solutions thicken or gel [15,40]. The phenomenon of gelation or phase separation resulting from the application of heat is known as thermogelation, in contrast to the thermomelting of other gelling systems described above. Thermogelation is also called inverse solubility-temperature behavior or thermophobic behavior [41]. Thermogelation is mainly due to the enhanced hydrophobic interaction between polymer chains at elevated temperatures.

5.1.2.1 HYDROPHOBIC INTERACTION

Even hydrophobic polymers can be dissolved in water to a certain extent as a result of hydrophobic hydration. Water molecules tend to form structured cages (also called clathrate) surrounding the hydrophobic solute [36]. The water molecules involved in the hydrophobic hydration are even more highly hydrogen bonded to each other than the water molecules in bulk water [41]. An increase in temperature results in a reduction in the total number of water molecules surrounding the hydrophobic solutes. Thus, temperature increase promotes hydrophobic interactions resulting in the association of hydrophobic polymer chains, and ultimately phase separation (effective aggregation of solutes) occurs [39,42–46]. This means that the driving force for the interaction between polymer chains is not an intrinsic attraction between them, but an increase in entropy of the water molecules released from the structured cages on the hydrophobic polymer chains [39]. Such behavior (i.e., exothermic heats of dilution) is not readily accounted for by the classical Flory-Huggins theory [39]. In essence, the formation of gel upon elevation of temperature is mainly due to the enhanced hydrophobic interaction among polymer chains at higher temperatures [38].

The gel temperature is affected by the molecular weight of a polymer, polymer concentration, and the presence of other dissolved species such as inorganic salts or organic additives [47]. Additives may either increase or decrease the temperature for thermogelation, since they affect the hydrophobic interaction between hydrophobic polymer chains. The lowering of gelation temperature is observed in the presence of co-solutes such as sucrose, glycerin, or ions [45]. The higher the charge on the ion, the greater the effect on thermal gel temperature [47]. Addition of non-electrolytes such as ethanol (20 percent), on the other hand, increased the thermal gel point of methylcellulose [48].

5.1.2.2 THERMOSHRINKING HYDROGELS

If physical gels are covalently crosslinked, they will not dissolve in water upon temperature changes. Rather, they will undergo dramatic changes in the extent of swelling. If a polymer is made of hydrophilic monomers, then the hydrogel will swell upon temperature increase and the hydrogel is called a thermoswelling hydrogel [46]. Most hydrogels belong to this category. On the other hand, some hydrogels made of relatively hydrophobic monomers undergo shrinkage upon temperature increase and are known as thermoshrinking gels [49]. Some hydrogels have both thermoswelling and thermoshrinking properties and are known as convexo hydrogels [50].

The thermoshrinking hydrogels undergo thermally reversible swelling and deswelling. Since the deswelling occurs rather dramatically with a minute temperature change, the phenomenon is often called gel collapse [49]. Gel collapse (i.e., volume phase transition of gels) is similar to the coil-globule transition of polymer chains in poor solvents [51,52]. The temperature which induces gel collapse corresponds to a lower critical solution temperature (LCST) of the uncrosslinked polymer [53]. Table 5.1 lists the phase transition temperatures of various polyacrylamide derivatives [54]. The structures of the polyacrylamide derivatives clearly show that the transition temperature becomes lower as the hydrophobicity of the side groups increases. The exact transition temperature of a given polymer may vary depending on the purity of the monomer used. For example, the coil-globule transition temperature of poly(*N*-isopropylacrylamide) ranges from 31 °C to 35 °C [46,52]. The transition temperature of linear polymer samples is independent of molecular weight [46]. Thus, it is expected to be the same as the transition temperature of gels. The two transition temperatures, however, may vary slightly depending on the nature of the crosslinking agent [53]. The transition temperature of poly(*N*-isopropylacrylamide) gel was always 1 ~ 2 °C higher than that of linear polymer solutions [46].

The transition temperature of thermoshrinking hydrogels or polymer solutions is lowered by the addition of low molecular weight additives, such as inorganic salts or alcohols [46]. These additives interact with bulk water and decrease entropy, which results in depression of the transition temperature [46]. These observations show that the hydrophobic interaction plays an important role in the volume phase transition [55].

Some crosslinked polypeptides also show inverse temperature transitions. Crosslinked sequential polypeptides such as poly(Val-Pro-Gly-Val-Gly), which is the main repeating unit of elastin, swell below 25 °C, but on raising the temperature they contract with the extrusion of water [56,57]. These bioelastic materials lose elastomeric force by means of thermal denaturation.

The thermoshrinking hydrogels have been examined extensively for their application in controlled release drug delivery [58–60] and bioassays [61,62].

5.2 THERMOREVERSIBLE GELLING POLYSACCHARIDES

Of the many polymers, polysaccharides present a rich source of physical gels. A large number of polysaccharides are able to form physical gels under various conditions. Polysaccharides can be classified according to structure, property, or origin. The structural classification is based on the monosaccharide composition of the main polymer chain. Polysaccharides are broadly classified into homo- and heteropolysaccharides. Homopolysaccharides are polymers formed from one monosaccharide only. Examples of

TABLE 5.1. Structures and Volume Transition Temperatures (T_{vt}) *of Thermoshrinking Polymers [54].*

Structure: $-[CH-C(R_1)]-$, with $C(R_1)$ bearing $C{=}O$, bonded to N, which carries R_2 and R_3

Polymer	R_1	R_2	R_3	T_{vt} (°C)
Poly(*N*-methyl-*N*-*n*-propylacrylamide)	$-H$	$-CH_3$	$-CH_2CH_2CH_3$	19.8
Poly(*N*-*n*-propylacrylamide)	$-H$	$-H$	$-CH_2CH_2CH_3$	21.5
Poly(*N*-methyl-*N*-isopropylacrylamide)	$-H$	$-CH_3$	$-CH(CH_3)_2$	22.3
Poly(*N*-*n*-propylmethacrylamide)	$-CH_3$	$-H$	$-CH_2CH_2CH_3$	28.0
Poly(*N*-isopropylacrylamide)	$-H$	$-H$	$-CH(CH_3)_2$	30.9
Poly(*N*,*n*-diethylacrylamide)	$-H$	$-CH_2CH_3$	$-CH_2CH_3$	32.0
Poly(*N*-isopropylmethacrylamide)	$-CH_3$	$-H$	$-CH(CH_3)_2$	44.0
Poly(*N*-cyclopropylacrylamide)	$-H$	$-H$	$-CH-CH_2$ (ring closed by CH_2)	45.5
Poly(*N*-ethylmethacrylamide)	$-CH_3$	$-H$	$-CH_2CH_3$	50.0
Poly(*N*-methyl-*N*-ethylacrylamide)	$-H$	$-CH_3$	$-CH_2CH_3$	56.0
Poly(*N*-cyclopropylmethacrylamide)	$-CH_3$	$-H$	$-CH-CH_2$ (ring closed by CH_2)	59.0
Poly(*N*-ethylacrylamide)	$-H$	$-H$	$-CH_2CH_3$	72.0

homopolysaccharides are starch (amylose and amylopectin), cellulose, dextrans, inulin, and chitin. Heteropolysaccharides are polymers made up of two or more different monosaccharide residues. Most polysaccharides belong to this group. The classification of polysaccharides based on function is rather difficult, since their functions are not fully understood [9]. Just for simplicity and convenience, polysaccharides are classified here based on their origin, such as algal, botanical, microbial, and animal (see Table 5.2).

All the polysaccharides in Table 5.2 are water-soluble except cellulose and chitin. The structures of cellulose, amylose, and chitin are shown in Figure 5.2. Cellulose is the main structural material of higher plant forms, e.g., wood, and is the most abundant naturally occurring organic substance on earth [9]. The repeating units, D-glucopyranose residues, are linked together by β-(1→4) bonds. These linkages are not flexible and the cellulose molecules tend to remain extended. The extended chains have strong intermolecular hydrogen bonds which make cellulose water-insoluble [6]. On the other hand, a polysaccharide of D-glucopyranose residues linked by α-(1→4) bonds (i.e., amylose, a component of starch) is flexible and water-soluble. Chitin is a structural polysaccharide of lower plant forms as well as of many insects and crustaceans. Chitosan is the second most abundant natural polysaccharide [63]. The structure of chitin is that of cellulose in which the hydroxyl groups on C-2 are replaced by acetylamino ($-NHCOCH_3$) groups [64].

Some polysaccharides are soluble in water, both cold and hot. Some are soluble only in hot water, while some are soluble only in cold water. Some are water-soluble even at very high concentrations. Some swell and absorb water to form gels at high concentrations. Some may form gels even at low concentrations. Some may form gels in the presence of inorganic salts.

Some polysaccharides, such as pectins and gums, are naturally present in foods. They are known as dietary fiber, although they may not be digested by the endogenous secretions of the human digestive tract [65]. Most plant gums are nontoxic to humans and have been used widely to produce edible gels and edible packings [66]. Plant gums are exuded as viscous fluids,

TABLE 5.2. Polysaccharides Derived from Different Sources [40,67,68].

Source	Polysaccharides
1. Algal	agar, furcelleran, alginate, carrageenan
2. Botanical	plant extracts: starch, pectin, cellulose exudate gums: gum arabic, tragacanth, karaya, ghatti seed gums: guar gum, locust bean gum
3. Microbial	xanthan, pullulan, scleroglucan, curdlan, dextran, gellan
4. Animal	chitin and chitosan, chondroitin sulfate, dermatan sulfate, heparin, keratan sulfate, hyaluronic acid

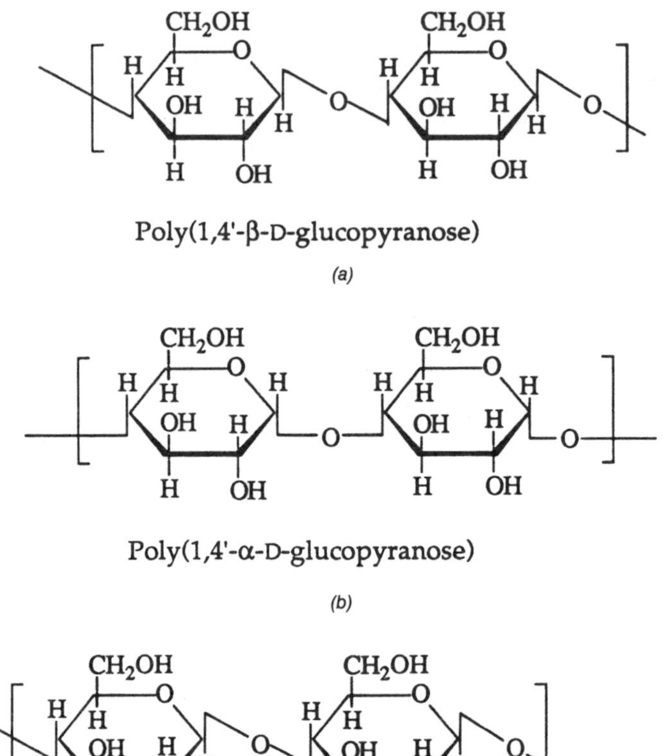

Poly(1,4'-β-D-glucopyranose)

(a)

Poly(1,4'-α-D-glucopyranose)

(b)

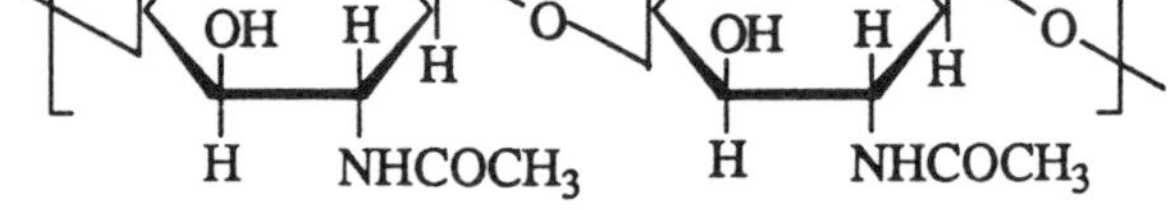

Poly(1,4'-β-N-acetyl-2-amino-2-deoxy-D-glucopyranose)

(c)

FIGURE 5.2. *Structures of cellulose (a), amylose (b), and chitin (c).*

either spontaneously or at the site of injury to protect the plant from further damage [9]. Edible gums are used as water binders, suspending agents, thickeners, and emulsion stabilizers in the food industry. They are essential in improving the "mouthfeel" of many products. Other gums find a variety of industrial applications in drilling fluids, sizings for textiles, paper coatings, paints, adhesives, and cosmetic lotions and creams [67].

5.2.1 Water-Soluble Polysaccharides

A number of polysaccharides are soluble in both cold and hot water. Upon addition to water, they tend to swell initially and are eventually converted to viscous or gel-like sols. Gum arabic, xanthan, and pullulan dissolve readily in water [68,69]. Gum arabic is highly water-soluble and forms more or less transparent, viscous, and adhesive solutions [40,70]. Gum ghatti is also water-soluble. Locust bean gum is partially hydrated in cold water, but fully hydrated when heated [70]. Individual solutions of these polysaccharides do not form gels by themselves, but the mixture of different polysaccharides may form gels. Such nonadditive behavior is known as synergism [71]. Xanthan forms cohesive, thermoreversible gels when mixed with locust bean gum, provided that the gum concentration exceeds 0.3 percent (w/v) [72,73]. Xanthan does not yield a gel when mixed with guar gum even though guar gum is also a galactomannan as is locust bean gum [74]. The galactose to mannose ratio is in the order of 1:2 for guar gum and 1:4 for locust bean gum [74].

Guar gum disperses and swells almost completely in cold or hot water to form a highly viscous sol [70]. Finely divided solid guar gum particles tend to take up water so rapidly that lumps or gel-like, semi-solid masses form [67]. Guar gum, in the presence of agar, forms extremely strong, rigid gels, resistant to heat and pH [70].

Semi-soluble gums consist of a mixture of water-insoluble and water-soluble polysaccharides. Gum tragacanth consists of a water-soluble fraction known as tragacanthin and a water-insoluble fraction known as bassorin [40]. It swells rapidly in either cold or hot water to give highly viscous colloidal sol or semi-gel [70]. Karaya gum partially dissolves in cold water and hydrates rapidly to form a swollen jelly which on the addition of sufficient water breaks down into a very thick, transparent solution [70,75,76].

Chitosan, chondroitin sulfate, dermatan sulfate, keratan sulfate, and heparin are water-soluble. Alginate, carrageenan, pectin, and gellan are all water-soluble, but they become a gel upon the addition of suitable inorganic metal salts (see Section 5.3).

5.2.2 Gel-Forming Polysaccharides

Many polysaccharides form gels at low temperatures, i.e., they form thermomelting gels. Agar forms strong gels even when very dilute [9]. Agar consists of two main components, agarose and agaropectin. Highly concentrated solutions of amylopectin (one of the starch components) can form gels [9]. Furcelleran forms strong but flexible clear gels [70]. The structure of furcelleran is similar to that of κ-carrageenan, but differs primarily in the amount of sulfate esters [40]. When a solution of scleroglucan is cooled, a weak gel is formed [77]. Dextrans are also known to form gels [9]. Curdlan is insoluble in cold water, but forms an elastic and firm irreversible gel when heated to above 54°C [68,78].

Hyaluronic acid (also called hyaluronate or hyaluronan) is capable of forming gels [9,79]. Hyaluronic acid molecules exist as individual molecules only at concentrations below 1 mg/ml. Above 1 mg/ml, hyaluronic acid adopts a three-dimensional network character due to entanglement of individual molecules [80–82]. The hyaluronic acid gels rapidly recover or "heal" from mechanical damage [11]. Hyaluronic acid is present in tissues, tissue fluids, and soft connective tissue. The concentration of hyaluronic acid in the human body ranges from less than 0.1 μg/ml in human blood plasma to more than 4 mg/ml in the umbilical cord [83]. As shown in Figure 5.3, hyaluronic acid is a linear polysaccharide containing alternating residues of glucuronic acid and *N*-acetylglucosamine. Hyaluronic acid plays many important roles such as lubrication of joints and regulation of the water balance of tissues [84]. Hyaluronic acid is removed from tissues either by local degradation by lysosomal hyaluronidase, β-glucuronidase, and β-*N*-acetylglucosaminidase or by lymph drainage. Recently, the moisturizing qualities of hyaluronic acid have been exploited by the cosmetics industry as an effective moisturizer for skincare. The films or microspheres made of hyaluronic acid were used to deliver steroids and peptide drugs [85,86].

5.3 PHYSICAL GELATION BY ION COMPLEXATION

Some water-soluble polymers containing complexing groups, whether neutral or charged, can be crosslinked to form gels in the presence of sufficient inorganic metal salts under appropriate conditions [87]. The bonding chemistry between the metal ions and the polymer functional groups is specific. Thus, each type of metal ion forms gels with different polymers under specific conditions of pH, ionic strength, and concentration of polymer [88–90].

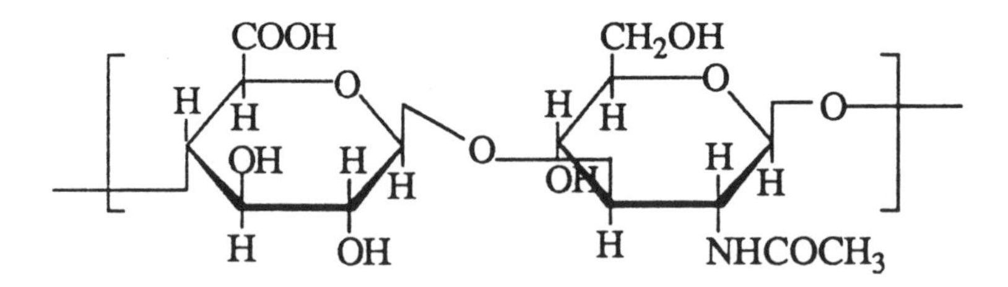

β-D-Glucuronic acid N-Acetyl-β-D-glucosamine

FIGURE 5.3. *Structure of hyaluronic acid.*

The ability of polyhydroxy compounds to complex with a variety of substances is well known [89,91]. Normally water-soluble, uncharged polyhydroxy compounds can be modified to render the compounds water-insoluble or less-soluble. For the association of charged polymers, counterions must be incorporated within the aggregate structure to preserve electrical neutrality and offset electrostatic repulsions between polyelectrolyte chains [45]. Polymers having 5–10 mol percent of acidic functional groups attached to their backbone chain can undergo reversible crosslinking. Upon neutralization with suitable counterions such as Ca^{++}, Mg^{++}, Zn^{++}, or others, pendant acid groups aggregate and form secondary bonds which crosslink the polymer chains [92]. Such crosslinking is temperature-dependent and becomes unstable upon heating.

5.3.1 Polymers with Hydroxyl Groups

Polyhydroxy compounds which contain accessible *cis*-hydroxyl groups form gels in the presence of anions such as borate, titanate, antimonate, vanadyl (VO^{++}), and permanganate (MnO_4-) ions [22,93–96]. Examples are guar gum, hydroxypropyl guar, scleroglucan, starch, and PVA. Each ion imposes its own conditions (concentration range, pH, ionic strength, oxidation state, etc.) for optimum crosslink formation. The pH of the solution must be maintained within limits to effect gelation. If the pH is too low, gelation will not occur or will occur too slowly. If the pH is too high, syneresis will result. The preferred pH range is between 4 and 11. The gel properties strongly depend on the nature of the binding ions and on the number of induced temporary links [87].

Guar gum is a high molecular weight polysaccharide made up of linear chains of β-D-mannopyranose joined by β-(1→4) linkage with α-D-galactopyranosyl units attached by 1,6′-links in the ratio of 1:2 (Figure 5.4). The structure is also called galactomannan. Guar gum and (hydroxypropyl)guar become gels in alkaline conditions by the addition of borate ions. Lowering the pH below 7, or heating, will cause collapse of the gel. Sodium tetraborate (borax) dissociates completely to form boric acid ($B(OH)_3$) and borate ions ($B(OH)_4^-$). The monoborate ions react with *cis*-hydroxyl groups in guar or (hydroxypropyl)guar to form crosslinks [97].

The tetrahedral borate ion is known to react with hydroxyl groups of PVA in *cis*-conformation [98]. PVA was crosslinked with boric acid in the presence of a small amount of calcium alginate to immobilize cells [99]. Scleroglucan in alkaline solution complexes with borate to form a stable gel [68]. Starch-borate matrices were further modified with calcium ions to control the release of low molecular weight drugs [100,101].

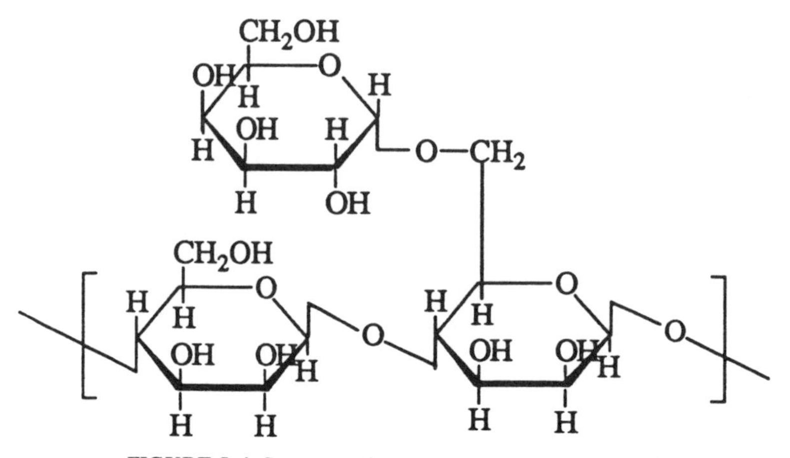

FIGURE 5.4. Structure of guar gum (from Reference [9]).

5.3.2 Carrageenan

Carrageenan is made up of a closely related mixture of sulfated, linear polysaccharides which have alternating 1,3′- and 1,4′-linked D-galactopyranose units. There are three different types of carrageenan: kappa, iota, and lambda. The type is determined by the number and position of the ester sulfate groups on the repeating galactose units. The degree of sulfation increases from kappa to iota to lambda [64,102]. The gel-forming property of carrageenan increases as the degree of sulfation decreases, i.e., as the proportion of anhydrogalactopyranose increases. Figure 5.5 shows the structure of κ-carrageenan which consists mainly of alternating 1,3′-linked β-D-galactopyranose-4-sulfate and 1,4′-linked 3,6-anhydro-α-D-galactopyranose units. The less common repeating unit of κ-carrageenan has α-D-galactopyranose-6-sulfate instead of 3,6-anhydro-α-D-galactopyranose [9].

Among the three types of carrageenan, only the κ-carrageenan exhibits significant gel-forming characteristics in water. At lower temperatures and higher salt concentrations, carrageenans in random coil conformation form double helices consisting of two parallel, staggered polysaccharide chains (Figure 5.6) [103]. At low salt content, electrostatic repulsive interactions increasingly disfavor the helix conformation as the helical content increases [104]. The sulfate groups are located at the outside of the helix, and the helix is stabilized by hydrogen bonding between the two chains [9,17]. The regular helical structure in Figure 5.6 is ended when the 3,6-anhydro-α-D-galactopyranose unit is replaced by a sulfated galactose unit.

The strength of the gel depends on the type of cations present. While Na^+ ions do not promote gel formation, K^+ ions greatly enhance the gel-forming

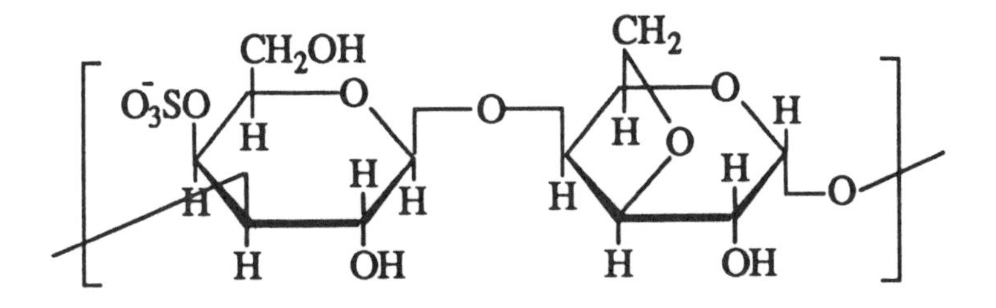

Main repeating unit

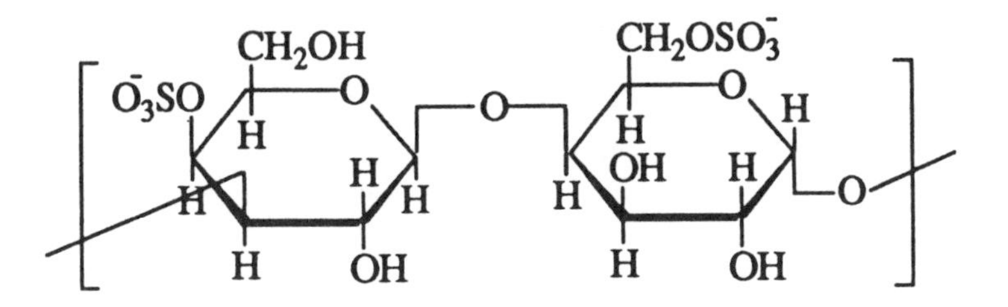

Less common unit

FIGURE 5.5. *Structures of main and less common repeating units of κ-carrageenan (from Reference [9]).*

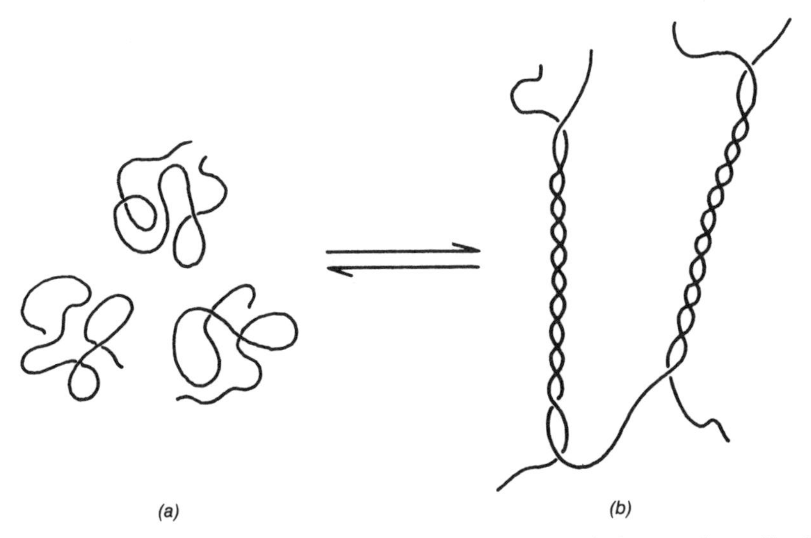

FIGURE 5.6. *Formation of double helical junction zones in gels from random coils of κ-carrageenan in solution (from Reference [103]).*

properties of $\varkappa$-carrageenan. The gelling temperature is about 25 °C or less depending on the concentration of K^+ ions, and the gel remelts at about 35 °C [90]. Ca^{++} ions cause aggregation of double helices of $\varkappa$-carrageenan and gel contraction which make the gel brittle. ι-Carrageenan forms elastic gels with Ca^{++}. λ-Carrageenan cannot form gels [70]. Recently, the gel forming ability of carrageenan was used to immobilize enzymes and cells on the surface of biosensor electrodes [105].

5.3.3 Gellan

Gellan produces gels on cooling a warm solution containing cations such as calcium or a very high concentration of monovalent cations [106]. The gel is formed by end-to-end association of gellan molecules through double helix formation [93]. The structure of gellan is shown in Figure 5.7. Approximately 25 percent of the repeating units have an O-acetyl group linked to C-6 of one of the β-D-glycopyranosyl residues [107]. The acetylated polysaccharide produces weak, very elastic gels whereas the deacetylated polysaccharide produces firm, brittle gels. Both types of gellan form thermoreversible gels. Unlike alginate and pectin which contain uronic acid, gellan gum does not have specificity among the alkaline-earth cations [107]. The ability of gellan to form gels in the presence of monovalent cations was used to deliver timolol to the eye [108]. The gel formation is expected to increase the residence time of the drug in the conjunctival sac. Bhakoo et al. [109] examined the drug release properties of gellan gum gels. Adriamycin, theophylline, ampicillin, amoxicillin, tetracycline, and erythromycin were loaded into the gel by imbibition for 24 hours. The initial release was the first order for all the drugs tested and more than 70 percent of the entrapped drugs were released within 24 hours.

5.3.4 Pectin

Pectins are colloidal polygalacturonic acids in which some of the carboxyl groups are esterified with methyl groups. The principal constituent of pectin is D-galacturonic acid (Figure 5.8). Other carbohydrate units, such as D-galactose, L-arabinose, and L-rhamnose, are also present either as side groups or in the main chain. Native pectins have methyl esterified carboxyl groups. Pectins are subdivided into pectic acid and pectinic acid depending on the extent of esterification. In pectinic acid about 50–80 percent of the carboxyl groups are esterified (high-ester pectin). Pectic acid is a low-methoxyl pectin in which only a few galacturonic acid units are esterified.

Pectinic acid can be induced to gel by adding a large amount of co-solute such as sucrose at low pH around 3 [9,110]. The addition of co-solute lowers the water activity and promotes polymer-polymer interaction rather than

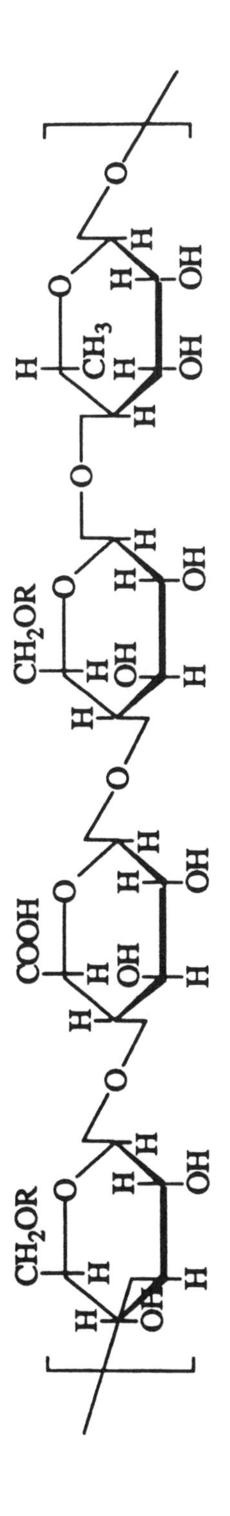

R = —H or —$COCH_3$

FIGURE 5.7. Structure of gellan (from Reference [9]).

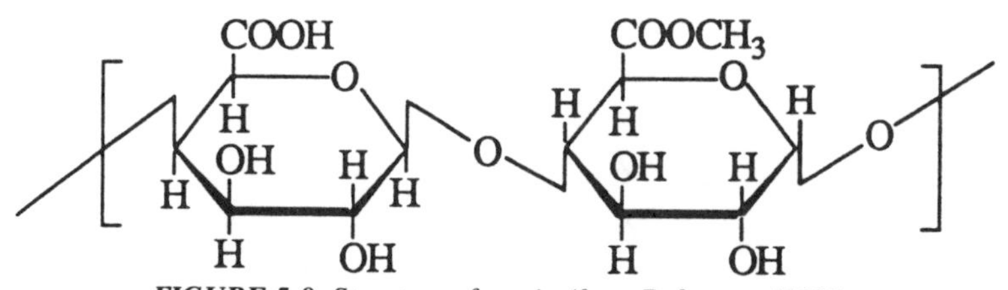

FIGURE 5.8. Structure of pectin (from Reference [102]).

polymer-solvent interaction [68,110]. Hydrophobic interaction between the ester methyl groups is known to be responsible for the gelation of pectinic acid [111,112]. Pectinic acid needs the presence of at least 55 percent soluble solids to gel [68].

Pectic acid, which is obtained by controlled deesterification, forms gel in the presence of divalent cations such as calcium ions. Calcium ions chelate in regular arrays of electronegative cavities formed by the galacturonic acid residues of pectic acid much the same way as an "egg box" [113]. Such chelation results in an initial dimerization followed by subsequent aggregation of the preformed dimers [8]. Pectic acid gels are colorless, coherent, and soft at low calcium concentrations. At high calcium concentrations, a brittle gel is formed with a strong tendency toward syneresis [90]. Pectic acid gels are heat-reversible whereas pectinic acid gels are not [68].

The ability of pectin solutions to form gels is determined by the number of successive negative cavities required to form a junction zone and the ratio of calcium to pectin in the system [114]. The pH and ionic strength of the solution also affect the extent of gel formation, since they affect the degree of ionization of carboxyl groups and the electrostatic interactions. The selectivity of cations to pectins is known to be $Ba^{2+} > Sr^{2+} > Ca^{2+}$. Mg^{2+} ions do not induce any chain association [114].

5.3.5 Alginate

Alginate is a linear polymer which has 1,4′-linked β-D-mannuronic acid and α-L-guluronic acid residues arranged as blocks of either type of unit or as a random distribution of each type (Figure 5.9). Poly(L-guluronic acid) sequences of alginate are rigid and buckle shaped. Alignment of two such sequences forms an array of cavities simulating the "egg box" which have carboxylate and oxygen atoms [24,68]. Poly(L-guluronic acid) sequences dimerize by selectively binding Ca^{2+} ions to adapt an ordered solution conformation (Figure 5.10). Monovalent cations and Mg^{2+} ions do not induce gelation [115]. Ba^{2+} and Sr^{2+} ions produce stronger alginate gels than Ca^{++} ions [2].

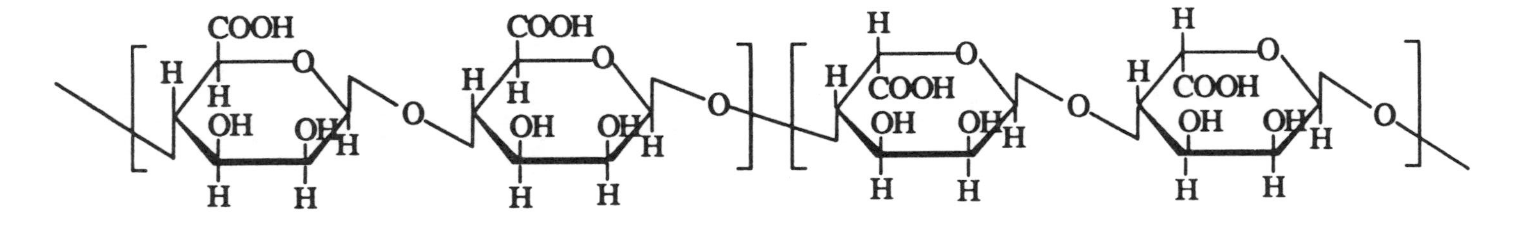

FIGURE 5.9. Structure of alginate (from Reference [9]).

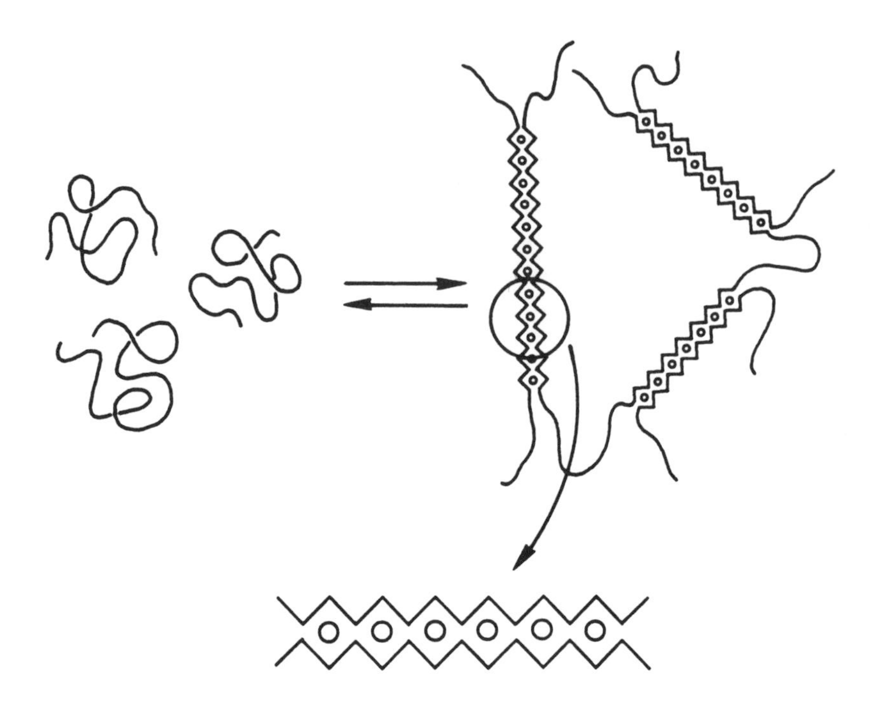

FIGURE 5.10. Schematic description of the calcium-mediated gelation of alginate by forming "egg-box" type associations of poly-L-guluronate sequences (from References [24] and [68]).

Since each alginate chain can form dimerization with many other chains, gel networks are formed rather than insoluble precipitates [103]. Since the chain sequences containing D-mannuronic acid do not dimerize, the content of D-mannuronic acid (or L-guluronic acid) in the alginate determines the properties of the alginate gels. Alginates with a high content of L-guluronic acids produce rigid and brittle gels that are subject to syneresis, while alginates with a low content of L-guluronic acid form more elastic gels [116]. Unlike pectic acid gels, alginate gels are not thermally reversible. Alginate generally forms thermostable gels over the range of 0–100°C, although the modulus of rigidity of the gels decreases with an increase in temperature [117]. Alginate produces clear, firm, quick-setting gels in hot or cold water [116].

Alginate has been used widely in the microencapsulation of living cells and microorganisms [118–121]. Alginate microgel capsules are generally made by spraying a sodium alginate solution containing cells into a calcium chloride solution. One of the disadvantages of the alginate gels is their low stability. Ca^{++} ions can be removed by chelating agents or by high concentrations of ions such as Na^+ or Mg^{2+} [115]. Alginate gel beads disintegrate and precipitate in 0.1 M phosphate buffer at pH 8.0 and become completely dissolved in 0.1 M sodium citrate buffer at pH 7.8 [122]. As Ca^{++} ions are removed from the gels, crosslinking diminishes and the gels are destabilized leading to leakage and loss of the entrapped material. Thus, it is necessary to enhance the stability of temporary calcium alginate gels. Polycations such as poly(L-lysine) and chitosan are commonly used to strengthen the calcium alginate gels and form permanent membranes [123–125].

Due to its simplicity of preparation, alginate gel has also been used to incorporate microspheres. Ethylcellulose microspheres containing drugs were dispersed into aqueous solutions of sodium alginate and alginate beads were formed by dropping them onto the $CaCl_2$ solution [126]. Drug containing liposomes were also embedded in an alginate matrix [127]. Aqueous solutions of sodium alginate and calcium alginate gel were also used in wound healing, since they form a protective film after air drying [128].

Recently, alginate was used to form a uniform, defect-free coating on solid dosage forms [129]. Calcium acetate tablets were added to an aqueous sodium alginate solution. As the tablet surface dissolved, the released calcium ions crosslink alginate to form a water-insoluble, calcium-alginate membrane. The thickness of the membrane can be controlled by the rate of diffusion of calcium ions from the tablet through the calcium alginate membrane already formed to the sodium alginate solution. The formed membrane can be further stabilized by treating the tablets with a calcium chloride solution.

The preparation of a membrane by alginate is better than other methods of microencapsulation which involve the use of an organic solvent and/or high temperature [123]. In a pH 7 phosphate buffer, the alginate gel swells, gradually disintegrates, and eventually disperses over time [130]. The fact that the alginate beads can be formed very easily in aqueous solution at room temperature and dissolve at a physiological condition makes the alginate gel very useful in the delivery of bioactive peptides and proteins [122,131,132].

5.3.6 Polyphosphazenes

Polyphosphazenes with carboxylic acid units in the para positions of aryloxy groups attached to a phosphazene chain, poly[bis(carboxylatophenoxy)phosphazene], are soluble in basic aqueous media [133]. Like poly(acrylic acid) [134], they crosslink in the presence of di- or trivalent cations (Figure 5.11). This process has been used to form hydrogels [133]. Polymeric phosphazenes bearing carboxylic acid form ionic crosslinks, or "salt bridges," to yield hydrogels when treated in water with salts of di- or trivalent cations, such as calcium chloride, copper bromide, copper sulfate,

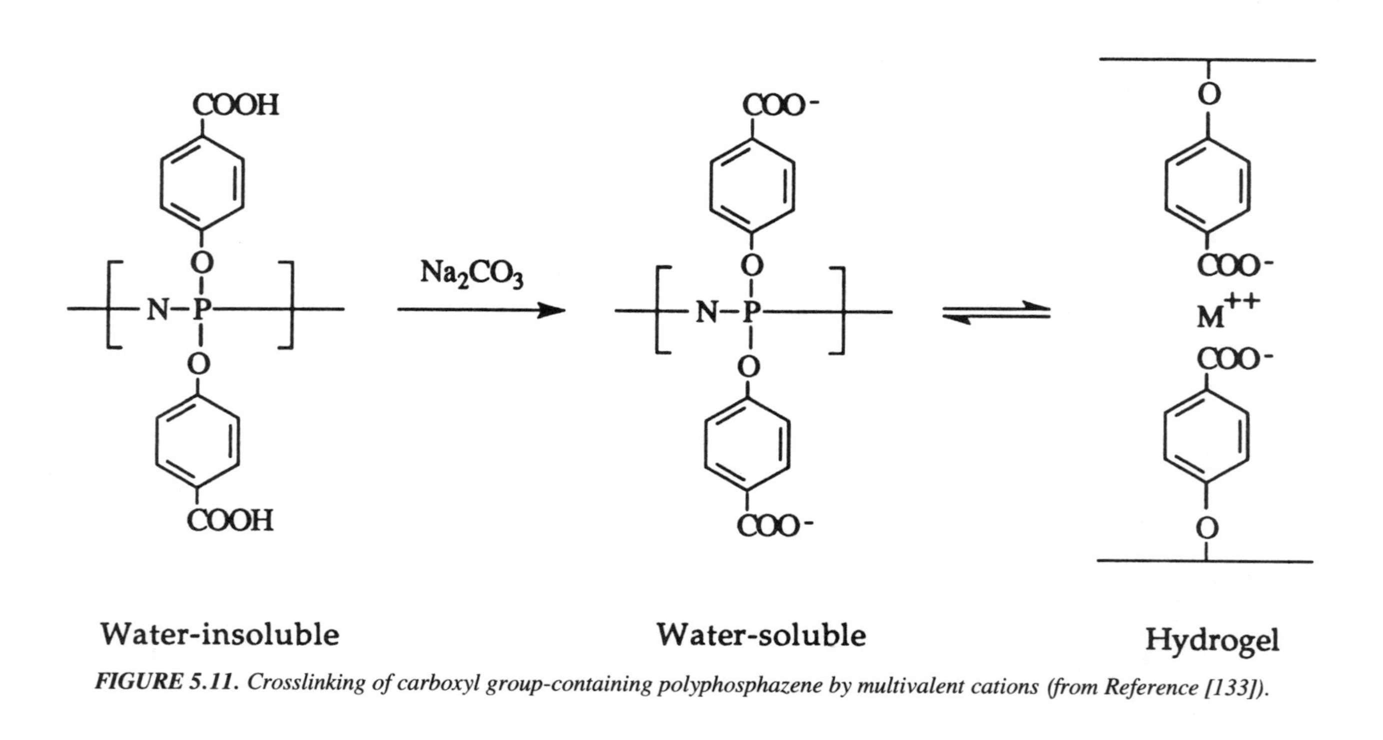

FIGURE 5.11. Crosslinking of carboxyl group-containing polyphosphazene by multivalent cations (from Reference [133]).

or aluminum acetate. Aluminum ions were more efficient in crosslinking than the divalent cations. The crosslinked gels were stable in neutral or strongly acidic aqueous media, but the crosslinking process was reversed in basic aqueous solutions of excess monovalent cations. This occurred at pH 7.5 for systems crosslinked by Ca^{2+} or Cu^{2+} ions. The system crosslinked by Al^{3+}, however, dissolved at pH 9 or above. Similar pH effects were observed with ionically crosslinked poly(acrylic acid) membranes [134]. As pH was increased, the carboxyl groups became ionized to increase the charge density, which resulted in the repulsion of the carboxylic groups. The swelling ratio of an aluminum ion crosslinked gel was close to 10. This polyphosphazene gel was used for microencapsulation of albumin [135] or hybridoma cells [136].

5.3.7 Other Polyanions

Water-soluble polymers containing chelating groups (or ligand residues) on the backbone chain can crosslink through chelation between two chelating groups (ligand residues) by the polyvalent metal ions such as Ca^{+2}, Mg^{+2}, Cu^{+2}, Al^{+3}, or Cr^{+}. Hyaluronic acid, carboxymethyldextran, dextran sulfate, carboxymethyl starch, and carboxymethylcellulose are examples of polymers that can form chelated hydrogels [45,68,92,137]. Chelate-forming functional groups can be introduced to neutral polymers such as dextran. Amino acids with functional side chains such as glutamic acid, aspartic acid, or cysteine, and α,ω-diaminodicarboxylic acids such as diaminopimelic acid can be incorporated into cyanogen bromide-activated dextran [22,138,139].

5.3.8 Chitosan

Chitosan is soluble in dilute acid, but precipitates at pH above 7. Chitosan forms a gel in the presence of anions such as phosphate [26]. Chitosan (1 percent, w/v in dilute acetic acid) beads were formed by dropping the bubble-free solution through a disposable syringe onto gently agitated solutions of the counterion (30 ml), tripolyphosphate. Chitosan beads disintegrated (dissolved) below pH 6, but they stayed intact in intestinal fluids [126]. The gel forming properties in the low pH range, and the antacid and anti-ulcer activities of chitosan may be used to prevent or weaken drug irritation to the stomach [140]. Chitosan has also been found to have pharmacological activity as an accelerator for wound healing, and activity as an immunoactive adjuvant [140].

5.4 PHYSICAL GELATION BY SPECIFIC INTERACTIONS

As described earlier, many water-soluble polysaccharides become gels by forming extended linear cooperative junction zones. Many proteins and glycoproteins can also form gels, but the mechanisms of gel formation are different from those open to polysaccharides. The association or gelation of protein molecules usually involves specific interactions which are often described as lock-and-key type interactions. A protein binds to another molecule only when the three-dimensional structures of both the molecules are complementary. Association of fibrin molecules into protofibrils and antigen-antibody interactions are good examples. Gel formation by glycoproteins such as mucin is not clearly understood mainly due to the poor characterization of the structures of glycoproteins. Protein molecules which have multibinding sites can function as crosslinking agents and form gels in the presence of suitable ligands. Examples are avidin, antibody, and lectin. Although these proteins form complexes with their ligands by physical association, the association is so tight that the complexes are essentially irreversible in physiological conditions.

5.4.1 Fibrin Gel

Fibrinogen is a blood protein which upon activation by thrombin causes clotting of blood. Fibrinogen has been studied extensively for its importance in thrombosis and hemostasis [141,142]. The thrombus formation on artificial materials exposed to blood is known to be the major deterrent in the development of biocompatible materials. The adsorption of fibrinogen on biomaterials is known to activate platelets and lead to surface-induced thrombus formation [143].

Fibrinogen is a dumbbell-shaped molecule consisting of three different types of chains (α, β, and γ), each having two identical halves, linked together by disulfide bonds. Thrombin, an enzyme, cleaves fibrinogen to release oligopeptides known as fibrinopeptides A and B from the N-terminal regions of the α and β chains. Upon removal of the fibrinopeptides, fibrinogen becomes a fibrin monomer which has two binding sites at the N-terminal domain in the center of the molecule. Fibrin monomers polymerize to form fibrin oligomers (protofibrils) by binding end-to-end with an overlap equal to half the monomer length (Figure 5.12). The protofibrils associate laterally to form fibrin polymers which leads to the formation of fibrin gel or clot. Fibrin clot formed from purified fibrinogen can be dissolved in a 3 M urea solution or 1 M NaBr solution at pH 5.3 [144]. Fibrin molecules in the protofibril can be crosslinked by transamidase (Factor XIIIa) in blood and that makes the fibrin gel more stable. The fibrin gel in the body is degraded by enzymes, mainly plasmin.

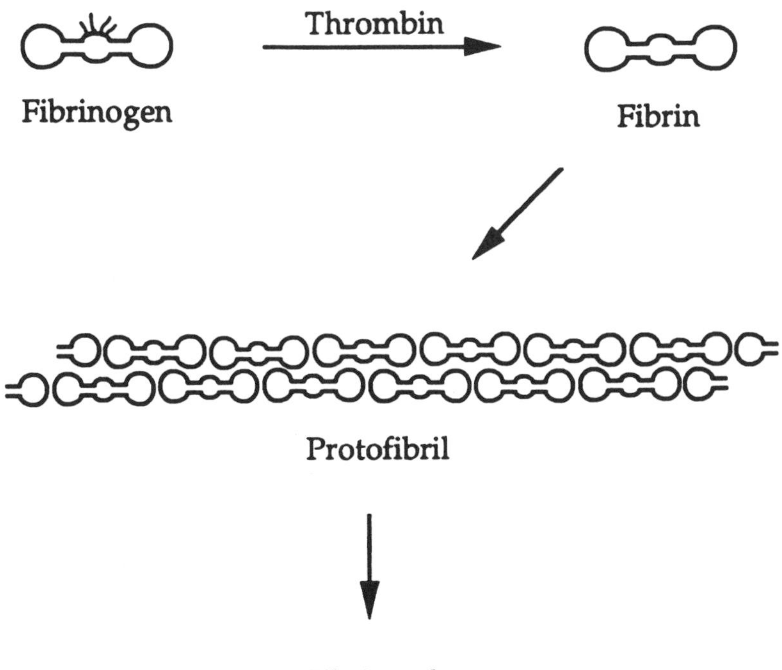

FIGURE 5.12. *Schematic representation of fibrin gel formation from fibrinogen.*

The fibrin gel *in vitro* can be stabilized by chemical crosslinking with various crosslinking agents. The chemically crosslinked gel can still be degraded by enzymes. The enzymatic degradation of the fibrin gel can be controlled by varying the extent of crosslinking. The fibrin gel was shown to be biodegraded and absorbed in rats in a relatively short period of time [145].

The fibrin gel is able to produce a stable linking between tissues and for this reason the fibrin gel is often called the fibrin glue or fibrin tissue adhesive. The fibrin glue has been used as an adhesive adjunct in a wide variety of surgical procedures such as wound closure after surgical injury [146,147]. In addition to wound sealing, the fibrin gel is also known to promote wound healing [148]. The ability of a fibrinogen solution to form a gel in any size and shape in the presence of other proteins is particularly useful for controlled drug delivery [149]. Drug molecules, regardless of their size and hydrophilicity, can be incorporated easily into the fibrin gel. It is a common practice to preseal synthetic, woven vascular grafts with fibrin gel to inhibit the leaking of blood through the pores of the grafts. During presealing, antibiotics can be incorporated into the fibrin gel [150].

One of the disadvantages of using fibrin gel as a drug delivery system is that most of the entrapped drug, whether hydrophilic or not, is released in less than several hours [151]. This is simply due to the large size pores present in the fibrin gel. The duration of drug release may be extended by increasing the concentration of the drug in the fibrin gel [152]. The fibrin gel is particularly useful in immobilizing living cells without affecting the bioactivity and handling the cells as a mass like a tissue [153]. This ability provides considerable advantages over other cell immobilization techniques.

5.4.2 Mucus Gel

Mucus is a translucent, viscid secretion which forms a thin, continuous gel blanket adherent to the mucosal epithelial surface in the eye, nose, respiratory tract, gastrointestinal tract, and female reproductive tract [154]. The mean thickness of the mucus layer varies from 50 μm to 450 μm in man and about half this in rats [155]. The primary functions of mucus are lubrication and protection of the underlying epithelial cells. Mucus has the same ability to recover rapidly from mechanical damage as hyaluronic acid does [11]. Continuous secretion of mucus from the goblet cells is necessary to compensate for the removal of the mucus layer due to digestion, bacterial degradation, and solubilization of mucin molecules. The turnover time of the intestinal mucus layer in rats was estimated to range from 47 to 270 minutes [156]. Soluble mucus may form temporary unstirred layers exterior to the adherent mucus gel [155]. The gastric mucus gel can be completely solubilized by guanidium chloride or by homogenation in water. This indicates that the mucus gel is stabilized by noncovalent interactions between the component glycoproteins [157].

The major constituents of mucus are high molecular weight glycoproteins capable of forming slimy, viscoelastic gels containing more than 95 percent water [154,158]. The exact composition of the mucus gel varies depending on the species, the anatomical location, and the pathophysiological state [159]. The structure of the mucin molecule is still a subject of controversy. The high molecular weight glycoproteins are known to form disulfide bonds as well as ionic bonds and physical entanglements. The molecular weight of glycoproteins varies from 2×10^6 to 14×10^6 daltons [154]. In general, a major portion of the peptide backbone is covered with carbohydrates grouped in various combinations. Over 70 percent of the dry weight of the mucin is accounted for by carbohydrate components [160–162]. Some of the polypeptide chain is not covered with carbohydrates and remains "naked." It is thought that this naked region interacts with similar regions on other molecules by noncovalent forces to produce a three-dimensional network [163].

Since the potential sites for drug absorption in the body are covered with a mucus layer, many attempts were made to localize the dosage forms on the surface of the mucus layer using mucoadhesive preparations. Research on mucoadhesive dosage forms is decades-old and continues to be active [164,165]. The development of effective mucoadhesives depends on understanding the nature of interactions between mucin and mucoadhesives. Further understanding of the structure of mucin and the mucus layer is necessary for the development of useful mucoadhesives.

5.4.3 Actin Gel

Actin is a globular protein with a molecular weight of about 42,000 daltons, found in muscle and nonmuscle cells [166]. It is one of the major components of the cytoskeleton which maintains the shape and internal organization of cells [167]. In addition, actin is known to be involved in the sol-gel transformation of the cytoplasm necessary for many cellular activities, such as pseudopod extension, cell spreading, and locomotion [166,168]. Thus, understanding of the sol-gel transformation of actin molecules is important in understanding the dynamical structural reorganizations of the cell structures. For this reason, many studies were done on the mathematical modeling of the nucleation and polymerization of actin molecules [169].

The actin monomers, known as globular actin monomers or G-actin, polymerize to form actin filaments (also called filamentous actin or F-actin) which form gels within the cells [170]. Actin filaments are long, linear, helical polymers with a diameter of 8 nm and a length of $5-10$ μm [171]. Conversion of actin filaments to actin monomers is reversible. Monomeric actin is easily polymerized *in vitro* to form gels in the presence of 100 mM K^+ or 1 mM Mg^{2+} or Ca^{2+} [3]. It is not known why the salt promotes polymerization of G-actin [170]. The assembly of actin into filaments and the crosslinking of the filaments into three-dimensional gels *in vitro* can be regulated by the actin-associated proteins [172,173].

5.4.4 Avidin-Biotin Mediated Gel

Avidin is found in hen egg white as a glycoprotein with a molecular weight of 67,000 daltons. Avidin is a tetramer containing four identical subunits and each subunit contains a high affinity binding site for biotin, a water-soluble vitamin. The dissociation constant between avidin and biotin is approximately 10^{-15} M [174,175]. The extremely strong binding ($K_D = 10^{-15}$ M) of four molecules of biotin to one of avidin allows the assembly of a molecular "sandwich," which serves to bind the two components together [176]. The avidin-biotin interaction is known to be the strongest noncovalent interaction between protein and ligand. This extraordinary affinity is rather surprising considering the fact that biotin is only a small molecule. The binding is undisturbed by extremes of pH, buffer salts, or even chaotropic agents, such as guanidine HCl (up to 3 M). Dissociation usually requires 6 M guanidine HCl, pH 1.5, an environment too extreme for many proteins. The biotin-avidin interactions can be reversed under milder conditions, e.g., pH 4, if a biotin analog, iminobiotin, is used [177].

The biotin molecule can be easily activated and coupled to any molecules. For example, biotinyl-*N*-hydroxysuccinimide ester (BNHS) was used to biotinylate proteins (as shown in Figure 5.13) [178,179]. BNHS reacts with ϵ-amino groups of lysine residues to form biotinyllysine. Instead of BNHS, longer chain homologs such as biotinyl-ϵ-aminocaproyl-*N*-hydroxysuccinimide ester or water-soluble analogs such as *N*-hydroxysulfosuccinimide derivatives of biotin can be used [178,180]. In addition, photoactivatable analogs of biotin called photobiotins can be used for biotinylation [181]. Since avidin has four binding sites for biotin, it can be used to crosslink biotin-containing molecules such as biotinylated proteins or polypeptides [175].

5.4.5 Antibody-Mediated Gel

Antibodies are a group of blood proteins known as immunoglobulins. They protect the body against invasion by foreign substances including proteins, polysaccharides, bacteria, viruses, or implanted organs. Monoclonal antibodies are homogeneous immunoglobulins directed against a single antigenic determinant. The affinity of monoclonal antibodies to antigens is generally lower than the affinity of polyclonal antibodies. The dissociation constant between antigen and antibody ranges from 10^{-5} to 10^{-12} [182,183]. Although the interaction between antigen and antibody is not as strong as that between avidin and biotin, the affinity of antigen to antibody is still so high that they separate only under rather drastic conditions, such as high concentrations of chaotropic ions, urea (3−8 M) or guanidine (6 M), and extremely acidic or alkaline pH values.

Each antibody molecule has two antigen binding sites, i.e., is bifunctional. If an antigen is large and possesses more than one antibody binding site, then a three-dimensional network of alternating antigen-antibody can be formed. The crosslinked network usually precipitates when it reaches sufficient size [184]. In theory, however, the antigen-antibody complex can form a three-dimensional network which spans the whole sample.

5.4.6 Lectin-Mediated Gel

Lectins are a class of proteins or glycoproteins of nonimmune origin that have specific and noncovalent binding sites for carbohydrates [185,186].

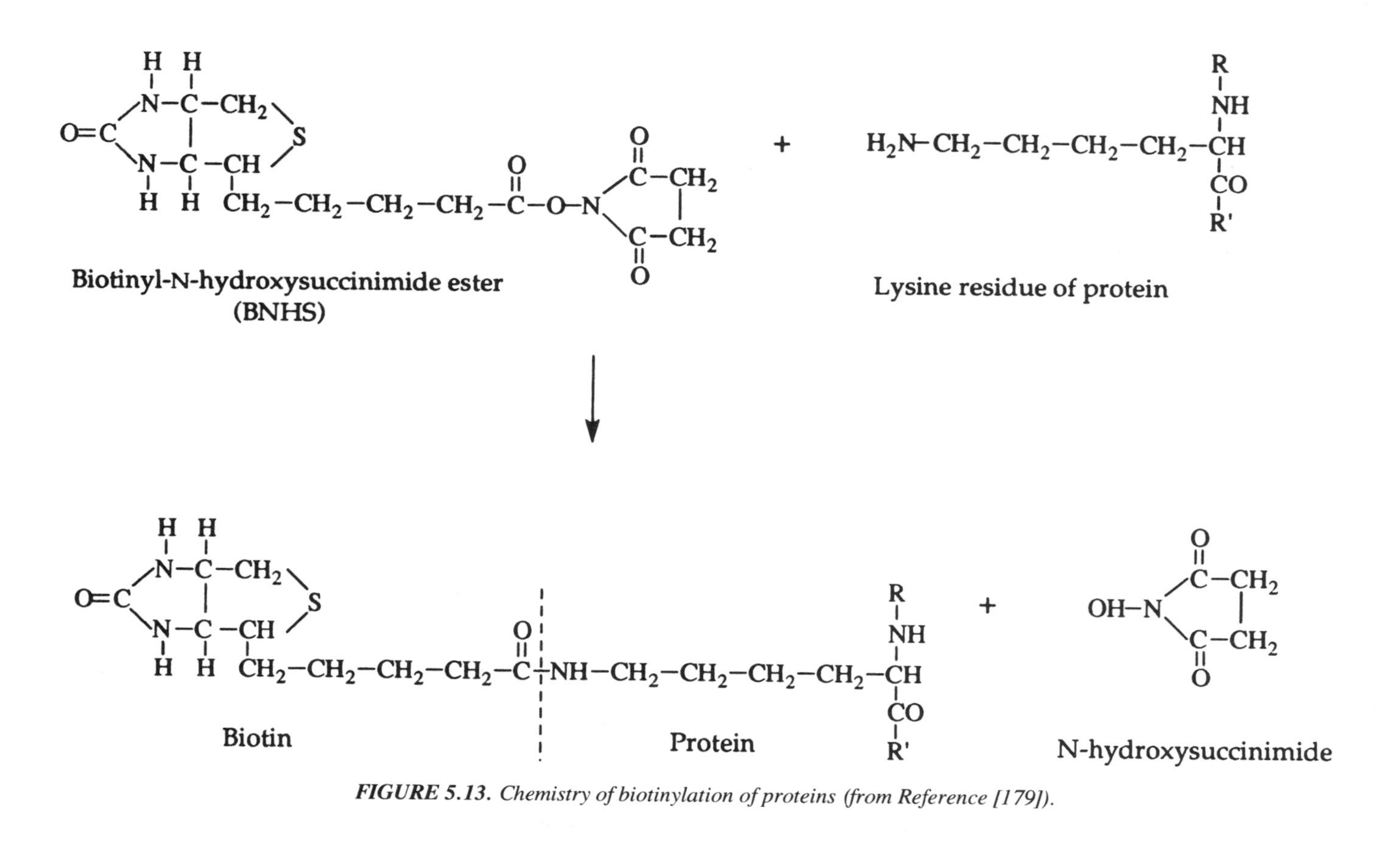

FIGURE 5.13. *Chemistry of biotinylation of proteins (from Reference [179]).*

The "nonimmune" origin means that lectins are different from anti-carbohydrate antibodies. Unlike carbohydrate-binding enzymes, lectins do not catalyze any reaction involving carbohydrates. Lectins are multivalent, i.e., they possess at least two sugar-binding sites that enable them to agglutinate cells bearing surface glycoproteins and/or precipitate glycoconjugates [187]. Carbohydrate specificities of various lectins are well documented [188,189]. Lectins, in the form of lectin affinity chromatography, were used extensively in the purification and characterization of various glycoconjugates [190].

The affinity between glycoconjugates and lectins is generally weak. This is probably due to the relatively weak polar interactions between hydroxyl groups of carbohydrates and the polar side chains of amino acid residues within the lectin's binding site [191]. The dissociation constant ranges from 10^{-2} to 10^{-4}. The glycoconjugates can be displaced readily from lectins by specific sugars at neutral pH [26]. This ability was used to develop self-regulated insulin delivery systems [192,193]. Carbohydrate-modified insulin molecules which are bound to lectins can be released after competitive interaction of glucose molecules. Due to their multivalent property, lectins can be used to crosslink carbohydrate-containing macromolecules.

5.5 DETERMINATION OF THE CROSSLINK DENSITY

While physical gels do not have well-defined, permanent crosslinks, the number of intermolecular interactions (e.g., hydrogen bonds) responsible for the formation of gels can be calculated from elastic moduli as described in Chapter 4. The rubber elasticity theory can be applied under the assumptions that the polymer chains retain sufficient flexibility for the gel network and the formation of junctions can be treated as an equilibrium process [194]. Obviously, the equilibrium swelling method cannot be used for many physical gels, since they will eventually be dissolved completely in water. The following is an example of measuring the number of intermolecular hydrogen bonds which are believed to be responsible for the gelation of gelatin [195].

According to the theory of rubber elasticity, Young's modulus (i.e., modulus of elasticity) E is related to the effective network chain concentration ν_e/V by the following equation [196].

$$E = 3(\nu_e/V)RT \tag{5.1}$$

where E is in the unit of dyne/cm^2, ν_e/V is in mol/cm^3, R is 8.314 x 10^7 dyne·cm/mol·K, and T is the absolute temperature in K. V is the volume of hydrogel when the gel is formed. Young's modulus E is a slope of the stress (τ)-strain (α-1) plot. A tetrafunctional crosslinking mode is assumed for gelatin. As described in Chapter 4, when the crosslink functionality is 4 [i.e., $\psi = 4$ in Equation (4.2)], the chain density ν_e/V is twice as large as the crosslink density μ/V. Thus, Equation (5.1) becomes

$$E = 6(\mu/V)RT \tag{5.2}$$

The crosslink density can be calculated from the Young's modulus E. The Young's modulus can be determined using the following equation from a simple experiment which measures the indentations of a heavy ball in the physical gel as shown in Figure 5.14.

$$E = 3(1 - \nu^2)\, F/(4h^{3/2}r^{1/2}) \tag{5.3}$$

where ν is the Poisson's ratio which is close to 0.5 for elastomers, F is the force of sphere against the physical gel surface (dyne), h is the depth of the indentation of sphere (cm), and r is the radius of the sphere (cm). This method is useful, since the idealized loading conditions can actually be realized in a practical situation [197].

The theory of rubber elasticity also provides that the effective network chain concentration is related to the shear modulus G by the following relationship:

$$G = (\nu_e/V)RT = (\varrho_2/M_c)RT = (c_2/M_c)RT \tag{5.4}$$

where ϱ_2 is the the density of the polymer, M_c is the average molecular weight between crosslinks, and c_2 is the weight concentration of polymer [194,198]. This relationship was used to estimate the size of junction zones in gelatin, pectin, and alginate gels.

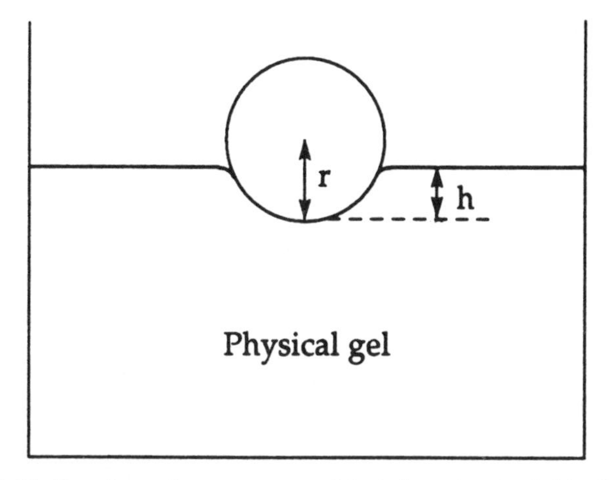

FIGURE 5.14. *Experimental measurement of the indentation depth* (h) *of a heavy ball in the physical gel (from Reference [195]).*

5.6 REFERENCES

1. Tanaka, F. and S. F. Edwards. 1992. "Viscoelastic Properties of Physically Cross-Linked Networks. Transient Network Theory," *Macromolecules*, 25:1516–1523.
2. Clark, A. H. and S. B. Ross-Murphy. 1987. "Structural and Mechanical Properties of Biopolymer Gels," *Adv. Polymer Sci.*, 83:57–192.
3. Hvidt, S. and K. Heller. 1990. "Viscoelastic Properties of Biological Networks and Gels," in *Physical Networks. Polymers and Gels*, W. Burchard and S. B. Ross-Murphy, eds., New York, NY: Elsevier, pp. 195–208.
4. Ross-Murphy, S. B. 1991. "Physical Gelation of Synthetic and Biological Macromolecules," in *Polymer Gels. Fundamentals and Biomedical Applications*, D. DeRossi, K. Kajiwara, Y. Osada and A. Yamauchi., eds., New York, NY: Plenum Press, pp. 21–39.
5. de Gennes, P.-G. 1979. *Scaling Concepts in Polymer Physics*. Ithaca, NY: Cornell University Press, p. 133.
6. Clark, A. H., S. B. Ross-Murphy, K. Nishinari and M. Watase. 1990. "Shear Modulus–Concentration Relationships for Biopolymer Gels. Comparison of Independent and Cooperative Crosslink Descriptions," in *Physical Networks. Polymers and Gels*, W. Burchard and S. B. Ross-Murphy, eds., New York, NY: Elsevier, pp. 209–229.
7. Nijenhuis, K. Te. 1990. "Viscoelastic Properties of Thermoreversible Gels," in *Physical Networks. Polymers and Gels*, W. Burchard and S. B. Ross-Murphy, eds., New York, NY: Elsevier, pp. 15–33.
8. Durand, D., C. Bertrand, J.-P. Busnel, J. R. Emery, M. A. V. Axelos, J. F. Thibault, J. Lefebvre, J. L. Doublier, A. H. Clark and A. Lips. 1990. "Physical Gelation Induced by Ionic Complexation: Pectin-Calcium Systems," in *Physical Networks. Polymers and Gels*, W. Burchard and S. B. Ross-Murphy, eds., New York, NY: Elsevier, pp. 283–300.
9. MacGreger, E. A and C. T. Greenwood. 1980. *Polymers in Nature*. New York, NY: John Wiley & Sons, Chapter 6.
10. Borchard, W. and B. Burg. 1989. "Investigations of the Complex Shear Modulus and the Optical Rotation in the System Gelatin-Water during the Thermoreversible Gelatin Process," in *Molecular Basis of Polymer Networks*, A. Baumgärtner and C. E. Picot, eds., Berlin: Spring-Verlag, pp. 162–168.
11. Morris, E. R. and D. A. Rees. 1978. "Principles of Biopolymer Gelation. Possible Models for Mucus Gel Structure," *Brit. Med. Bull.*, 34:49–53.
12. Egelandsdal, B. 1984. "A Comparison between Ovalbumin Gels Formed by Heat and by Guanidinium Hydrochloride Denaturation," *J. Food Sci.*, 49:1099–1102.
13. Tombs, M. P. 1974. "Gelation of Globular Proteins," *Faraday Discuss. Chem. Soc.*, 57:158–164.
14. Clark, A. H. and S. B. Ross-Murphy. 1985. "The Concentration Dependence of Biopolymer Gel Modulus," *Brit. Polymer J.*, 17:164–168.
15. Borchard, W. 1983. "Thermoreversible Gelation," in *Chemistry and Technology of Water-Soluble Polymers*, C. A. Finch, ed., New York, NY: Plenum Press, pp. 113–124.
16. Cohen, Y., Y. Talmon and E. L. Thomas. 1990. "On the Structure of Poly(γ-benzyl-L-glutamate) (PBLG) Gels," in *Physical Networks. Polymers and Gels*, W. Burchard and S. B. Ross-Murphy, eds., New York, NY: Elsevier, pp. 147–158.

17. Hermansson, A.-M. 1990. ''Structure and Rheological Properties of Kappa-Carrageenan Gels,'' in *Physical Networks. Polymers and Gels*, W. Burchard and S. B. Ross-Murphy, eds., New York, NY: Elsevier, pp. 271–282.

18. Rogovina, L. Z. 1990. ''Comparison of the Formation and Properties of Physical and Chemical Networks Prepared in the Swollen State,'' in *Physical Networks. Polymers and Gels*, W. Burchard and S. B. Ross-Murphy, eds., New York, NY: Elsevier, pp. 133–145.

19. Flory, P. J. 1953. *Principles of Polymer Chemistry*. Ithaca, NY: Cornell University Press, p. 47.

20. Richtering, H. W., K. D. Gagnon, R. W. Lenz, R. C. Fuller and H. H. Winter. 1992. ''Physical Gelation of a Bacterial Thermoplastic Elastomer,'' *Macromolecules*, 25:2429–2433.

21. Pritchard, J. G. 1970. *Poly(vinyl Alcohol). Basic Properties and Uses*. New York, NY: Gordon and Breach Science Publishers, Chapter 4.

22. Finch, C. A. 1983. ''Chemical Modification and Some Cross-Linking Reactions of Water-Soluble Polymers,'' in *Chemistry and Technology of Water-Soluble Polymers*, C. A. Finch, ed., New York, NY: Plenum Press, pp. 81–111.

23. Burchard, W. and S. B. Ross-Murphy. 1990. ''Introduction: Physical Gels from Synthetic and Biological Macromolecules,'' in *Physical Networks. Polymers and Gels*, W. Burchard and S. B. Ross-Murphy, eds., New York, NY: Elsevier, pp. 1–14.

24. Higgs, P. G. and R. C. Ball. 1990. ''A 'Reel-Chain' Model for the Elasticity of Biopolymer Gels, and Its Relationship to Slip-Link Treatments of Entanglements,'' in *Physical Networks. Polymers and Gels*, W. Burchard and S. B. Ross-Murphy, eds., New York, NY: Elsevier, pp. 185–194.

25. Djadourov, M. and J. Leblond. 1987. ''Thermally Reversible Gelatin of the Gelatin-Water System,'' Chapter 14 in *Reversible Polymeric Gels and Related Systems*, P. S. Russo, ed., Washington, DC: American Chemical Society.

26. Sturgeon, C. M. 1988. ''The Synthesis of Polysaccharide Derivatives,'' Chapter 13 in *Carbohydrate Chemistry*, J. F. Kennedy, ed., New York, NY: Oxford University Press.

27. Arnott, S., A. Fulmer, W. E. Scott, I. C. M. Dea, R. Moorhouse and D. A. Rees. 1974. ''Agarose Double Helix and Its Function in Agarose Gel Structure,'' *J. Mol. Biol.*, 90:269–284.

28. 1982. *The Agarose Monograph*. Rockland, MA: FMC Corp., Marine Colloids Division, p. 21.

29. Iwata, H., T. Takagi, H. Amemiya, H. Shimizu, K. Yamashita, K. Kobayashi and T. Akutsu. 1992. ''Agarose for a Bioartificial Pancreas,'' *J. Biomed. Mater. Res.*, 26:967–977.

30. Noguchi, T., T. Yamamuro, M. Oka, P. Kumar, Y. Kotoura, S.-H. Hyon and Y. Ikada. 1991. ''Poly(vinyl Alcohol) Hydrogel as an Artificial Articular Cartilage: Evaluation of Biocompatibility,'' *J. Appl. Biomater.*, 2:101–107.

31. Peppas, N. A. and S. R. Stauffer. 1991. ''Reinforced Uncrosslinked Poly(vinyl Alcohol) Gels Produced by Cyclic Freeze-Thawing Processes: A Short Review,'' *J. Controlled Rel.*, 16:305–310.

32. Vinogradov, S. N. and R. H. Linnell. 1971. *Hydrogen Bonding*. New York, NY: Van Nostrand Reinhold Company, Chapter 9.

33. Tsuchida, E. and K. Abe. 1982. ''Interactions between Macromolecules in Solution and Intermacromolecular Complexes,'' *Adv. Polymer Sci.*, 45:1–119.

34. Walker, J. S. and C. A Vause. 1987. "Reappearing Phases," *Sci. Amer.*, 256:98–105.

35. Leach, H. W. 1965. "Gelatinization of Starch," Chapter 12 in *Starch: Chemistry and Technology*, R. L. Whistler and E. F. Paschall, eds., New York, NY: Academic Press.

36. Israelachvili, J. N. 1985. *Intermolecular and Surface Forces*. New York, NY: Academic Press, Chapter 8.

37. Bednar, B., Z. Li, Y. Huang, L.-C. P. Chang and H. Morawetz. 1985. "Fluorescence Study of Factors Affecting the Complexation of Poly(acrylic Acid) with Polyoxyethylene," *Macromolecules*, 18:1829–1833.

38. Bekturov, E. A. and L. A. Bimendina. 1981. "Interpolymer Complexes," *Adv. Polymer Sci.*, 41:99–147.

39. Franks, F. 1975. "The Hydrophobic Interaction," Chapter 1 in *A Comprehensive Treatise, Vol. 4,* F. Franks, ed., New York, NY: Plenum Press.

40. Franz, G. 1986. "Polysaccharides in Pharmacy," *Adv. Polymer Sci.*, 76:1–30.

41. Molyneux, P. 1983. "Water-Soluble Synthetic Polymers," Chapter 1 in *Water Soluble Synthetic Polymers: Properties and Behavior, Vol. 1.* Boca Raton, FL: CRC Press.

42. Tanford, C. 1980. *The Hydrophobic Effect: Formation of Micelles and Biological Membranes, Second Edition.* New York, NY: John Wiley & Sons, Chapter 4.

43. Ben-Naim, A. 1980. *Hydrophobic Interactions.* New York, NY: Plenum Press, Chapter 5.

44. Nakai, S. and E. Li-Chan. 1988. *Hydrophobic Interactions in Food Systems.* Boca Raton, FL: CRC Press, Chapter 1.

45. Morris, E. R. and I. T Norton. 1983. "Polysacchraide Aggregation in Solutions and Gels," Chapter 19 in *Aggregation Processes in Solution*, E. Wyn-Jones and J. Gormally, eds., New York, NY: Elsevier Scientific Pub. Co.

46. Otake, K., H. Inomato, M. Konno and S. Saito. 1990. "Thermal Analysis of the Volume Phase Transition with *N*-Isopropylacrylamide Gels," *Macromolecules*, 23:283–289.

47. Doelker, E. 1986. "Water-Swollen Cellulose Derivatives in Pharmacy," Chapter 5 in *Hydrogels in Medicine and Pharmacy, Vol. II*, N. A. Peppas, ed., Boca Raton, FL: CRC Press.

48. Product information from the Dow Chemical Company, Midland, Michigan 48640.

49. Hirokawa, Y. and T. Tanaka. 1984. "Volume Phase Transition in a Nonionic Gel," *J. Chem. Phys.*, 81:6379–6380.

50. Katayama, S., Y. Hirokawa and T. Tanaka. 1984. "Reentrant Phase Transition in Acrylamide Derivative Copolymer Gels," *Macromolecules*, 17:2641–2643.

51. Dusek, K. and D. Patterson. 1968. "Transition in Swollen Networks Induced by Intramolecular Condensation," *J. Polymer Sci.: Part A-2*, 6:1209–1216.

52. Meewes, M., J. Riĉka, M. de Silva, R. Nyffenegger and Th. Binkert. 1991. "Coil-Globule Transition of Poly(*N*-isopropylacrylamide). A Study of Surfactant Effects by Light Scattering," *Macromolecules*, 24:5811–5816.

53. Bae, Y. H., T. Okano and S.W. Kim. 1990. "Temperature Dependence of Swelling of Crosslinked Poly(*N,N'*-alkyl Substituted Acrylamides) in Water," *J. Polymer Sci.: Part B: Polym. Phys.*, 28:923–936.

54. Hirasa, O., S. Ito, A. Yamauchi, S. Fujishige and H. Ichijo. 1991. "Thermoresponsive Polymer Hydrogel," in *Polymer Gels. Fundamentals and Biomedical Applica-*

tions, D. DeRossi, K. Kajiwara, Y. Osada and A. Yamauchi., eds., New York, NY: Plenum Press, pp. 247–256.

55. Inomata, H., S. Goto, K. Otaka and S. Saito. 1992. "Effect of Additives on Phase Transition of *N*-Isopropylacrylamide Gels," *Langmuir*, 8:687–690.

56. Urry, D. W. 1990. "Preprogrammed Drug Delivery Systems Using Chemical Triggers for Drug Release by Mechanochemical Coupling," *Proc. ACS Div. Polym. Mat. Sci. Eng.*, 63:329–336.

57. Urry, D. W. 1988. "Entropic Elastic Processes in Protein Mechanisms. II. Simple (Passive) and Coupled (Active) Development of Elastic Forces," *J. Protein Chem.*, 7:81–114.

58. Bae, Y. H., T. Okano and S. W. Kim. 1991. "Insulin Permeation through Thermo-Sensitive Hydrogels," *J. Controlled Rel.*, 9:271–279.

59. Bae, Y. H., T. Okano and S. W. Kim. 1991. " 'On-Off' Thermocontrol of Solute Transport. II. Solute Release from Thermosensitive Hydrogels," *Pharm. Res.*, 8:624–628.

60. Dong, L. C. and A. S. Hoffman. 1990. "Synthesis and Application of Thermally Reversible Heterogels for Drug Delivery," *J. Controlled Rel.*, 13:21–31.

61. Cole, C.-A., S. M. Scheriner, J. H. Priest, N. Monji and A. S. Hoffman. 1987. "*N*-Isopropylacrylamide and *N*-Acryloxysuccinimide Copolymer. A Thermally Reversible, Water-Soluble, Activated Polymer for Protein Conjugation," Chapter 17 in *Reversible Polymeric Gels and Related Systems*, P. S. Russo, ed., Washington, DC: American Chemical Society.

62. Park, T. G. and A. S. Hoffman. 1990. "Immobilization and Characterization of B-Galactosidase in Thermally Reversible Hydrogel Beads," *J. Biomed. Mater. Res.*, 24:21–38.

63. Claesson, P. M. and B. W. Ninham. 1992. "pH-Dependent Interactions between Adsorbed Chitosan Layers," *Langmuir*, 8:1406–1412.

64. Alger, M. S. M. 1989. *Polymer Science Dictionary*. New York, NY: Elsevier Applied Science, p. 70.

65. Connell, A. M. 1981. "Dietary Fiber," Chapter 51 in *Physiology of the Gastrointestinal Tract*, L. R. Johnson, ed., New York, NY: Raven Press.

66. Daniel, R. 1973. *Edible Coatings and Soluble Packaging*. Park Ridge, NJ: Noyes Data Corp.

67. Lawrence, A. A. 1976. *Natural Gums for Edible Purposes*. Park Ridge, NJ: Noyes Data Corp., pp. 17–19.

68. Griffiths, A. J. and J. F. Kennedy. 1988. "Biotechnology of Polysaccharides," Chapter 14 in *Carbohydrate Chemistry*, J. F. Kennedy, ed., New York, NY: Oxford University Press.

69. Paul, F., A. Morin and P. Monsan. 1986. "Microbial Polysaccharides with Actual Potential Industrial Applications," *Biotech. Adv.*, 4:245–259.

70. Flick, E. W. 1986. *Water-Soluble Resins. An Industrial Guide*. Park Ridge, NJ: Noyes Data Corp., Section III.

71. Morris, E. R. 1990. "Mixed Polymer Gels," Chapter 8 in *Food Gels*, P. Hans, ed., New York, NY: Elsevier.

72. Morris, E. R. and D. A. Rees. 1980. "Competitive Inhibition of Interchain Interactions in Polysaccharide Systems," *J. Mol. Biol.*, 138:363–374.

73. Sandford, P. A. and J. Baird. 1983. "Industrial Utilization of Polysaccharides," in *The Polysaccharides, Vol. 2,* G. O. Aspinall, ed., New York, NY: Academic Press, pp. 411–490.

74. Doublier, J.-L. and G. Llamas. 1990. "Flow and Viscoelastic Properties of Mixed Xanthan Gum + Galactomannan Systems," in *Food, Polymers, Gels, and Colloids*, E. Dickinson, ed., Norwich, England: The Royal Society of Chemistry, pp. 349 –356.

75. Mantell, C. L. 1947. *The Water-Soluble Gums*. New York, NY: Reinhold Publishing Corp., Chapter 1.

76. Howes, F. N. 1949. *Vegetable Gums and Resins*. Waltham, MA: Chronica Botanica Co., Chapter 1.

77. Rinaudo, M. and M. Milas. 1987. "On the Properties of Polysaccharides. Relation between Chemical Structure and Physical Properties," in *Industrial Polysaccharides. Progress in Biotechnology, Vol. 3,* M. Yalpani, ed., Amsterdam, The Netherlands: Elsevier, pp. 217–223.

78. Harada, T. 1974. "Production, Properties and Application of Curdlan," in *Extracellular Microbial Polysaccharides*, P. A. Sandford and A. Laskin, eds., Washington, DC: American Chemical Society, pp. 265–283.

79. Swann, D. A. and J. Kuo. 1991. "Hyaluronic Acid," Chapter 6 in *Biomaterials. Novel Materials from Biological Sources*, D. Byrom, ed., New York, NY: Stockton Press.

80. Bother, H. and O. Wik. 1987. "Rheology of Hyaluronate," *Acta Otolaryngol.*, Suppl. 442:25–30.

81. Drobnik, J. 1991. "Hyaluronan in Drug Delivery," *Adv. Drug Delivery Rev.*, 7:295–308.

82. Shah, C. B. and S. M. Barnett. 1992. "Hyaluronic Acid Gels," *ACS Symp. Ser.*, 480:116–130.

83. Laurent, T. C. and J. R. E. Fraser. 1986. "The Properties and Turnover of Hyaluronan," in *Functions of the Proteoglycans, Ciba Foundation Symposium 143*, D. Evered and J. Whelan, eds., Chichester, England: John Wiley & Sons, p. 9.

84. Laurent, T. C. and J. R. E. Fraser. 1991. "Catabolism of Hyaluronan," Chapter 16 in *Degradation of Bioactive Substances: Physiology and Pathophysiology*, J. H. Henriksen, ed., Boca Raton, Florida: CRC Press.

85. Hunt, J. A., H. N. Joshi, V. J. Stella and E. M. Topp. 1990. "Diffusion and Drug Release in Polymer Films Prepared from Ester Derivatives of Hyaluronic Acid," *J. Controlled Release*, 12:159–169.

86. Papini, D., S. Hejri, V. J. Stella and E. M. Topp. 1992. "Diffusion of Peptides in Polymer Films Prepared from Ester Derivatives of Hyaluronic Acid," *Proceed. Intern. Symp. Control. Rel. Bioact. Mater.*, 18:611–622.

87. Pezron, E., A. Richard, F. Lafuma and R. Audebert. 1988. "Reversible Gel Formation Induced by Ion Complexation. 1. Borax-Galactomannan Interactions," *Macromolecules*, 21:1121–1125.

88. Bucci, S., G. Gallino and T. P. Lockhart. 1991. "B +3/Polysaccharide Gels: Influence of Crosslinking Chemistry on Rheological Properties," *Polymer Preprints*, 32:457–458.

89. BeMiller, J. N. 1965. "Organic Complexes and Coordination Compounds of Carbohydrates," Chapter 13 in *Starch: Chemistry and Technology*, R. L. Whistler and E. F. Paschall, eds., New York, NY:Academic Press.

90. Johnson, J. C., ed. 1983. *Food Additives. Recent Developments*. Park Ridge, NJ: Noyes Data Corp., pp. 90–132.

91. Deuel, H. and H. Neukom. 1949. "The Reaction of Boric Acid and Borax with Polysaccharides and Other High Molecular Polyhydroxy Compounds," *Makromol. Chem.*, 3:13–30.

92. Lazár, M., T. Bleha and J. Rychly. 1989. *Chemical Reactions of Natural and Synthetic Polymers*. West Sussex, England: Ellis Horwood Ltd.

93. Linton, J. D., S. G. Ash and L. Huybrechts. 1991. "Microbial Polysaccharides," Chapter 4 in *Biomaterials. Novel Materials from Biological Sources*, D. Byrom, ed., New York, NY: Stockton Press.

94. Chrisp, J. D. U.S. patent 3,518,242, June 30, 1970.

95. Meltzer, Y. L. 1976. *Water-Soluble Resins and Polymers*. Park Ridge, NJ: Noyes Data Corp., pp. 229–230.

96. Pezron, E., L. Leibler, A. Richard and R. Audebert. 1988. "Reversible Gel Formation Induced by Ion Complexation. 2. Phase Diagrams," *Macromolecules*, 21:1126–1131.

97. Kesavan, S. and R. K. Prud'homme. 1992. "Rheology of Guar and HPG Cross-Linked by Borate," *Macromolecules*, 25:2026–2032.

98. Ochiai, H., S. Shimizu, Y. Tadokoro and I. Murakami. 1981. "Complex Formation between Poly(vinyl Alcohol) and Borate Ion," *Polymer*, 22:1456–1458.

99. Wu, K.-Y. A. and K. D. Wisecarver. 1992. "Cell Immobilization Using PVA Crosslinked with Boric Acid," *Biotechnol. Bioeng.*, 39:447–449.

100. Kost, J. and S. Shefer. 1990. "Chemically-Modified Polysaccharides for Enzymatically-Controlled Oral Drug Delivery," *Biomaterials*, 11:695–698.

101. Wing, R. E., S. Maiti and W. M. Doane. 1987. "Factors Affecting Release of Butylate from Calcium Ion-Modified Starch-Borate Matrices," *J. Controlled Rel.*, 5:79–89.

102. Whistler, R. L. and J. R. Daniel. 1985. "Carbohydrates," Chapter 3 in *Food Chemistry, 2nd Edition*, O. R. Fennema, ed., New York, NY: Marcel Dekker, Inc.

103. Rees, D. A. and E. J. Welsh. 1977. "Secondary and Tertiary Structure of Polysaccharides in Solutions and Gels," *Angew. Chem. Int. Ed. Engl.*, 16:214–224.

104. Nilsson, S., L. Piculell and B. Jönsson. 1989. "Salt Effect on Gel-Forming Ionic Polysaccharides," in *Molecular Basis of Polymer Networks*, A. Baumgärtner and C. E. Picot, eds., Berlin: Spring-Verlag, pp. 157–161.

105. Crumbliss, A. L., R. W. Henkens, S. C. Perine, K. R. Tubergen, B. S. Kitchell and J. Stonehuerner. 1990. "Amperometric Glucose Sensor Fabricated from Glucose Oxidase and a Mediator Coimmobilized on a Colloidal Gold Hydrogel Electrode," in *Biosensor Technology. Fundamental and Applications*, R. P. Buck, W. E. Hatfield, M. Umana and E. F. Bowden, eds., New York, NY: Marcel Dekker, Inc.

106. Moorhouse, R., G. T. Colegrove, P. A. Sanford, J. K Baird and K. S. Kang. 1981. "PS-60: A New Gel-Forming Polysaccharide," *Amer. Chem. Soc. Symp. Ser.*, 150:111–124.

107. Grasdalen, H. and O. Smidsrød. 1987. "Gelation of Gellan Gum," *Carbohydrate Polymers*, 7:371–393.

108. Rozier, A., C. Mazuel, J. Grove and B. Plazonnet. 1989. "Gelrite: A Novel, Ion-Activated, *in situ* Gelling Polymer for Opthalmic Vehicles. Effect on Bioavailability of Timolol," *Int. J. Pharm.*, 57:163–168.

109. Bhakoo, M., S. Woerly and R. Duncan. 1991. "Release of Antibiotics and Antitumor Agents from Alginate and Gellan Gum Gels," *Proceed. Intern. Symp. Control. Rel. Bioact. Mater.*, 18:441–442.

110. Rolin, C. and J. De Vries. 1990. "Pectin," Chapter 10 in *Food Gels*, P. Hans, ed., New York, NY: Elsevier.

111. Oakenful, D. and A. Scott. 1984. "Hydrophobic Interaction in the Gelation of High Methoxyl Pectins," *J. Food Sci.*, 49:1093–1098.

112. Burchard, W. 1985. "Networks in Nature," *Brit. Polymer J.*, 17:154–163.

113. Grant, G. T., E. R. Morris, D. A. Rees, P. J. C. Smith and D. Thom. 1973. "Biological Interactions between Polysaccharides and Divalent Cations: The Egg-Box Model," *FEBS Lett.*, 32:195–198.

114. Thibault, J.-F. and M. Rinaudo. 1985. "Gelation of Pectinic Acids in the Presence of Calcium Counterions," *Brit. Polymer J.*, 17:181–184.

115. Sutherland, I. W. 1991. "Alginates," Chapter 7 in *Biomaterials. Novel Materials from Biological Sources*, D. Byrom, ed., New York, NY: Stockton Press.

116. Chapman, V. J. and D. J. Chapman. 1980. *Seaweeds and Their Uses*. New York, NY: Chapman and Hall, Chapter 6.

117. Gacesa, P. 1988. "Alginates," *Carbohydrate Polymers*, 8:161–182.

118. Goosen, M. F. A., G. M. O'Shea, H. M. Gharapetian, S. Chou and A. M. Sun. 1985. "Optimization of Microencapsulation Parameters: Semipermeable Microcapsules as a Bioartificial Pancreas," *Biotech. Bioeng.*, 27:146–150.

119. Chang, T. M. S. 1992. "Hybrid Artificial Cells: Microencapsulation of Living Cells," *ASAIO Journal*, 38:128–130.

120. Gòdia, F., C. Casas and C. Solà. 1991. "Application of Immobilized Yeast Cells to Sparkling Wine Fermentation," *Biotechnol. Prog.*, 7:468–470.

121. Tomkins, R. G., E. A. Carter, J. D. Carlson and M. L. Yamush. 1988. "Enzymatic Function of Alginate Immobilized Rat Hepatocytes," *Biotechnol. Bioeng.*, 31:11–18.

122. Gray, C. J. and J. Dowsett. 1988. "Retention of Insulin in Alginate Gel Beads," *Biotechnol. Bioeng.*, 31:607–612.

123. Kwok, K. K., M. J. Groves and D. J. Burgess. 1991. "Production of 5–15 μm Diameter Alginate-Polylysine Microcapsules by an Air-Atomization Technique," *Pharm. Res.*, 8:341–344.

124. Posillico, E. G. 1986. "Microencapsulation Technology for Large-Scale Antibody Production," *Bio/Technology*, 4:114–117.

125. McNight, C. A., A. Ku, M. F. A. Goosen, D. Sun and C. Penny. 1988. "Synthesis of Chitosan-Alginate Microcapsule Membranes," *J. Bioact. Compat. Polymers*, 3:334–355.

126. Bodmeier, R., H. Chen and O. Paeratakul. 1989. "A Novel Approach to the Oral Delivery of Micro- or Nanoparticles," *Pharm. Res.*, 6:413–417.

127. Igari, Y., P. G. Kibat and R. Langer. 1990. "Optimization of a Microencapsulated Liposome System for Enzymatically Controlled Release of Macromolecules," *J. Controlled Rel.*, 14:263–267.

128. Gilchrist, T. and A. M. Martin. 1983. "Wound Treatment with Sorbsan–An Alginate Fibre Dressing," *Biomaterials*, 4:317–320.

129. Bhagat, H. R., R. W. Mendes, E. Mathiowitz and N. Bhargava, N. 1991. "A Novel, Self-Correcting Membrane Coating Technique," *Pharm. Res.*, 8:576–583.

130. Yotsuyanagi, T., T. Ohkubo, T. Ohhashi and K. Ikeda. 1987. "Calcium-Induced Gelation of Alginic Acid and pH-Sensitive Reswelling of Dried Gels," *Chem. Pharm. Bull.*, 35:1555–1563.

131. Husain, Q., J. Iqbal and M. Saleemuddin. 1985. ''Entrapment of Concanavalin A-Glycoenzyme Complexes in Calcium Alginate Gels,'' *Biotechnol. Bioeng.*, 27:1102–1107.

132. Wheatley, M. A., M. Chang, E. Park and R. Langer. 1991. ''Coated Alginate Microspheres: Factors Influencing the Controlled Delivery of Macromolecules,'' *J. Appl. Polymer Sci.*, 43:2123–2135.

133. Allcock, H. R. and S. Kwon. 1989. ''An Ionically Cross-Linkable Polyphosphazene: Poly[bis(carboxyatophenoxy)phosphazene] and Its Hydrogels and Membranes,'' *Macromolecules*, 22:75–79.

134. Habert, A. C., C. M. Burns and R. Y. M. Huang. 1979. ''Ionically Crosslinked Poly-(acrylic Acid) Membranes. II. Dry Technique,'' *J. Appl. Polymer Sci.*, 24:801–809.

135. Cohen, S., M. C. Bano, K. B. Visscher, M. Chow, H. R. Allcock and R. Langer. 1990. ''Ionically Cross-Linkable Polyphosphazene: A Novel Polymer for Microencapsulation,'' *J. Am. Chem. Soc.*, 112:7832–7833.

136. Bano, M. C., S. Cohen, K. B. Visscher, H. R. Allcock and R. Langer. 1991. ''A Novel Synthetic Method for Hybridoma Cell Encapsulation,'' *Bio/Technology*, 9:468–471.

137. Mortimer, D. A. 1991. ''Synthetic Polyelectrolytes–A Review,'' *Polymer International*, 25:29–41.

138. Shimizu, H. and H. Ozawa. U.S patent 4,024,073, May 17, 1977.

139. Johnson, J. C., ed. 1980. *Sustained Release Medications*. Park Ridge, NJ: Noyes Data Corp., pp. 10–12.

140. Karlson, J. and O. Skaugrud. 1991. ''Excipient Properties of Chitosan,'' *Manuf. Chem.*, 62:18–19.

141. Mosesson, M. W. and R. F. Doolittle, eds. 1983. *Molecular Biology of Fibrinogen and Fibrin, Ann. N.Y. Acad. Sci., Vol. 408.* New York, NY: The New York Academy of Sciences.

142. Sherry, S. 1992. *Fibrinolysis, Thrombosis, and Hemostasis. Concepts, Perspectives, and Clinical Applications*. New York, NY: Marcel Dekker, Inc., pp. 187–193.

143. Park, K., F. W. Mao and H. Park. 1991. ''The Minimum Surface Fibrinogen Concentration Necessary for Platelet Activation on Dimethyldichlorosilane-Coated Glass,'' *J. Biomed. Mater. Res.*, 25:407–420.

144. Donnelly, T. H., M. Laskowski, Jr., N. Notley and H. A. Sheraga. 1955. ''Equilibria in the Fibrinogen-Fibrin Conversion. II. Reversibility of the Polymerization Steps,'' *Arch. Biochem. Biophys.*, 56:369–387.

145. Kerenyi, G. 1980. ''Properties and Applications of Bioplast, an Absorbable Surgical Implant Material from Fibrin,'' *Biomaterials*, 1:30–32.

146. Scheele, J., H. H. Gentsch and E. Matteson. 1984. ''Splenic Repair by Fibrin Tissue Adhesive and Collagen Fleece,'' *Surgery*, 95:6–13.

147. Sierra, D. H., D. S. Feldman, R. Saltz and S. Huang. 1992. ''A Method to Determine Shear Adhesive Strength of Fibrin Sealants,'' *J. Appl. Biomater.*, 3:147–151.

148. Siedentop, K. H., D. M. Harris and A. Loewy. 1983. ''Experimental Use of Fibrin Tissue Adhesive in Middle Ear Surgery,'' *Laryngoscope*, 93:1310–1313.

149. Senderoff, R. I., M. T. Sheu and T. D. Sokoloski. 1991. ''Fibrin Based Drug Delivery Systems,'' *J. Parenter. Sci. Technol.*, 45:2–6.

150. Sakurai, T., N. Nishikimi, K. Yamamura and S. Shionoya. 1992. ''Controlled Release of Sisomicin from Fibrin Glue,'' *J. Controlled Rel.*, 18:39–44.

151. Miyazaki, S. and T. Nadai. 1980. "Use of Fibrin Film as a Carrier for Drug Delivery: *In vitro* Drug Permeabilities of Fibrin Film," *Chem. Pharm. Bull.*, 28:2261–2264.
152. Redl, H., G. Schlag, G. Stanek, A. Hirschl and T. Seelich. 1983. "*In vitro* Properties of Mixtures of Fibrin Seal and Antibiotics," *Biomaterials*, 4:29–32.
153. Inada, Y., S. Hirose, A. Matsushima, H. Mihama and Y. Hiramoto. 1977. "Fibrin Membrane Endowed with Biological Function. III. Fixing Living Cells in Fibrin Gel without Impairing Their Functions," *Experientia*, 33:1257–1259.
154. Marriott, C. and N. P. Gregory. 1990. "Mucus Physiology and Pathology," in *Bioadhesive Drug Delivery Systems*, V. Lenaerts and R. Gurny, eds., Boca Raton, FL: CRC Press, pp. 1–24.
155. Allen, A. and N. J. H. Carroll. 1985. "Adherent and Soluble Mucus in the Stomach," *Digestive Disease and Sciences*, 30:555–625.
156. Lehr, C.-M., F. G. J. Poelma, H. E. Junginger and J. J. Tukker. 1991. "An Estimate of Turnover Time of Intestinal Mucus Gel Layer in the Rat *in situ* Loop," *Int. J. Pharm.*, 70:235–240.
157. Allen, A., R. H. Pain and T. R. Robson. 1976. "Model for the Structure of the Gastric Mucous Gel," *Nature*, 264:88–89.
158. Allen, A.1981. "The Stucture and Function of Gastrointestinal Mucus," in *Basic Mechanisms of Gastrointestinal Mucosal Cell Injury and Protection*, J. W. Harmon, ed., Baltimore, MD: Williams and Wilkins, pp. 351–367.
159. Gandhi, R. B. and J. R. Robinson. 1988. "Bioadhesion in Drug Delivery," *Indian J. Pharm. Sci.*, 50:145–156.
160. Beeley, J. G. 1985. *Glycoprotein and Proteoglycan Techniques*. New York, NY: Elsevier, Chapter 2.
161. Marriott, C. and D. R. L. Hughes. 1989. "Mucus Physiology and Pathology," in *Bioadhesion–Possibilities and Future Trends*, R. Gurny and H. E. Junginger, eds., Stuttgart: Wissenschaftliche Verlagsgesellschaft mbH, pp. 29–43.
162. Allen, A., W. J. Cunliffe, J. P. Pearson, L. A. Sellers and R. Ward. 1984. "Studies on Gastrointestinal Mucus," *Scand. J. Gastroenterol*, Suppl., 19:101–113.
163. Roberts, G. P. 1976. "The Role of Disulfide Bonds in Maintaining the Gel Structure of Bronchial Mucus," *Arch. Biochem. Biophys.*, 173:528–537.
164. Lenaerts, V. and R. Gurny, eds. 1990. *Bioadhesive Drug Delivery Systems*. Boca Raton, FL: CRC Press.
165. Kamath, K. R. and K. Park. In press. "Mucosal Adhesive Preparations," in *Encyclopedia of Pharmaceutical Technology*, J. Swarbrick and J. C. Boylan, eds., New York, NY: Marcel Dekker, Inc.
166. Fulton, A. 1984. *The Cytoskeleton: Cellular Architecture and Choreography*. New York, NY: Chapman and Hall, Chapter 1.
167. Spudich, J. A. and S. Watts. 1971. "The Regulation of Rabbit Skeletal Muscle Contraction. 1. Biochemical Studies of the Interaction of the Tropomyosin/Troponin Complex with Actin and the Proteolytic Fragments of Myosin," *J. Biol. Chem.*, 246:4866–4871.
168. Cunningham, C. C., J. B. Gorlin, D. J. Kwiatkowski, J. H. Hartwig, P. A. Janmey, H. R. Byers and T. P. Stossel. 1992. "Actin-Binding Protein Requirement for Cortical Stability and Efficient Locomotion," *Science*, 255:325–327.
169. Nossal, R. 1987. "*In vitro* Polymerization of Complex Cytoplasmic Gels," Chapter 15 in *Reversible Polymeric Gels and Related Systems*, P. S. Russo, ed., Washington, DC: American Chemical Society.

170. Bershadsky, A. D. and J. M. Vasiliev. 1988. *Cytoskeleton*. New York, NY: Plenum Press, Chapter 1.

171. Dirlikov, S. 1992. "Living Polymerization of Proteins: Actin and Tubulin. A Review," *Proc. ACS Div. Polym. Mat. Sci. Eng.*, 66:393–340.

172. Stossel, T. P. 1989. "From Signal to Pseudopod. How Cells Control Cytoplasmic Actin Assembly," *J. Biol. Chem.*, 264:18261–18264.

173. Pollard, T. D. and J. A. Cooper. 1986. "Actin and Actin-Binding Proteins. A Critical Evaluation of Mechanisms and Functions," *Annu. Rev. Biochem.*, 55:987–1035.

174. Richards, F. M. 1990. "Reflections," *Meth. Enzymol.*, 184:3–5.

175. Wilchek, M. and E. A. Bayer. 1990. "Introduction to Avidin-Biotin Technology," *Meth. Enzymol.*, 184:5–13.

176. Pantano, P., T. H. Morton and W. G. Kuhr. 1991. "Enzyme-Modified Carbon-Fiber Microelectrodes with Millisecond Response Times," *J. Am. Chem. Soc.*, 113:1832–1833.

177. Green, N. M. 1966. "Thermodynamics of the Binding of Biotin and Some Analogs by Avidin," *Biochem. J.*, 101:774–780.

178. Bayer, E. A. and M. Wilchek. 1990. "Protein Biotinylation," *Meth. Enzymol.*, 184: 138–160.

179. Boyce, S. T., B. E. Stompro and J. E. Hansbrough. 1992. "Biotinylation of Implantable Collagen for Drug Delivery," *J. Biomed. Mater. Res.*, 26:547–553.

180. Billingsley, M. L., K. R. Pennypacker, C. G. Hoover and R. L. Kincaid. 1987. "Biotinylated Proteins as Probes of Protein Structure and Protein-Protein Interactions," *BioTechniques*, 5:22–31.

181. Mclnnes, J. L., A. C. Foster, D. C. Skingle and R. H. Symons. 1990. "Preparation and Uses of Photobiotin," *Methods Enzymol.*, 184:588–600.

182. Steward, M. W. 1981. "Affinity of the Antibody-Antigen Reaction and Its Biological Significance," Chapter 7 in *Structure and Function of Antibodies*, L. E. Glynn and M. W. Steward, eds., New York, NY: John Wiley & Sons.

183. Pecht, I. 1982. "Dynamic Aspects of Antibody Function," Chapter 1 in *The Antigens, Vol. VI*, M. Sela, ed., New York, NY: Academic Press.

184. MacGreger, E. A and C. T. Greenwood. 1980. *Polymers in Nature*. New York, NY: John Wiley & Sons, Chapter 4.

185. Sharon, N. and H. Lis. 1989. "Lectins as Cell Recognition Molecules," *Science*, 246:227–234.

186. Quiocho, F. A. 1986. "Carbohydrate-Binding Proteins: Tertiary Structures and Protein-Sugar Interactions," *Ann. Rev. Biochem.*, 55:287–315.

187. Benhamou, N. 1989. "Preparation and Application of Lectin-Gold Complexes," Chapter 4 in *Colloidal Gold: Principles, Methods, and Applications*, *Vol. 1*, M. A. Hayat, ed., New York, NY: Academic Press.

188. Clarke, A. E. and R. M. Hoggart. 1982. "The Use of Lectins in the Study of Glycoproteins," Chapter 11 in *Antibody as a Tool*, J. J. Marchalonis and G. W. Warr, eds., New York, NY: John Wiley & Sons.

189. Wu, A. M., S. Sugii and A. Herp. 1988. "A Table of Lectin Carbohydrate Specificities," in *Lectins–Biology, Biochemistry, Clinical Biochemistry, Vol. 6*, T. C. Bog-Hansen and D. L. J. Freed, eds., St. Louis, MO: Sigma Chemical Company, pp. 723–740.

190. Lakhtin, V. M. and I. A. Yamskov. 1991. "Lectins in the Investigation of Receptors," *Russian Chemical Reviews*, 60:903–923.

191. Goldstein, I. J. and R. D. Poretz. 1986. "Isolation, Physico-Chemical Characterization and Carbohydrate-Binding Specificity of Lectin," in *The Lectins: Properties, Functions and Applications in Biology and Medicine*, I. E. Liener, N. Sharon and I. J. Goldstein, eds., New York, NY: Academic Press, pp. 33–50.

192. Brownlee, M. and A. Cerami. 1979. "A Glucose-Controlled Insulin-Delivery System: Semisynthetic Insulin Bound to Lectin," *Science*, 206:1190–1191.

193. Jeong, S. Y., S. W. Kim, D. L. Holmberg and J. C. McRea. 1985. "Self-Regulating Insulin Delivery Systems," *J. Controlled Rel.*, 2:143–152.

194. Oakenfull, F. 1984. "A Method for Using Measurements of Shear Modulus to Estimate the Size and Thermodynamic Stability of Junction Zones in Noncovalently Cross-Linked Gels," *J. Food Sci.*, 49:1103–1110.

195. Henderson, G. V. S., Jr., D. O. Campbell, V. Kuzmicz and L. H. Sperling. 1985. "Gelatin as a Physically Crosslinked Elastomer," *J. Chemical Education*, 62:269–270.

196. Rodriguez, F. 1982. *Principles of Polymer Systems, 2nd Ed*. New York, NY: McGraw-Hill Book Company, p. 221.

197. Arridge, R. G. C. and P. J. Barham. 1982. "Polymer Elasticity. Discrete and Continuum Models," *Adv. Polymer Sci.*, 46:67–117.

198. Munk, P. 1989. *Introduction to Macromolecular Science*. New York, NY: John Wiley & Sons, pp. 423–431.

CHAPTER 6

Chemically-Induced Degradation

A variety of biodegradable polymers have been synthesized for biomedical and pharmaceutical applications. The majority of them are water-insoluble and degrade by hydrolysis [1]. Although most synthetic, biodegradable polymers are not water-soluble, they can be used to prepare hydrogels by making copolymers, polymer blends, and interpenetrating polymer networks with water-soluble polymers. The water-insoluble fraction provides structural rigidity as well as degradability, while the hydrophilic polymer fraction provides properties of the hydrogel. Of course, the hydrophilic fraction can also be degraded.

There are several classes of synthetic polymers which are susceptible to chemically-induced hydrolysis. A variety of hydrolytically-labile polymers can be developed using ester, anhydride, carbonate, amide, urethane, orthoester, and acetal linkages. Table 6.1 provides relative hydrolysis rates for different types of biodegradable polymers. Polyanhydride, polyketal, and polyorthoester degrade quite rapidly, while polyurea, polycarbonate, polyurethane, and polyamide degrade too slowly. Polyacetal and polyester appear to have reasonable degradation rates in our time scale [2–7]. Polyesters appear to be particularly useful in the preparation of biodegradable hydrogels. For this reason, the *in vivo* and *in vitro* degradation of several polyesters are briefly reviewed in this chapter.

6.1 CHEMICALLY-INDUCED HYDROLYSIS IN THE BODY

6.1.1 Distribution and Elimination

In the development of biodegradable polymers and hydrogels for drug delivery, one should consider the route of administration and subsequent distribution in the body, the degradation behavior, and the extent of elimination from the body. Hydrogels entering the body by parenteral administration will likely be taken up by cells of the reticuloendothelial system by the process known as endocytosis. Water-soluble products released from the

TABLE 6.1. Hydrolysis Rates for Several Classes of Biodegradable Polymers.

Class	Hydrolysis Rate
Polyanhydride	0.1 hours*
Polyketal	3 hours
Polyorthoester	4 hours
Polyacetal	0.8 years
Polyester	3.3 years
Polyurea	33 years
Polycarbonate	42,000 years
Polyurethane	42,000 years
Polyamide	83,000 years

*Time required for 50 percent hydrolysis at pH 7 and 25°C [2].

degrading hydrogels may also be eliminated by the kidneys [1]. The disposition of water-soluble polymers will vary depending on the size of the fragments and the chemical nature of the polymers. Orally administered polymeric microspheres, in the size of 10 μm or less, may enter the body via phagocytosis by the cells of the Peyer's patches in the small intestine [8,9].

The distribution and elimination kinetics of biodegradable polymers and their degradation products may be treated by standard pharmacokinetic models. Drobnik and Rypacek used a multicompartmental model to treat polymer disposition in the body [1]. Elimination of polymers from the body may occur via metabolism or by excretion depending on the properties of the degradation products. Metabolism is mainly a function of the liver, but may occur to some extent in the lungs [10]. Excretion, on the other hand, is a renal function associated with glomerular filtration, tubular secretion, and tubular reabsorption. The low molecular weight, water-soluble degradation products may be eliminated by glomerular filtration. Renal metabolism and reabsorption are known to play some role in the elimination of natural oligomers [11]. Their role in the metabolism of synthetic oligomers, however, has yet to be determined. The overall advantage of compartmental modeling is that the kinetics of absorption, distribution, and elimination can be determined from plasma levels and/or urinary excretion data. The disadvantage of this model is that it does not relate to specific anatomical structures or physiological barriers [12].

6.1.2 Factors Influencing Chemically-Induced Hydrolysis

Chemically-induced hydrolysis can be broadly classified as main chain or side chain scission. The hydrolysis reaction can be either reversible or irreversible, and may or may not depend on the reactive state of neighboring groups [2]. In addition, scission can occur randomly along the chain (random chain scission) or preferentially at the chain ends (depolymerization) [3]. Macroscopically, degradation events are observed as homogeneous or heterogeneous. In general, homogeneous hydrolysis (or bulk degradation) arises when the diffusion of water into the polymer matrix is faster than the overall rate of hydrolysis. Heterogeneous hydrolysis (or surface degradation), however, occurs when the diffusion of water is slower than the rate of hydrolysis. The extent of heterogeneous hydrolysis is mainly determined by the surface area of the polymeric system. Thus, any micro- or macroscopic parameters that change the surface area of the device will likely alter the kinetics of surface degradation.

As shown in Table 6.1, there are several types of synthetic polymers which are susceptible to hydrolysis. One important parameter which determines the rate and extent of hydrolysis is the hydrophilicity of the polymer. The more hydrophilic the polymer, the more extensive hydrolysis is. Polymers with high molecular weight, extensive branching, high glass transition temperature, or high crystallinity are generally resistant to hydrolysis due to the restricted polymer chain flexibility which reduces the diffusion of water. The addition of hydrophilic components by copolymerization with hydrophilic monomers or blending with hydrophilic polymers will enhance hydrolysis [5,13]. Other factors which affect hydrolysis are the presence of excipients or impurities in the polymer matrix, the properties of the hydrolyzed products, and the reactive state of neighboring groups. The last example is of particular importance in side chain hydrolysis [4]. Structural modification has been used to alter the rate of chemically-induced hydrolysis. For example, the rate of ester hydrolysis can be increased by incorporating electron withdrawing groups near the ester bond [14,15] or by treatment with γ-irradiation [16]. The hydrolysis rate can be increased substantially by catalysts, such as acid, base, nucleophiles other than hydroxyl ions [17], or metal ions [18,19].

6.2 CHEMICALLY-INDUCED HYDROLYSIS OF POLYESTERS

6.2.1 Poly(glycolic Acid)

Poly(glycolic acid) (PGA) is the most hydrophilic polyester commercially available to date. PGA is a semi-crystalline polymer with a normal crystallinity of 50 percent [10]. The hydrolysis at the ester bond of PGA yields an acid and an alcohol. Water penetration into PGA fibers occurs readily within a few minutes due to the hydrophilicity of PGA [13]. The initial water content in PGA at 37°C is approximately 14 percent and increases during hydrolysis to as much as 42 percent. The *in vivo* hydrolysis of PGA leads to the formation of glycolic acid [20].

PGA is hydrolyzed by a two stage mechanism. In the first stage of hydrolysis the random chain scission occurs preferentially in the amorphous phase, because water cannot penetrate into the crystalline region [13]. During the first stage, crystallinity of the polymer steadily increases and reaches a maximum before the beginning of the second stage of hydrolysis [21,22]. Such changes in crystallinity are thought to result from the reorganization of chains to the more ordered crystalline state following hydrolysis [21]. The second stage of hydrolysis occurs in the crystalline region and is characterized by a steady decrease in crystallinity, a sharp decrease in pH, and a significant increase in the glycolic acid concentration [21]. Any treatment which alters the ratio of amorphous to crystalline regions will significantly affect the rate of hydrolysis. For example, annealing of PGA fibers in different conditions is known to alter the ratio and thus the hydrolysis rate [23,24].

The rate of PGA hydrolysis is pH-dependent. As the pH increases from 5.25 to 7.4, there is very little change in the hydrolysis rate [22,25,26]. As the pH increases to 10.6, however, the rate of hydrolysis increases significantly [22,26]. The rate of PGA hydrolysis is also affected by the presence of buffer [26]. The PGA fibers lost tensile strength faster in a phosphate buffered saline solution (pH 7.4) than in an unbuffered saline solution (pH 5.0). In the absence of buffer, the pH of the solution reduced as degradation proceeded, and as a result, the rate of hydrolysis decreased.

Exposure of PGA to γ-irradiation was shown to alter the hydrolytic and the morphological properties of PGA fibers in aqueous solutions [16,27]. Samples exposed to γ-irradiation underwent a more rapid drop in pH and tensile strength when immersed in water [16]. It was postulated that γ-irradiation led to the predominance of main chain scission in the amorphous regions of the PGA fibers and such scission shortened the duration of the first hydrolysis stage by facilitating chain mobility.

6.2.2 Poly(lactic Acid)

Poly(lactic acid) (PLA) is commonly synthesized from dilactide by stannous-octoate-catalyzed ring opening polymerization [28]. Because of its chirality, there are three polymers that can be synthesized; poly(D-lactic acid) (P(D-LA)), poly(D,L-lactic acid) (P(D,L-LA)), and poly(L-lactic acid) (P(L-LA)). P(L-LA) and P(D-LA) are semi-crystalline, while P(D,L-LA) is amorphous. P(L-LA) is preferred over P(D-LA) in bioapplications, since D-lactic acid is not readily metabolized by the body. Water uptake by PLA is restricted to 2 percent by weight due to the hydrophobic nature of a methyl group in the polymer backbone. The hydrophobicity and high crystallinity make P(L-LA) very resistant to hydrolysis. P(D,L-LA), however, is more susceptible to hydrolysis due to its amorphous structure [29].

The hydrolysis of P(L-LA) and P(D,L-LA) was observed to occur in two stages. Stage one hydrolysis was characterized by random chain scission of ester groups which resulted in a reduction in molecular weight and tensile strength, but only a minimal loss in mass [30]. Stage two hydrolysis was distinguished by a significant loss in weight and mechanical strength [31]. For semi-crystalline P(L-LA), stage one hydrolysis was restricted to the amorphous regions of the sample and stage two hydrolysis was associated with the remaining crystalline regions [32]. Hydrolysis of the crystalline region occurred at a much slower rate than that of the amorphous region. For amorphous P(D,L-LA), however, stage two hydrolysis was marked by an enhanced rate of chain scission [33].

The hydrolysis rate of P(L-LA) can be altered by blending with other biodegradable polymers, such as poly(ϵ-caprolactone) and/or poly(glycolic acid-*co*-lactic acid) [34,35]. Blending of P(L-LA) with other polymers resulted in increased hydrolysis. Base-catalyzed hydrolysis of P(L-LA) has been demonstrated under both *in vitro* and *in vivo* conditions [36,37]. Hydrolysis of P(L-LA) microspheres was catalyzed by the incorporation of tertiary amine-containing drugs. It was found that the incorporation of methadone, promethazine, and meperidine increased the rate of microsphere hydrolysis in phosphate buffer. Meperidine produced the greatest increase in hydrolysis followed by methadone, and promethazine. Naltrexone did not have any catalytic effects. No particular relationship was observed between the base-catalyzed hydrolysis and the pK_a of the tertiary amines or changes in glass transition temperature by drug incorporation.

The results of *in vivo* studies with ^{14}C-labeled P(L-LA) and P(D,L-LA) suggested that PLA was degraded and eliminated from the body mainly as CO_2 via the respiratory route [38,39]. When microspheres, which were prepared from low molecular weight P(L-LA) and P(D,L-LA), were added to mouse peritoneal macrophages, they were phagocytosed within six hours and degraded completely over a seven-day period [40].

6.2.3 Polycaprolactone

One of the main applications of polycaprolactone (PCL) is the degradation-controlled drug delivery [41]. PCL is a semi-crystalline polymer with a melting point of 63°C and a glass transition temperature ranging from −60°C to −70°C [42]. The hydrolysis of PCL also proceeds in two stages. Stage one hydrolysis is restricted to the amorphous regions of the polymer and is very similar to that observed with other polyesters [43]. During stage one hydrolysis, the molecular weight decreases by a first-order kinetics and the crystallinity of the polymer increases [42,44]. Stage two hydrolysis is characterized by a dramatic decrease in molecular weight, mechanical strength, and sample weight. The loss of sample weight during stage two hydrolysis was attributed to the diffusion of small PCL fragments from the sample.

When implanted in rats, the PCL powders were fragmented and phagocytosed by neutrophils after three days [45]. The PCL fragmentation further elicited a vascular foreign body response containing macrophages, giant cells, and fibroblasts. Over time, PCL samples ranging from less than 10 μm to 80 μm were phagocytosed by macrophages. The size of PCL particles in phagosomes decreased as compared to the initial sample size. This observation suggested that intracellular degradation took place.

The hydrolysis rate of PCL can be increased by copolymerizing ϵ-caprolactone with D,L-lactide, γ-valerolactone, or ϵ-decalactone [42,46,47]. The effect of comonomers on the hydrolysis rate was dependent on the reduction in crystallinity of the sample by the comonomers. It was also shown that PCL hydrolysis is susceptible to specific acid and base catalysis [48]. The rate of PCL hydrolysis increases dramatically in the presence of oleic acid or decylamine and to a moderate extent in the presence of tributylamine.

6.2.4 Poly(ethylene Terephthalate)

Poly(ethylene terephthalate) (PET) is a semi-crystalline polymer having a crystallinity of around 40 percent. The degradation of PET by acid-catalyzed hydrolysis proceeds by a bulk degradation mechanism and is restricted to the amorphous regions of the polymer, while the base-catalyzed hydrolysis is characterized by surface degradation [13]. *In vivo* degradation of PET is very slow. The mechanical strength of PET decreased to about half of its original strength 10 years after implantation [49,50]. Furthermore, only a moderate decline in molecular weight was observed with no loss in sample weight. The mean absorption time for PET in the body was estimated to be 30 years.

6.2.5 Poly(β-hydroxybutyrate)

Poly(β-hydroxybutyrate) (PHB) is a highly crystalline, thermoplastic polyester which is found in a variety of microorganisms [51–54]. Bacteria can synthesize a range of copolymers from 3-hydroxybutyrate and 3-hydroxyvaleric acid (HV) [55]. The crystallinity and melting point of PHB can be reduced by preparing copolymers of HB and HV (P(HB-HV)).

The hydrolytic degradation of PHB and P(HB-HV) occurred in two stages. The first stage was characterized by a moderate loss in weight and an increase in water uptake. The second stage was marked by a more dramatic change in weight and water uptake. The degradation of PHB and P(HB-HV) after subcutaneous implantation in rats was faster than that in buffer solution [56]. Thus, it was assumed that enzyme-catalyzed hydrolysis played some role during *in vivo* degradation. The rate of hydrolysis of P(HB-HV) was inversely related to the degree of crystallinity [57].

The rate of P(HB-HV) hydrolysis can also be changed by blending with various polysaccharides [58,59]. P(HB-HV) samples were melt blended with amylose, dextran, dextrin, or sodium alginate. The content of HV in the copolymer was either 12 or 20 percent while the polysaccharide comprised 10 or 30 percent (w/w) of the blend. The weight loss was greater with the blended than with the pure P(HB-HV). As shown in Table 6.2, the weight loss increased as the content of polysaccharide in the blend increased. At pH 7.4, the hydrolysis of P(HB-HV)/alginate blends was fastest. The increase in the HV content in the copolymer increased the hydrolysis rate due to a decrease in the crystallinity of the sample [56]. In general, the blending was believed to increase hydrolysis rates by increasing the matrix porosity resulting from the dissolution and release of the blended polysaccharide. Increased porosity facilitated fluid penetration and enabled both low and high molecular weight fragments to diffuse more readily from the matrix following chain scission.

TABLE 6.2. Time for 10 Percent Weight Loss ($t_{0.9}$) of P(HB-HV)/Polysaccharide Blends after Immersion in a pH 7.4 Buffer at 37°C [59].

	$t_{0.9}$ (days)	
Sample	12% HV	20% HV
Unblended	>600	>600
Amylose		
10%	484	431
30%	240	230
Dextrin		
10%	462	410
30%	84	14
Dextran		
10%	312	76
30%	25	9
Na-Alginate		
10%	122	44
30%	5	2

6.2.6 Polydioxanone

Polydioxanone (PDO) has been used traditionally as a monofilament suture or as a biodegradable ligating clip [60,61]. When PDO was implanted in rats, the loss in tensile strength and molecular weight of PDO implanted in rats was comparable to that of PDO in a buffer solution [60]. This indicated that the PDO degrades in the body by a nonenzymatic hydrolysis mechanism. The study with ^{14}C-labelled PDO showed that PDO was not accumulated in any major tissues. The hydrolysis occurred faster when copolymers of PDO were made with glycolic acid, lactic acid, or 2,5-morpholinedione [62–64].

6.3 POLYESTER HYDROGELS

As described in Chapter 3, Section 3.1, polyesters were used in the preparation of biodegradable hydrogels in different forms. Since many polyesters are known to be biocompatible and chemically well characterized, they are expected to be used widely in the preparation of new biodegradable hydrogels in the near future. Recently, Pathak et al. prepared biodegradable hydrogels for the prevention of postoperative adhesions using copolymers of poly(ethylene glycol) (PEO) and poly(glycolic acid) [65]. In this work, the number of glycolidyl residues on each end of PEO was kept short, only about 10. Thus, the hydrogel maintained the properties of PEO while becoming biodegradable. Slight variations of the existing hydrogel systems with polyesters will provide biodegradable hydrogels without altering the properties of the native hydrogels significantly.

6.4 REFERENCES

1. Drobnik, J. and F. Rypacek. 1984. "Soluble Synthetic Polymers in Biological Systems," *Adv. Polym. Sci.*, 57:1–50.
2. St. Pierre, T. and E. Chiellini. 1986. "Biodegradability of Synthetic Polymers Used for Medical and Pharmaceutical Applications: Part 1–Principles of Hydrolysis Mechanisms," *J. Bioact. Compatible Polym.*, 1:467–497.
3. St. Pierre, T. and E. Chiellini. 1987. "Biodegradability of Synthetic Polymers for Medical and Pharmaceutical Applications: Part 2–Backbone Hydrolysis," *J. Bioact. Compatible Polym.*, 2:4–30.
4. St. Pierre, T. and E. Chiellini. 1987. "Biodegradability of Synthetic Polymers for Medical and Pharmaceutical Applications: Part 3–Pendant Group Hydrolysis and General Conclusions," *J. Bioact. Compatible Polym.*, 2:238–257.
5. Holland, S. J. and B. J. Tighe. 1992. "Biodegradable Polymers," in *Advances in Pharmaceutical Sciences, Vol. 6,* D. Ganderton and T. Jones, eds., San Diego, CA: Academic Press Inc., pp. 101–164.

6. Holland, S. J., B. J. Tighe and P. L. Gould. 1986. "Polymers for Biodegradable Medical Devices. 1. The Potential of Polyesters as Controlled Macromolecular Release Systems," *J. Controlled Release*, 4:155–180.
7. Shalaby, S. W. 1988. "Bioabsorbable Polymers," in *Encyclopedia of Pharmaceutical Technology, Vol. 1,* J. Swarbrick and J. C. Boylan, eds., New York, NY: Marcel Dekker, Inc., pp. 465–475.
8. Eldridge, J. H., C. J. Hammond, J. A. Meulbroek, J. K. Staas, R. M. Gilley and T. R. Tice. 1990. "Controlled Vaccine Release in the Gut-Associated Lymphoid Tissues. I. Orally Administered Biodegradable Microspheres Target the Peyer's Patches," *J. Controlled Release*, 11:205–214.
9. O'Hagan, D. T., K. Palin, S. S. Davis, P. Artursson and I. Sjoholm. 1989. "Microparticles as Potentially Orally Active Immunological Adjuvants," *Vaccine*, 7:421–424.
10. Gilding, D. K. 1981. "Biodegradable Polymers," in *Biocompatibility of Clinical Implant Materials, Vol. 2,* D. F. Williams, ed., Boca Raton, Florida: CRC Press, Inc., pp. 209–232.
11. Carone, F. R., D. R. Peterson and G. Flouret. 1982. "Renal Tubular Processing of Small Peptide Hormones," *J. Lab. Clin. Med.*, 100(1):1–14.
12. Gibaldi, M. and D. Perrier. 1982. *Pharmacokinetics, Second Edition*. New York, NY: Marcel Dekker, Inc., pp. 355–384.
13. Zaikov, G. E. 1985. "Quantitative Aspects of Polymer Degradation in the Living Body," *J. Macromol. Sci.–Rev. Macromol. Chem. Phys.*, C25(4):551–597.
14. Euranto, E. K. 1969. "Esterification and Ester Hydrolysis," in *The Chemistry of Carboxylic Acid and Esters*. Sussex, England: John Wiley & Sons, Ltd., pp. 505–588.
15. Heller, J., R. F. Helwing, R. W. Baker and M. E. Tuttle. 1983. "Controlled Release of Water-Soluble Macromolecules from Bioerodible Hydrogels," *Biomaterials*, 4:262–266.
16. Chu, C. C. 1985. "Degradation Phenomena of Two Linear Aliphatic Polyester Fibres Used in Medicine and Surgery," *Polymer,* 26:591–594.
17. Fife, T. H. and J. E. C. Hutchins. 1981. "Effect of Nucleophile Basicity on Intramolecular Nucleophilic Aminolysis Reactions of Carbonate Diesters," *J. Am. Chem. Soc.*, 103:4194–4199.
18. Fife, T. H. and T. J. Przystas. 1985. "Divalent Metal Ion Catalysis in the Hydrolysis of Esters of Picolinic Acid. Metal Ion Promoted Hydroxide Ion and Water Catalyzed Reactions," *J. Am. Chem. Soc.*, 107:1041–1047.
19. Sayre, L. M. 1986. "Metal Ion Catalysis of Amide Hydrolysis," *J. Am. Chem. Soc.*, 108:1632–1635.
20. Frazza, E. J. and E. E. Schmitt. 1971. "A New Absorbable Suture," *J. Biomed. Mater. Res. Symposium*, 1:43–58.
21. Chu, C. C. 1981. "Hydrolytic Degradation of Polyglycolic Acid: Tensile Strength and Crystallinity Study," *J. Appl. Polym. Sci.*, 26:1727–1734.
22. Ginde, R. M. and R. K. Gupta. 1987. "*In vitro* Chemical Degradation of Poly(glycolic Acid) Pellets and Fibres," *J. Appl. Polym. Sci.*, 33:2411–2429.
23. Browning, A. and C. C. Chu. 1986. "The Annealing Treatments on the Tensile Properties and Hydrolytic Degradative Properties of Polyglycolic Acid Sutures," *J. Biomed. Mater. Res.*, 20:613–632.
24. Chu, C. C. and A. Browning. 1988. "The Study of Thermal and Gross Morphological Properties of Polyglycolic Acid upon Annealing and Degradation Treatments," *J. Biomed. Mater. Res.*, 22:699–712.

25. Williams, D. F. 1980. "The Effect of Bacteria on Absorbable Sutures," *J. Biomed. Mater. Res.*, 14:329–338.
26. Chu, C. C. 1981. "The *in-vitro* Degradation of Poly(glycolic Acid) Sutures–Effect of pH," *J. Biomed. Mater. Res.*, 15:795–804.
27. Chu, C. C. and N. D. Campbell. 1982. "Scanning Electron Microscopic Study of the Hydrolytic Degradation of Poly(glycolic Acid) Suture," *J. Biomed. Mater. Res.*, 16:417–430.
28. Huang, S. J. 1985. "Biodegradable Polymers," in *Encyclopedia of Polymer Science and Engineering*, *Vol. 2*, H. F. Mark, N. M. Bikales, C. G. Overberger and G. Menges, eds., New York, NY: John Wiley & Sons, pp. 220–243.
29. Kulkarni, R. K., E. G. Moore, A. F. Hegyeli and F. Leonard. 1971. "Biodegradable Poly(lactic Acid) Polymers," *J. Biomed. Mater. Res.*, 5:169–181.
30. Reed, A. M. and D. K. Gilding. 1981. "Biodegradable Polymers for Use in Surgery–Poly(glycolic/lactic Acid) Homo and Copolymers: 2. *In vitro* Degradation," *Polymer*, 22:494–498.
31. Christel, P., F. Chabot and M. Vert. 1984. "*In vivo* Fate of Bioresorbable Bone Plate in Long-Lasting Poly(L-lactic Acid)," *2nd World Congr. Biomater.*, *Washington, DC*, p. 279.
32. Leenslag, J. W., A. J. Pennings, R. R. M. Bos, F. R. Rozema and F. Boering. 1987. "Resorbable Materials of Poly(L-lactide)," *Biomaterials*, 8:311–314.
33. Pitt, C. G., M. M. Gratzl, G. L. Kimmel, J. Surles and A. Schindler. 1981. "Aliphatic Polyesters II. The Degradation of Poly(DL-lactide), Poly(ϵ-caprolactone) and Their Copolymers *in Vivo*," *Biomaterials*, 2:215–220.
34. Cha, Y. and C. G. Pitt. 1990. "The Biodegradability of Polyester Blends," *Biomaterials*, 11:108–112.
35. Coombes, A. G. A. and J. D. Heckman. 1992. "Gel Casting of Resorbable Polymers, 2. *In vitro* Degradation of Bone Graft Substitutes," *Biomaterials*, 13(5):297–307.
36. Cha, Y. and C. G. Pitt. 1988. "A One-Week Subdermal Delivery System for L-Methadone Based on Biodegradable Microcapsules," *J. Controlled Release*, 7:69–78.
37. Cha, Y. and C. G. Pitt. 1989. "The Acceleration of Degradation-Controlled Drug Delivery from Polyester Microspheres," *J. Controlled Release*, 8:259–265.
38. Kulkarni, R. K., K. C. Pani, C. Neuman and F. Leonard. 1966. "Polylactic Acid for Surgical Implants," *Arch. Surg.*, 93:839–843.
39. Brady, J. M., D. E. Cutright, R. A. Miller, G. C. Battistone and E. E. Hunsuck. 1973. "Resorption Rate, Route of Elimination, and Ultrastructure of the Implant Site of Polylactic Acid in the Abdominal Wall of the Rat," *J. Biomed. Mater. Res.*, 7:155–166.
40. Tabata, Y. and Y. Ikada. 1988. "Macrophage Phagocytosis of Biodegradable Microspheres Composed of L-Lactic Acid/Glycolic Acid Homo- and Copolymers," *J. Biomed. Mater. Res.*, 22:837–858.
41. Pitt, C. G., T. A. Marks and A. Schindler. 1980. "Biodegradable Drug Delivery Systems Based on Aliphatic Polyesters: Application to Contraceptives and Narcotic Antagonists," in *Controlled Release of Bioactive Materials*, R. W. Baker, ed., New York, NY: Academic Press, pp. 19–43.
42. Schindler, A., R. Jeffcoat, G. L. Kimmel, C. G. Pitt, M. E. Wall and R. Zweidinger. 1977. "Biodegradable Polymers for Sustained Drug Delivery," in *Contemporary Topics in Polymer Science*, *Vol. 2,* E. M. Pearce and J. R. Schaefgen, eds., New York, NY: Plenum Press, pp. 251–286.

43. Jarrett, P., C. Benedict, J. P. Bell, J. A. Cameron and S. J. Huang. 1983. "Mechanism of the Biodegradation of Polycaprolactone," *Polym. Prepr.*, 24(1):32–33.

44. Pitt, C. G., F. I. Chasalow, Y. M. Hibionada, D. M. Klimas and A. Schindler. 1981. "Aliphatic Polyesters. I. The Degradation of Poly(ϵ-caprolactone) *in Vivo*," *J. Appl. Polym. Sci.*, 26:3779–3787.

45. Woodward, S. C., P. S. Brewer, F. Moatamed, A. Schindler and C. G. Pitt. 1985. "The Intracellular Degradation of Poly(ϵ-caprolactone)," *J. Biomed. Mater. Res.*, 19:437–444.

46. Pitt, C. G., M. M. Gratzl, G. L. Kimmel, J. Surles and A. Schindler. 1981. "Aliphatic Polyesters II. The Degradation of Poly(DL-lactide), Poly(ϵ-caprolactone), and Their Copolymers *in Vivo*," *Biomaterials*, 2:215–220.

47. Gilbert, R. D., V. Stannett, C. G. Pitt and A. Schindler. 1982. "The Design of Biodegradable Polymers: Two Approaches," in *Developments in Polymer Degradation, Vol. 4,* N. Grassie, ed., London: Applied Science Publishers LTD, pp. 259–293.

48. Pitt, C. G. and Z. Gu. 1987. "Modification of the Rates of Chain Cleavage of Poly(ϵ-caprolactone) and Related Polyesters in the Solid State," *J. Controlled Release*, 4:283–292.

49. Rudakova, T. E., G. E. Zaikov, O. S. Voronkova, T. T. Daurova and S. M. Degtyareva. 1979. "The Kinetic Specificity of Polyethylene Terephthalate Degradation in the Living Body," *J. Polym. Sci. Polym. Symp.*, 66:277–281.

50. Gumargalieva, K. Z., Y. V. Moiseev, T. T. Daurova and O. S. Voronkova. 1982. "The Effect of Infections on the Degradation of Polyethylene Terphthalate Implants," *Biomaterials*, 3:177–180.

51. Homles, P. A. 1985. "Application of PHB–A Microbially Produced Biodegradable Thermoplastic," *Phys. Technol.*, 16:32–36.

52. Williamson, D. H. and J. F. Wilkinson. 1958. "The Isolation and Estimation of Poly-β-hydroxybutyrate Inclusions of *Bacillus* Species," *J. Gen. Microbiol.*, 19:198–209.

53. Macrae, R. M. and J. F. Wilkinson. 1958. "Poly-β-hydroxybutyrate Metabolism in Washed Suspensions of *Bacillus cereus* and *Bacillus megaterium*," *J. Gen. Microbiol.*, 19:210–222.

54. Alper, R., D. G. Lundgren, R. H. Marchessault and W. A. Cote. 1963. "Properties of Poly-β-hydroxybutyrate. I. General Considerations Concerning the Naturally Occurring Polymer," *Biopolymers*, 1:545–556.

55. Holmes, P. A., L. F. Wright and S. H. Collins. European patent 52459, 1981.

56. Miller, N. D. and D. F. Williams. 1987. "On the Biodegradation of Poly-β-hydroxybutyrate (PHB) Homopolymer and Poly-β-hydroxybutyrate-hydroxyvalerate Copolymers," *Biomaterials*, 8:129–137.

57. Yasin, M., S. J. Holland and B. J. Tighe. 1990. "Polymers for Biodegradable Medical Devices. V. Hydroxybutyrate-Hydroxyvalerate Copolymers: Effects of Polymer Processing on Hydrolytic Degradation," *Biomaterials*, 11:451–454.

58. Yasin, M., S. J. Holland, A. M. Jolly and B. J. Tighe. 1989. "Polymers for Biodegradable Medical Devices. VI. Hydroxybutyrate-Hydroxyvalerate Copolymers: Accelerated Degradation of Blends with Polysaccharides," *Biomaterials*, 10:400–412.

59. Holland, S. J., M. Yasin and B. J. Tighe. 1990. "Polymers for Biodegradable Medical Devices. VII. Hydroxybutyrate-Hydroxyvalerate Copolymers: Degradation of Copolymers and Their Blends with Polysaccharides under *in vitro* Physiological Conditions," *Biomaterials*, 11:206–215.

60. Ray, J. A., N. Doddi, D. Regula, J. A. Williams and A. Melveger. 1981. "Polydioxanon (PDS), a Novel Monofilament Synthetic Absorbable Suture," *Surg. Gynecol. Obstet.*, 153:497–507.

61. Schaefer, C. J., P. M. Colombani and G. W. Geelhoed. 1982. "Absorbable Ligating Clips," *Surg. Gynecol. Obstet.*, 154:513–516.

62. Bezwada, R. S., S. W. Shalaby and H. D. Newman. U.S. patent 4,653,497, 1987.

63. Bezwada, R. S., S. W. Shalaby, H. D. Newman and A. Kafrawy. U.S. patent 4,643,191, 1987.

64. Shalaby, S. W. and D. F. Koelmel. U.S. patent 4,441,496, 1984.

65. Pathak, C. P., A. S. Sawhney, R. C. Dunn and J. A. Hubbell. 1992. "Photopolymerizable Biodegradable Hdyrogels for the Prevention of Postoperative Adhesions," *Transactions of 4th World Biomaterials Congress*, p. 231.

CHAPTER 7

Enzyme-Catalyzed Degradation

7.1 PROPERTIES OF ENZYMES

Enzymes are proteins that are highly specific in the reactions they catalyze as well as in their choice of reactants, which are called substrates [1]. Enzymes can accelerate reactions by forming an enzyme-substrate complex. The active site of an enzyme is complementary to the structure of the reaction's transition state [2]. The active site is made up of catalytic and noncatalytic amino acid residues. The catalytic groups induce changes in electron density on the substrate and such changes promote the formation or cleavage of chemical bonds. Enzymes are also capable of catalyzing reactions which involve different substrates providing that the electron density changes resemble those of the native substrates [3]. The noncatalytic groups in the active site lie in close proximity to the site of bond cleavage or formation. The main function of the noncatalytic groups is to interact with the substrate and lower the activation energy of the reaction [4,5].

Enzymes are divided into six classes: oxidoreductases, transferases, hydrolases, lyases, isomerases, and ligases [6]. Hydrolases are of particular relevance to enzyme-degradable hydrogels since they catalyze the hydrolysis of C–O, C–N, and C–C bonds. Hydrolases that act on proteins are divided into peptidases (exopeptidases) and proteinases (endopeptidases) [6]. Aminopeptidases and carboxypeptidases catalyze the hydrolysis of the N-terminal and C-terminal ends of polypeptide chains, respectively. Different peptidases require different amino acid residues and/or transition metals to be present at the active site. Carboxypeptidases are subdivided into serine carboxypeptidases, cysteine carboxypeptidases, and metallocarboxypeptidases. Proteinases catalyze the hydrolysis of peptide bonds within the polypeptide chain. Like peptidases, proteinases are further subdivided into serine proteinases, cysteine proteinases, aspartic proteinases, and metalloproteinases depending on the requirements of the active site.

Enzymes that catalyze hydrolysis of polysaccharides are called glycosidases. There are three types of glycosidases that hydrolyze either *O*-glycosyl, *N*-glycosyl, or *S*-glycosyl bonds in a polysaccharide chain. The largest class of glycosidases is the *O*-glycosidases. Enzyme-catalyzed hydrolysis of *O*-glycosyl bonds is carried out by nucleophilic substitution at the anomeric carbon [7]. The reaction results in either retention or inversion of the configuration at the anomeric carbon [8]. Thus, glycosidases are classified by the overall stereochemistry of the reaction. *O*-glycosidases that act on pyranosides are divided into 4 classes: (1) those that produce retention of equatorial configuration, (2) those that produce retention of axial configuration, (3) those that produce equatorial to axial inversion of configuration, and (4) those that produce axial to equatorial inversion of configuration. *O*-glycosidases that act on furanosides are also classified by the retention or inversion of the configuration following hydrolysis. The properties of some protein and polysaccharide hydrolases are presented in Table 7.1.

As mentioned above, the main function of the noncatalytic groups at the active site of an enzyme is to lower the activation energy of the reaction. Their contribution to enzyme-catalyzed hydrolysis has been studied extensively [9–13]. Schechter and Berger [9] divided the active site of an enzyme into subsites in which each subsite (S) can accommodate one amino acid residue of the substrate (P). Subsites are designated by their relative position from the cleavage site as shown in Figure 7.1. The rate of enzyme-catalyzed hydrolysis is dependent on the favorability of the subsite-substrate (S-P) interactions. For example, collagenase from *Clostridium histolyticum* contains six subsites in its active site [12,13]. Reactivity is determined by the S_3-P_3, S_2-P_2, S_1-P_1, S_1'-P_1', S_2'-P_2', and S_3'-P_3' interactions. It has been shown that the best substrates for collagenase have glycine at P_3 and P_1', proline or alanine at P_2 and P_2', hydroxyproline, arginine, or alanine at P_3', and a large hydrophobic residue at P_1. Thus, any chemical modification near the substrate's cleavage site may alter the rate of enzyme-catalyzed hydrolysis depending on the changes it caused in subsite-substrate interactions. The importance of subsite-substrate interactions in controlling enzyme-catalyzed hydrolysis of modified proteins and polysaccharides is discussed in Sections 7.4 and 7.5.

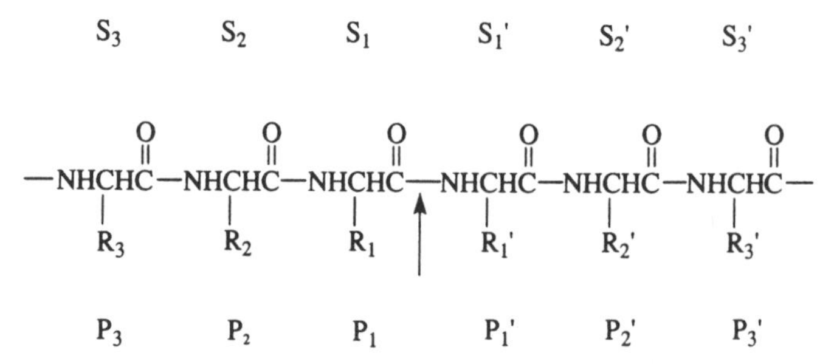

FIGURE 7.1. Subsite-substrate interactions between a hydrolase and a polypeptide. The arrow denotes the site of cleavage.

TABLE 7.1. Enzymes That Hydrolyze Proteins and Polysaccharides [6].

Enzymes	Type	Other Names	Preferential Cleavage Sites
I: Protein Hydrolases			
Alanine carboxypeptidase	metallocarboxypeptidase	–	peptidyl-L-alanine
Arginine carboxypeptidase	metallocarboxypeptidase	carboxypeptidase N	peptidyl-L-arginine
Aspartate carboxypeptidase	metallocarboxypeptidase	–	peptidyl-L-asparate
Bromelain	cysteine proteinase	–	Lys-, Ala-, Tyr-, Gly-
Carboxypeptidase A	metallocarboxypeptidase	carboxypolypeptidase	peptidyl-L-amino acid
Carboxypeptidase B	metallocarboxypeptidase	protaminase	peptidyl-L-lysine (-L-arginine)
Cathepsin B	cysteine proteinase	cathepsin B_1	Arg-, Lys-, Phe-X-aa (carbonyl side of amino acid residue next to Phe)
Cathepsin D	aspartic proteinase	–	Phe-, Leu-
Cathepsin G	serine proteinase	–	Tyr-, Trp-, Phe-, Leu-
Cathepsin H	cysteine proteinase	cathepsin B_3, cathepsin Ba, *N*- benzoylarginine-β-naphthylamide hydrolase	hydrolysis of proteins and hydrolysis of peptide substrates with free N-terminal group
Cathepsin L	cysteine proteinase	–	hydrolysis of proteins with no action on acylamino acid esters
Chymosin	aspartic proteinase	rennin	cleaves a single bond in casein κ
Chymotrypsin	serine proteinase	chymotrypsin A and B	Tyr-, Trp-, Phe-, Leu-
Coagulation factor Xa	serine proteinase	thromokinase, prothrombase, prothrombinase	Arg-Ile, Arg-Gly; activates prothrombin to thrombin

(continued)

TABLE 7.1. (continued).

Enzymes	Type	Other Names	Preferential Cleavage Sites
I: Protein Hydrolases			
Ficin	cysteine proteinase	–	Lys-, Ala-, Tyr-, Gly-, Asn-, Leu-, Val-
Glycine carboxypeptidase	metallocarboxypeptidase	yeast carboxypeptidase	peptidyl-L-glycine
Lysosomal carboxypeptidase B	cysteine carboxypeptidase	cathepsin B_2, cathepsin IV	peptidyl-L-amino acid
Pancreatic elastase	serine proteinase	pancreatopeptidase E, pancreatic elastase I	cleaves bonds at the carboxyl side of Gly, Ala, Val, Leu, Ile in elastin
Papain	cysteine proteinase	papaya peptidase I	Arg-, Lys-, Phe-X-aa (carbonyl side of amino acid residue next to Phe)
Pepsin A	aspartic proteinase	pepsin	Phe-, Leu-
Plasmin	serine proteinase	fibrinase, fibrinolysin	Arg-, Lys-
Plasminogen activator	serine proteinase	urokinase	Arg-Val in plasminogen
Proline carboxypeptidase	serine carboxypeptidase	angiotensinase C, lysosomal carboxypeptidase C	peptidylprolyl-L-amino acid
Serine carboxypeptidase	serine carboxypeptidase	–	peptidyl-L-amino acid
Thrombin	serine proteinase	fibrinogenase	Arg-; activates fibrinogen to fibrin
Trypsin	serine proteinase	α- and β-trypsin	Arg-, Lys-
Tyrosine carboxypeptidase	serine carboxypeptidase	thyroid peptide carboxypeptidase	peptidyl-L-tyrosine
Vertebrate collagenase	metalloproteinase	–	cleaves a single bond in native collagen leaving an N-terminal (75 percent) and a C-terminal (25 percent) fragment

TABLE 7.1. (continued).

Enzymes	Type	Other Names	Preferential Cleavage Sites
II: Glycosidases			
α-*N*-Acetylgalactosaminidase	*O*-glycosidase	–	hydrolysis of terminal, nonreducing *N*-acetyl-D-galactosamine residues in *N*-acetyl-α-D-galactosaminides
α-*N*-Acetylglucosaminidase	*O*-glycosidase	–	hydrolysis of terminal, nonreducing *N*-acetyl-D-galactosamine residues in *N*-acetyl-β-D-galactosaminides
β-*N*-Acetylglucosaminidase	*O*-glycosidase	–	hydrolysis of terminal, nonreducing *N*-acetyl-D-glucosamine residues in *N*-acetyl-α-D-glucosaminides
Agarase	*O*-glycosidase	–	hydrolysis of 1,3-β-D-galactosidic linkages in agarose
α-Amylase	*O*-glycosidase	glycogenase	endohydrolysis of 1,4-α-D-glucosidic linkages containing three or more 1,4-α-linked D-glucose units
β-Amylase	*O*-glycosidase	saccharogen amylase, glycogenase	hydrolysis of 1,4-α-D-glucosidic linkages from nonreducing ends of chains
ϰ-Carrageenase	*O*-glycosidase	–	hydrolysis of 1,4-β-D-linkages between D-galactose-4-sulfate and 3,6-anhydro-D-galactose in carrageenans

(continued)

TABLE 7.1. (continued).

Enzymes	Type	Other Names	Preferential Cleavage Sites
II: Glycosidases			
Cellulase	*O*-glycosidase	endo-1,4-β-glucanase	endohydrolysis of 1,4-β-D-glucosidic linkages in cellulose
Chitinase	*O*-glycosidase	chitodextrinase, 1,4-β-poly-*N*-acetylglucosaminidase, poly-β-glucosaminidase	random hydrolysis of *N*-acetyl- β-D-glucosaminide 1,4-β-linkages in chitin and chitodextrins
Dextranase	*O*-glycosidase	—	endohydrolysis of 1,6-α-D-glucosidic linkages in dextran
α-L-Fucosidase	*O*-glycosidase	—	α-L-fucoside + H_2O yields an alcohol + L-fucose
α-Galactosidase	*O*-glycosidase	melibiase	hydrolysis of terminal, nonreducing α-D-galactosidase residues in α-D-galactosides
β-Galactosidase	*O*-glycosidase	lactase	hydrolysis of terminal, nonreducing β-D-galactose residues in β-D-galactosides
α-Glucosidase	*O*-glycosidase	maltase, glucoinvertase, glucosidosucrase, maltase-glucoamylase	hydrolysis of terminal, nonreducing 1,4-linked α-D-glucose residues
β-Glucosidase	*O*-glycosidase	gentiobiase, cellobiase, amygdalase	hydrolysis of terminal, nonreducing β-D-glucose residues
β-Glucuronidase	*O*-glycosidase	—	β-D-glucuronoside + H_2O yields an alcohol + D-glucuronate

TABLE 7.1. (continued).

Enzymes	Type	Other Names	Preferential Cleavage Sites
II: Glycosidases			
Hyaluronoglucosaminidase	*O*-glycosidase	hyaluronidase	random hydrolysis of 1,4-linkages between *N*-acetyl-β-D-glucosamine and D-glucuronate residues in hyaluronate
Hyaluronoglucuronidase	*O*-glycosidase	hyaluronidase	random hyrolysis of 1,3-linkages between β-D-glucuronate and *N*-acetyl-D-glucosamine residues in hyaluronate
α-L-Iduronidase	*O*-glycosidase	—	hydrolysis of α-L-iduronosidic linkages in desulfated dermatan
Inulinase	*O*-glycosidase	inulase	endohydrolysis of 2,1-β-D-fructosidic linkages in inulin
Isoamylase	*O*-glycosidase	debranching enzyme	hydrolysis of 1,6-α-D-glucosidic branch linkages in glycogen, amylopectin, and their β-limit dextrins
Lysozyme	*O*-glycosidase	muramidase	hydrolysis of 1,4-β-linkages between *N*-acetylmuramic acid and *N*-acetyl-D-glucosamine residues in peptidoglycan and between *N*-acetyl-D-glucosamine residues in chitodextrin

(continued)

TABLE 7.1. (continued).

Enzymes	Type	Other Names	Preferential Cleavage Sites
II: Glycosidases			
α-Mannosidase	*O*-glycosidase	—	hydrolysis of terminal, nonreducing α-D-mannose residues in α-D-mannosides
β-Mannosidase	*O*-glycosidase	mannanase, mannase	hydrolysis of terminal, nonreducing β-D-mannose residues in β-D-mannosides
Oligo-1,6-glucosidase	*O*-glycosidase	α-limit dextrinase, isomaltase, sucrase-isomaltase	hydrolysis of 1,6-α-D-glucosidic linkages in isomaltose and dextrins produced from starch and glycogen α-amylase
Sialidase	*O*-glycosidase	neuraminidase	hydrolysis of 2,3-, 2,6-, and 2,8-glycosidic linkages joining terminal, nonreducing *N*- or *O*-acylneuraminyl residues to galactose, *N*-acetylhexosamine, or *N*- or *O*-acylated neuraminyl residues in oligosaccharides, glycoproteins, glycolipids, or colominic acid
Sucrose α-glucosidase	*O*-glycosidase	sucrase, sucrose α-glucohydrolase, sucrase-isomaltase	hydrolysis of sucrose and maltose by an α-D-glucosidase-type action

7.2 HYDROLYTIC ENZYMES IN THE BODY

7.2.1 Gastrointestinal Tract

Enzymes of the gastrointestinal (GI) tract function in the digestion of proteins, polysaccharides, triglycerides, and nucleic acids to absorbable metabolites. Hydrolysis occurs either by enzymes secreted into the luminal contents or by membrane-bound enzymes that line the small intestine. Luminal enzymes are secreted by the glands in the mouth, stomach, and small intestine. The function of luminal enzymes is to hydrolyze large macromolecules to di-, tri-, and oligomeric constituents. Membrane-bound enzymes of the small intestine provide additional hydrolytic activity which is critical to the absorption of nutrients. A list of the luminal and surface bound enzymes of the GI tract is presented in Table 7.2.

TABLE 7.2. Hydrolytic Enzymes of the Gastrointestinal Tract.

	Mouth		
	α-Amylase	Lingual lipase	
	Stomach		
	Pepsin	Gelatinase	Gastric amylase
	Gastric lipase		
	Small Intestine		
1.	Pancreatic		
	Chymotrypsin	Trypsin	Pancreatic elastase
	Carboxypeptidase A	Carboxypeptidase B	α-Amylase
	Pancreatic lipase	Cholesterol esterase	Phospholipase A_2
2.	Brush border bound		
	α-Limit Dextrinase	Maltase	Lactase
	Sucrase	Aminooligopeptidase	Aminopeptidase A
	Dipeptidase I	Dipeptidase III	
	Dipeptidyl aminopeptidase IV		
3.	Cytoplasm of mucosal cells		
	Dipeptidase	Aminotripeptidase	Proline dipeptidase
	Colon		
	β-Glucuronidase	β-Galactosidase	β-Glucosidase
	Dextranase	Urease	Azoreductase
	Cholanoylglycine hydrolase	Hydroxy-steroid oxido-reductase	
	Hydroxycholanoyl-dehydroxylases		

The salivary glands in the mouth secrete α-amylase and ptyalin [14,15]. They catalyze the endohydrolysis of 1,4-α-D-glucosidic linkages in starch. The products of hydrolysis range from maltose, a disaccharide, to various oligosaccharides that contain up to nine glucose units. Because of the brief residence time in the mouth and the inactivation of α-amylase under gastric conditions, only 30 percent to 40 percent of all ingested starch is hydrolyzed by salivary enzymes. Triglycerides are partially hydrolyzed in the mouth by lingual lipase. Although the residence time in the mouth can be brief, lingual lipase remains active in the stomach and can hydrolyze up to 30 percent of all dietary triglycerides [14].

The major hydrolase found in the stomach is pepsin. Its proenzymes pepsinogen I and II are secreted from the oxyntic and pyloric glands of the stomach. Pepsin preferentially cleaves proteins near aromatic amino acids. One of the most important roles of pepsin is in the digestion of collagen. It is estimated that pepsin contributes between 10 percent and 30 percent of protein digestion in the GI tract [16]. Optimal pepsin activity occurs between pH 1.6 and pH 2.3. Thus, the gastric emptying of pepsin into the duodenum where the pH ranges from 2.0 to 4.0 will lead to its inactivation. In addition to pepsinogens I and II, small quantities of gastric amylase, gastric lipase, and gelatinase are also secreted in the gastric juices [14]. These enzymes, however, play a minor role in the digestion process.

The most extensive enzyme-catalyzed hydrolysis occurs in the lumen and at the brush border membrane of mucosal cells in the small intestine. Enzymes which are present in the luminal contents are secreted from the pancreas via the pancreatic duct. Polysaccharides which are not broken down by salivary α-amylase are almost completely hydrolyzed by pancreatic α-amylase [15]. As mentioned, these enzymes endohydrolyze 1,4-α-D-glucosidic linkages but spare 1,6-α-linkages, terminal 1,4-α-linkages, and 1,4-α-linkages near branch points. Pancreatic lipase is responsible for the hydrolysis of triglycerides in the small intestine [17]. Lipase-catalyzed hydrolysis leads to the formation of monoglycerides and fatty acids. Cholesterol esterase and phospholipase A_2 are two other lipases secreted from the pancreas [17]. Their function is to hydrolyze cholesterol esters and phospholipids. The main site for protein digestion is in the small intestine. Luminal proteinases and peptidases are secreted by the pancreas. They include chymotrypsin, trypsin, elastase, and carboxypeptidases A and B [16].

The microvilli that comprise the brush border of the small intestine contain membrane-bound enzymes that hydrolyze disaccharides, dipeptides, oligosaccharides, and oligopeptides. Glycosidic enzymes that hydrolyze α-limit dextrins, maltotriose, maltose, lactose, and sucrose are shown in Table 7.2 [18]. These enzymes are primarily located in the upper and mid-jejunum [15]. Hydrolysis by membrane-bound glycosidases leads to the formation of glucose, galactose, or fructose. Oligopeptides and dipeptides are hydrolyzed to amino acids by brush border-bound aminopeptidases and dipeptidases (Table 7.2) [19]. Aminopeptidases and dipeptidases are also found in the cytoplasmic fraction of the mucosal cells (Table 7.2) [19]. They function in the hydrolysis of di-, tri-, and oligopeptides which are actively transported from the lumen [20–24].

The GI tract contains a variety of aerobic and anaerobic bacterial flora that are necessary for normal metabolic activity. The greatest concentrations of bacteria are found in the distal ileum and the colon [25]. The environment of the distal ileum is both aerobic and anaerobic. In man, the distal ileum contains enterobacteria, streptococci, staphylococci, lactobacilli, bacteroides, bifidobacteria, clostridia, and fungi [25–33]. The colon contains many of the same bacterial flora which are present in the distal ileum, but in much higher concentrations. The colonic environment, however, is predominantly anaerobic [25]. Proteins and polysaccharides are digested by bacterial flora as a result of secreted or cell-bound hydrolases. The bacterial flora of the colon metabolize disaccharides, oligosaccharides, and polysaccharides by fermentative digestion [34–38]. The major bacterial glycosidases are β-glucuronidase, β-galactosidase, β-glucosidase, and dextranase as listed in Table 7.2 [39,40]. Substrates for these enzymes include mucopolysaccharides, dietary fibers, and partially digested carbohydrates. In some instances, colonic bacteria can utilize proteins for nutrition [41]. Recent findings suggest that colonic bacteria may play an important role in the degradation of pancreatic enzymes such as leucine aminopeptidase, trypsin, and chymotrypsin [42]. Another unique enzyme secreted by bacterial flora in the colon is azoreductase. Because of its activity and location in the GI tract, the hydrolysis of compounds containing azo groups has been utilized in the development of antibiotics and colon-specific drug delivery systems as described in Chapter 3, Section 3.2.1.2 [43,44].

7.2.2 Reticuloendothelial System

The reticuloendothelial system (RES) is comprised of macrophages which either circulate in the blood or reside in specific tissues of the body. Macrophages in the lymph nodes, spleen, and bone marrow are known as tissue macrophages. Macrophages in the liver, lung alveoli, subcutaneous tissue, and brain are called Kupffer cells, alveolar macrophages, tissue histiocytes, and microglia, respectively [45]. Macrophages of the RES are capable of intracellular and extracellular digestion. Intracellular digestion begins with a process known as endocytosis. Endocytosis is an intracellular transport mechanism by which solids and macromolecules are incorporated into intracellular vesicles or endosomes. The term phagocytosis is used to describe the endocytosis of solid material while endocytosis of fluids is called pinocytosis. Macrophages provide phagocytic activity toward various pathogens, foreign substances, and particulate matter that enter the body. They also function in the removal of endogenous materials such as plasma proteins and necrotic tissue [45]. Following endocytosis, the formed endosome fuses with cytoplasmic organelles known as lysosomes [46]. Lysosomes contain a variety of hydrolases that function in the degradation of proteins, polysaccharides, nucleic acids, and lipids into low molecular weight products [47–50]. The types of hydrolases generally found in lysosomes are presented in Table 7.3 [51].

Macrophages of the RES also secrete lysosomal enzymes for extracellular digestion. The release of lysosomal enzymes occurs in response to phagocytosis [52], lymphokines [53,54], components of the complement system [55,56], IgE aggregates [57,58], and weak bases [59–61]. Enzymes that are released include lysozyme, plasminogen activator, elastase, collagenase, proteoglycan degrading enzyme, angiotensinase, arginase, and acid hydrolase [62]. Proteinases are secreted after an inflammatory, immune, or endocytic stimuli. Acid hydrolases are released in response to a variety of stimuli such as digestible and nonparticulate substances [63,64]. The secretion of lysozyme, however, is independent of cell stimulation [65]. Monocytes and macrophages also possess surface-bound peptidases which may play some digestive role [66].

7.2.3 Others

The inflammatory response is a sequence of events that arises from injury to soft tissues by bacteria, trauma, chemical irritants, endogenous inflammatory molecules, or heat. Phagocytosis by macrophages of the RES provides the first line of defense during the early stages of inflammation. Within hours, neutrophils move into the inflamed area to provide the major source of phagocytic action. Under a variety of stimuli, neutrophils may release lysosomal enzymes, such as elastase, cathepsin G, collagenase, and gelatinase, into the extracellular space [67,68]. In the latter stages of inflammation, a large population of blood-born monocytes infiltrate the inflamed area and mature into macrophages, which provide long-term phagocytic action [45]. The inflammatory response will also activate mast cells that secrete chymase and tryptase which have chymotrypsin-like and trypsin-like properties, respectively [69–72].

Many of the hydrolytic enzymes found in plasma are associated with the regulation of blood pressure, circulation, clotting, and anticlotting. Because of their specificity, these particular enzymes are unlikely to contribute to the biodegradation of hydrogels. It is likely that macrophages of the RES will provide the major contribution to biodegradation. Hydrolases of lysosomal

TABLE 7.3. Lysosomal Hydrolases [51].

Proteolytic			
	Carboxypeptidase A	Cathepsin H	Dipeptidyl peptidase I
	Carboxypeptidase B	Cathepsin L	Dipeptidyl peptidase II
	Cathepsin B	Dipeptidase I	Proline carboxypeptidase
	Cathepsin D	Dipeptidase II	Tripeptidyl peptidase
			Tyrosine carboxypeptidase
Glycosidic			
	α-*N*-Acetylgalactosaminidase	α-Galactosidase	α-L-Iduronidase
	α-*N*-Acetylglucosaminidase	β-Galactosidase	α-Mannosidase
	β-*N*-Acetylglucosaminidase	α-Glucosidase	β-Mannosidase
	β-Aspartylglucosylaminase	β-Glucosidase	Sialidase
	Chondroitin-6-sulfatase	β-Glucuronidase	Sulfatase A
	α-L-Fucosidase	Heparin sulfamatase	Sulfatase B
Lipases			
	Triacylglycerol lipase		
	Phospholipase A_1		
	Phospholipase A_2		

origin are released in small amounts into the plasma [73]. Recently, two membrane-bound, multicatalytic proteinases have been isolated from human erythrocytes [74]. Both proteinases demonstrated trypsin- and chymotrypsin-like activities.

7.3 FACTORS INFLUENCING ENZYME-CATALYZED REACTIONS

7.3.1 Conformation of Enzymes and Substrates

Changes in the conformation of enzymes or substrates will have a dramatic effect on the rate of enzyme catalysis. Conformational changes to the catalytic and noncatalytic residues in the active site of the enzyme will impair substrate recognition and binding. The conformational changes may occur by altering temperature, pH, ionic strength, or solvent type [75]. These factors will either impair or enhance the formation of enzyme-substrate complexes. For example, the activity of proteolytic enzymes found in the gastrointestinal tract are pH-dependent. Trypsin and chymotrypsin show the greatest activity in the small intestine while the activity of pepsin is greatest in the stomach. The conformational and configurational properties of the substrate will also influence the rate of enzyme-catalyzed reactions. In some instances, the inherent three-dimensional configuration of the substrate may sterically block close encounters by enzymes [76–78]. It was reported that denaturation of proteins made the protein more susceptible to enzyme-catalyzed hydrolysis [75].

Quite often, the kinetics of enzyme-catalyzed hydrolysis can be altered by chemical modification of the substrate. The influence of chemical modification on the formation of enzyme-substrate complexes has been studied extensively with gastrointestinal, lysosomal, and blood proteases [75,77,79–81]. Rejmanová, Duncan, Ringsdorf, Řihová, and Kopeček have developed site-directed lysosomotropic drug delivery systems which are composed of synthetic polymers containing drug-bound oligopeptide side chains [82–87]. The drug release was dependent on the enzyme-catalyzed hydrolysis of the side chain. The drug release rates increased when the structure of the side chain favored subsite-substrate interactions. The amino acid sequence of the oligopeptide side chains can be manipulated in such a way that the side chains are resistant to hydrolysis by blood proteases but susceptible to lysosomal proteases. Park and his coworkers examined the enzymatic degradation of human serum albumin after modification with glycidyl acrylate [88,89]. The pepsin-catalyzed hydrolysis became slower as the exent of the modification increased. The introduction of vinyl pendant groups to albumin may have impaired subsite-substrate interactions. The notion of steric hindrances at the active site may be supported by the fact that in the primary structure of human serum albumin, more than 30 percent of the lysine residues which react with glycidyl acrylate are within three amino acid residues from the major or minor cleavage sites by pepsin [90].

7.3.2 Presence of Polymer Molecules

The presence of polymers in the reaction medium can have a significant influence on the rate of enzyme-catalyzed hydrolysis. Polymer chains in solution increase the viscosity of the medium and thus lower the diffusion coefficients of both enzyme and substrate. As a result, the enzyme-substrate collision frequency is reduced and the overall reaction rate is decreased. This phenomenon is known as the sieving effect [91]. If substrates are associated with polymer chains, the formation of enzyme-substrate complexes will be restricted [92,93]. Polymer chains are expected to restrict the position and orientation of enzymes near the clevage sites on substrates. This steric constraint on enzyme activity becomes prominent if the substrate is chemically bound to polymer chains. In the degradation of oligopeptide chains, which are attached to the polymer backbone, steric constraints from polymer chains decreased as the site of cleavage moved away from the polymer backbone [94,95]. Examples of the degradation of oligopeptide side chains are described in Chapter 3, Section 3.2.2. The presence of polymer chains near the substrate generally decreases the rate of enzyme-catalyzed hydrolysis.

7.4 ENZYME-CATALYZED DEGRADATION OF MODIFIED PROTEINS

Chemical modification of proteins may alter a protein's susceptibility to enzyme-catalyzed degradation by either enhancing or impairing subsite-substrate interactions. Glycoproteins, which are known to be resistant to trypsin-catalyzed hydrolysis, became highly susceptible to degradation after periodate oxidation or acid hydrolysis [96]. Trypsin-catalyzed degradation of bovine serum albumin was enhanced following metholation with thionyl chloride and methanol [97]. Increased hydrolysis by trypsin was thought to be due to protein denaturation which made the protein more accessible by trypsin. On the other hand, human and bovine serum albumin showed decreased hydrolysis by both trypsin and pepsin when modified with glycidyl acrylate [88,89]. Clearly, the effect of chemical modification on the rate of enzyme-catalyzed degradation is complex. The enzymatic degradation of some modified proteins is described in this section.

7.4.1 Albumin

7.4.1.1 ALBUMIN-CROSSLINKED HYDROGELS

Enzyme-degradable hydrogels have been prepared by copolymerizing vinyl monomers with glycidyl acrylate-modified albumin as a crosslinking agent (see Chapter 3, Section 3.2). Figure 7.2 shows the swelling and degradation of an albumin-crosslinked hydrogel in the absence and presence of pepsin in the simulated gastric fluid. In the absence of enzymes, dried gels swelled in buffer solutions to reach equilibrium states. In the presence of enzymes, however, gels swelled faster and to a greater extent than in the absence of enzymes. The presence of enzymes resulted in the disruption of the gel structure and the eventual complete dissolution of the gels. The digestion of hydrogels by pepsin is obvious as shown in Figure 7.2(d).

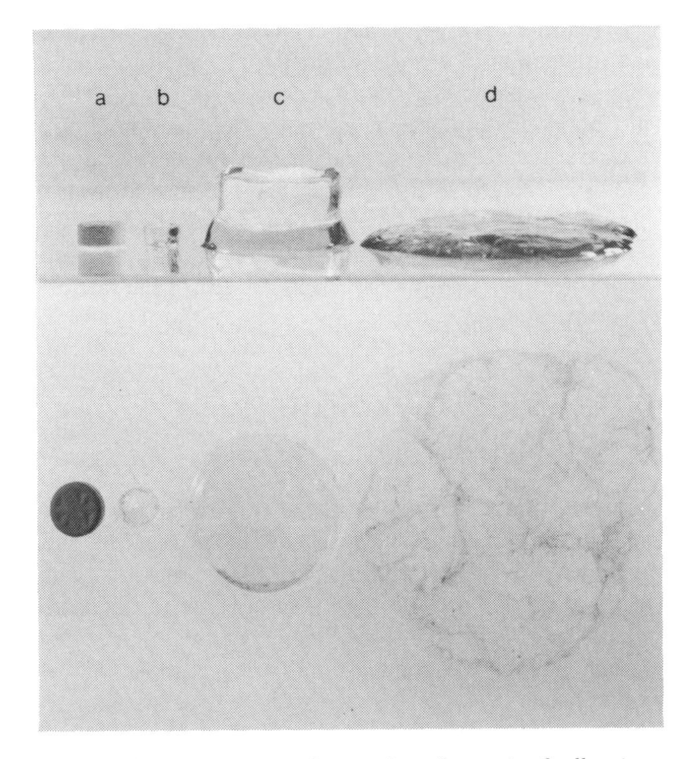

FIGURE 7.2. *Side view (top) and top view (bottom) of albumin-crosslinked polyacrylamide gels in various states. Extra-strength Tylenol tablet as a size indicator (a), dried polyacrylamide gel (b), polyacrylamide gel swollen in the simulated gastric juice without pepsin at 37° C (c), and polyacrylamide gel swollen in the simulated gastric juice with pepsin (570 unit/ml) at 37° C for 24 hours (d). The concentration of acrylamide was 15 percent (w/w) and that of albumin was 1 percent (w/w) of the monomer (from Reference [88]).*

The mode (surface vs. bulk) of hydrogel degradation was controlled by the extent of albumin modification and the polymerization condition such as the initiator concentration [88,89,98]. The rate and mode of hydrogel degradation is thought to depend largely on the degree of albumin incorporation within the polymer network and the steric constraints imposed by the polymer chains. As the number of polymer chains crosslinked by each albumin molecule increases, the size and mobility of the oligopeptide segments between polymer chains is reduced. This will restrict the formation of enzyme-substrate complexes. Bulk degradation results when albumin incorporation is high enough to slow down the rate of enzymatic degradation, and thus allow swelling-controlled penetration of the enzyme into the hydrogel. Surface degradation results when the degree of albumin incorporation is very low. Since the size and mobility of oligopeptide segments between crosslinks is large, enzyme-catalyzed degradation is rapid. It was postulated that surface degradation occurred due to restricted enzyme penetration during gel swelling from the glassy, dry state and the rapid hydrolysis of albumin crosslinks followed by loss of polymer chains from the rubbery phase of the gel.

7.4.1.2 ALBUMIN MICROSPHERES

Albumin microspheres have been prepared as an injectable delivery system for a variety of drugs [99]. It has been shown that albumin microspheres can be degraded by collagenase, papain, protease, and trypsin [99 – 101]. Lee et al. studied the enzyme-degradable properties of glutaraldehyde-crosslinked microspheres loaded with progesterone [102]. The microspheres were prepared by crosslinking emulsified serum albumin with glutaraldehyde in the presence of progesterone. The crosslinking density of the microsphere was altered by varying the glutaraldehyde concentration. The number of modified lysine residues increased from 21 to 47 as the concentration of glutaraldehyde increased from 1 percent to 4 percent (v/v) of the albumin solution. Microspheres prepared with 1 percent glutaraldehyde were the only samples degradable by chymotrypsin. When injected intramuscularly into a rabbit, all the albumin microspheres degraded two months after administration.

Willmott et al. studied release and degradation of adriamycin-loaded, glutaraldehyde-crosslinked microspheres [103]. Again, the properties of the microspheres were altered by varying the glutaraldehyde concentration from 0.25 percent to 2 percent (v/v) of the albumin solution. The residence time of the intravenously injected microspheres in the lungs was studied by examining treated tissue samples with fluorescence microscopy. The residence time of microspheres entrapped in the lungs was dependent on the concentration of glutaraldehyde used for crosslinking. Two days after administration, samples crosslinked with 0.3 percent glutaraldehyde were nearly all degraded whereas samples crosslinked with 0.5 percent and 1.0 percent gluteraldehyde underwent 87 percent and 57 percent degradation, respectively. In more recent work, Willmott et al. [100] studied the biodegradability of ^{125}I-labeled, glutaraldehyde-crosslinked albumin microspheres which were entrapped in the capillary beds of the lungs, liver, and kidneys of rats. The rate of biodegradation was measured by the loss in radioactivity in each respective tissue. It took only 2 days to lose 50 percent of the radioactivity in the lungs while it took 3.6 days in the liver. The degradation of microspheres was also observed in the kidneys. When the same microspheres were incubated in serum, however, only a slight loss in radioactivity was observed over a 9-day period. Similar findings were also reported for doxorubicin-loaded albumin microspheres in human plasma [104]. Because of the slow degradation observed in plasma, the authors suggested that other enzymes, such as those involved in the inflammatory response, might be responsible for regulating biodegradation of crosslinked albumin.

7.4.2 Gelatin

7.4.2.1 GELATIN HYDROGELS

Gelatin hydrogels were prepared by exposing solutions of the glycidyl acrylate-modified gelatin to γ-rays [105]. In pepsin-free simulated gastric fluid, the swelling ratio of the hydrogels decreased by 30 percent as the γ-irradiation dose increased from 0.16 Mrad to 0.48 Mrad. The rate of degradation in the presence of pepsin decreased as the γ-irradiation dose was increased, i.e., as the crosslinking density increased. The swelling ratio measured over time, in the presence of pepsin, passed a transient maximum, which indicates the bulk degradation of the hydrogel. The time to reach the maximum swelling was prolonged as the γ-irradiation dose was increased. Once again it was observed that the crosslinking density of the network significantly influences the rate of enzyme-catalyzed hydrolysis.

7.4.2.2 GELATIN MICROSPHERES

The degradation properties associated with interferon-loaded, glutaraldehyde-crosslinked gelatin microspheres was studied by Tabata et al. [106]. Gelatin microspheres were prepared by crosslinking with glutaraldehyde, either in the absence of, or in the presence of ^{125}I-labeled interferon. The resistance to collagenase-catalyzed hydrolysis increased as the reaction time increased and as the concentration of glutaraldehyde increased. As the resistance to enzymatic degradation increased, the release rate of ^{125}I-labeled interferon decreased. The release of hydrolyzed gelatin was monitered using the ninhydrin method [107]. The phagocytosis of gelatin microspheres containing ^{125}I-labeled interferon was studied using mouse peritoneal macrophages. The rate of the microsphere degradation, and the subsequent release of ^{125}I-labeled interferon, was reduced as the glutaraldehyde content increased. The crosslinking density of the network profoundly influenced the rate of enzyme-catalyzed hydrolysis presumably by limiting the penetration of enzymes into the network and by restricting subsite-substrate interactions. Glutaraldehyde-crosslinked gelatin microspheres labeled with ^{131}I have been used as a carrier for mitomycin C, a chemotherapeutic agent [108]. The gelatin microspheres achieved sufficient peripheral embolizaton in the hepatic artery and could be detected in the liver for as long as 28 days. The half-life of the microspheres was between 7 and 14 days. Degradation was believed to result from lysosomal breakdown following phagocytosis by the surrounding cells.

7.4.3 Modified Enzymes

Recently, polymer-grafted enzymes have been used in various applications. The pharmacological properties of the polymer-enzyme conjugates and other polymer-protein conjugates have been recently reviewed by many investigators [109 – 112]. The main attributes of the polymer-protein conjugates is their ability to minimize immunological side effects as compared with the native proteins and to prolong the circulation half-life. The dramatic increase in the circulation half-life is still not fully understood, but the steric repulsion of other blood proteins by the grafted water-soluble polymer is expected to play a major role. Polymer-protein conjugates generally undergo slow degradation by proteolytic enzymes in plasma and/or by intracellular lysosomal enzymes.

7.4.3.1 POLYETHYLENE GLYCOL CONJUGATES

Monomethoxypolyethylene glycol [MPEG] has been used most widely in the preparation of polymer-enzyme conjugates. Polymer-enzyme conjugates based on polyvinylpyrrolidone have also been prepared [113 – 115]. MPEG is generally activated with either cyanuric chloride or carbonyldiimidazole [116,117]. Abuchowski et al. coupled triazine-activated MPEG to liver catalase [118]. The degree of enzyme modification, as measured by the loss in primary amine groups, ranged from 13 percent to 37 percent depending on the amount of activated MPEG added to the reaction mixture. Native catalase was completely inactivated by trypsin within 40 minutes, while MPEG-catalase retained 90 percent of its original activity even after 150 minutes. A similar yet slower rate of degradation was observed with chymotrypsin. After intravenous administration, native catalase lost nearly 80 percent of its activity within 10 hours whereas MPEG-catalase remained active for over 50 hours. Triazine-activated MPEG was also coupled to trypsin [119]. The degree of trypsin modification varied from 24 percent to 59 percent. The 24 percent-modified MPEG-trypsin was completely resistant to inactivation by native trypsin. Resistance to inactivation by trypsin was thought to be due to the steric hindrances imparted by the MPEG chains and restricted subsite-substrate interactions due to the modification of lysine residues.

Triazine-activated MPEG and its derivatives have also been coupled to other enzymes such as phenylalanine ammonia-lyase [120], asparaginase [121,122], superoxide dismutase [123], and acyl-plasmin-streptokinase complex [124]. MPEG conjugates of phenylalanine ammonia-lyase and asparaginase elicited a marked resistance to trypsin-catalyzed degradation as observed with other MPEG conjugates. The circulation half-lives of the native and MPEG-modified phenylalanine ammonia-lyases in rabbits were 6 hours and 20 hours, respectively. In rats, the circulation half-life of the MPEG-asparaginase was 56 hours as compared with 2.9 hours for the native asparaginase. The degree and type of MPEG modification on superoxide dismutase conjugates was shown to have significant effects on its circulation half-life in rats. As the degree of modification increased, the circulation half-life of the MPEG-superoxide dismutase was prolonged from 3 hours to 25 hours. At the same degree of modification, however, the circulation half-life increased from 1.5 hours to 3.0 hours as the molecular weight of the conjugated MPEG increased from 1,900 to 5,000 daltons. Increased half-lives for MPEG-superoxide dismutase conjugates have also been observed by Veronese et al. [125].

Katre et al. examined the circulation half-life of MPEG-conjugated recombinant interleukin-2 (MPEG-rIL-2) [126]. MPEG was activated by acylating the MPEG with glutaric anhydride followed by succinimidation to form *N*-hydroxysuccinimide esters. The plasma level of MPEG-rIL-2 was nearly 14 fold higher than that of the native lymphokine when the degree of MPEG modification was low. The plasma level was increased more than 140 fold when the degree of MPEG modification was high. It was suggested that the prolonged level of MPEG-rIL-2 in the circulation was due to the steric hindrances from the MPEG chains and/or by size-dependent differences in glomerular filtration.

7.4.3.2 DEXTRAN CONJUGATES

Dextran-enzyme conjugates are generally prepared with cyanogen bromide-activated dextran or periodate-activated dextran. Cyanogen bromide-activated dextran has been conjugated with lysozyme, chymotrypsin, and β-glucosidase by Vegarud et al. [127]. Dextran-lysozyme conjugates showed significant resistance to inactivation by chymotrypsin. While the native lysozyme was completely inactivated within two days, the dextran-lysozyme conjugates retained 40 percent activity after the same time period.

α-Amylase and catalase have also been conjugated with cyanogen bromide-activated dextran [128,129]. The molecular weight of the conjugated dextran ranged from 60,000 to 90,000 daltons. The activity of both dextran-α-amylase and dextran-catalase conjugates was prolonged as compared to the native enzymes in mice and rats. Two hours after intravenous injection, dextran-α-amylase retained 75 percent of its original activity while native α-amylase retained only 16 percent of its original activity. Similarly, dextran-catalase retained 70 percent of its original activity while native catalase retained only 7 percent of its original activity. It was postulated that the increased circulation half-life was related to shifts in the enzyme's isoelectric point which may have impaired cellular recognition and subsequent uptake. In the case of dextran-α-amylase, however, reduction in glomerular filtration may have also contributed to the prolonged circulation half-life. Steric constraints from the dextran chains are also expected to restrict inactivation by the enzymes in the plasma [130].

Conjugates of carboxypeptidase G_2 and dextran were studied by Melton et al. [131,132]. The molecular weight of cyanogen bromide-activated dextran ranged from 40,000 to 150,000 daltons. The activity of dextran-carboxypeptidase G_2 was not affected by trypsin or chymotrypsin for 18 hours, whereas native carboxypeptidase G_2 was completely inactivated in less than 3 hours. The circulation half-life of carboxypeptidase G_2 was prolonged from 3 hours to 14 hours by conjugation with dextran. The circulation half-life was increased up to 46 hours as the molecular weight of dextran increased. The tissue distribution studies showed that non-degraded, fully active dextran-carboxypeptidase G_2 was preferentially taken up by the liver. In contrast, native carboxypeptidase G_2 showed no preferential tissue uptake and was rapidly cleared in the urine as low molecular weight fragments. The degradation site of native carboxypeptidase G_2, however, could not be determined. Apparently, conjugation with dextran alters the uptake and subsequent degradation of carboxypeptidase G_2 by parenchymal cells. Furthermore, the extent of uptake by macrophage appears to decrease as the molecular weight of dextran increases.

L-Asparaginase has been conjugated to periodate-activated dextran for the treatment of lymphoblastic leukemia [133]. Dextran-L-asparaginase conjugates showed a marked resistance to inactivation by trypsin or chymotrypsin and were able to prolong their bioactivity in the plasma. Native L-asparaginase had a half-life of 12 hours in man while dextran-L-asparaginase conjugate displayed a prolonged half-life of 11 days [134]. As the molecular weight of dextran increased from 10,000 daltons to 70,000 daltons, the circulation half-life in rabbits increased nearly seven fold [135]. The pharmacological properties of dextran-asparaginase conjugates have recently been reviewed by Wileman [136]. Although the degradation mechanisms of polymer-protein conjugates are not fully understood, it is clear that the circulation half-life of therapeutically active proteins can be significantly prolonged by conjugation with water-soluble polymers. This approach may profoundly enhance the therapeutic efficacy of certain drugs.

7.5 ENZYME-CATALYZED DEGRADATION OF MODIFIED POLYSACCHARIDES

7.5.1 Dextran

7.5.1.1 DEXTRAN DRUG CARRIERS

The biodegradable properties of modified dextrans were studied by Schacht et al. [137,138] and Vercauteren et al. [139]. Dextran was modified by three methods: (1) periodate-activated oxidation followed by reduction with sodium borohydride to yield reduced aldehyde dextrans, (2) acylation with succinic anhydride to form dextran monosuccinate derivatives, and (3) acylation with 4-nitrophenyl chloroformate followed by subsequent reactions with amines resulting in dextran carbamate derivatives. Samples of modified dextrans were incubated with either penicillum dextranase or lysosomal enzymes isolated from rat liver. Dextranase-catalyzed degradation was monitored by gel permeation chromatography. In all samples, as the degree of modification increased, the size of the hydrolyzed fragments increased, i.e., the rate of degradation decreased. The authors contended that the increased degree of modification impaired subsite-substrate interactions. The type of chemical modification on dextran had a minor influence on the rate of degradation.

Chaves et al. modified dextran with ethyl and butyl chloroformate [140]. Such modification led to the formation of cyclic carbonates and interchain carbonates (i.e., crosslinked dextran). The crosslinked, water-insoluble dextran was resistant to hydrolysis by dextranase. The hydrolysis rates of the water-soluble ethyl and butyl carbonate derivatives, however, were dependent on the degree of substitution and the concentration of dextranase. The modified dextran degraded more rapidly as the degree of dextran substitution decreased or as the dextranase concentration increased.

The importance of chemical structure on subsite-substrate interactions between dextranase and modified dextran was studied by Crepon et al. [141]. Dextran modification was carried out by carboxymethylation followed by benzylamine coupling to a fraction of the carboxyl groups to form carboxymethylbenzylamine dextran. Additional modification was carried out by sulfonating a fraction of the benzylamine groups to form carboxymethylbenzylamine sulfonated dextran. The degree of each type of substitution was determined by potentiometric titration and elemental analysis. The extent of dextranase-catalyzed degradation was determined by measuring the changes in molecular weight. The extent of degradation by dextranase decreased as the degree of carboxylic functionality increased. Dextran derivatives were completely resistant to degradation when the carboxylic functionality was 54 percent and the benzylamine-sulfonate functionality was 19.5 percent. When the overall degree of modification was nearly equivalent, the benzylamine and benzylamine-sulfonate groups had a stronger inhibitory effect on dextranase than the carboxylic group did. It was suggested that the bulky size of the benzylamine and benzylamine-sulfonate side chains was most effective in hindering the formation of enzyme-substrate complexes.

7.5.1.2 DEXTRAN MICROSPHERES

Edman et al. have studied the use of modified dextran microspheres as lysosome-directed drug delivery systems [142–144]. The objective of their work was to deliver agents to cells of the reticuloendothelial system using microspheres which were susceptible to degradation by lysosomal enzymes. To prepare the microspheres, samples of dextran were first modified with glycidyl acrylate. ^{14}C-labeled microspheres were prepared by emulsion polymerization where the degree of crosslinking was increased by using dextran derivatives with higher degrees of modification. After intravenous injection in rats, it was found that the microspheres were phagocytized predominantly by the liver, spleen, and bone marrow. The half-life of the microspheres in the liver and spleen was directly related to degree of dextran modification. It was shown that particles containing a smaller number of acrylic groups were eliminated more rapidly. *In vitro* degradation studies with dextranase also showed that as the degree of dextran modification increased, the rate of microsphere degradation was reduced.

7.5.2 Starch

7.5.2.1 STARCH HYDROGELS

Heller et al. developed enzyme-degradable starch hydrogels as a part of a "triggered" drug delivery system [145]. Soluble starch was functionalized with glycidyl methacrylate under alkaline conditions. The degree of modification was controlled by altering the amount of glycidyl methacrylate or the reaction time. As the degree of modification increased, the prepared hydrogels exhibited increased firmness indicating higher crosslinking density. All hydrogels tested were susceptible to hydrolysis by α-amylase as measured by gel swelling and dissolution. As expected, the rate of hydrogel degradation decreased as the degree of starch modification increased. The hydrolysis rate of the modified, water-soluble starch, however, increased as the degree of modification increased. The results suggest that the formation of a three-dimensional network imparts significantly larger steric constraints on the enzyme and such constraints increase with higher degrees of crosslinking density (i.e., increased starch modification).

7.5.2.2 STARCH MICROSPHERES

Artursson et al. [146] and Laakso et al. [147,148] studied the degradation properties of modified starch microspheres which were prepared as lysosomotropic drug delivery systems. Soluble starch was modified by alkylation using glycidyl acrylate or by acylation using acryloyl chloride. Starch microspheres were prepared by free radical emulsion polymerization. As the degree of starch modification increased, the crosslinking density of the microsphere increased. The degradability of the microsphere by mouse serum, α-amylase, amyloglucosidase, or lysosomal enzymes decreased as the degree of modification increased. When the modified starch was copolymerized with methylenebisacrylamide (BIS), the degradability of microspheres increased as compared to the modified starch controls. The copolymerization with BIS resulted in the formation of larger pore sizes in microspheres. The increased degradation was thought to be due to the enhanced penetration of enzymes into the microspheres. The delivery of high [149] and low molecular weight [150,151] agents to tissues of the reticuloendothelial system using the glycidyl acrylate-modified starch microspheres has been shown to be quite promising.

7.5.3 Amylose

The enzyme-degradable properties of epichlorohydrin-crosslinked amylose were studied by Mateescu et al. [152]. The rate of enzyme-catalyzed hydrolysis was studied by an iodometric method. The iodine formed inclusion complexes with the hydrolyzed amylose chains which were released from the network. Thus, the rate of hydrolysis was determined by monitoring the change in color of the supernatant. It was found that the rate of hydrolysis by α-amylase was inversely related to the crosslinking density of the amylose. The authors postulate that the increase in crosslinking density limited penetration of enzymes into the network, and thus restricted the enzymes to the surface of the network. As the crosslinking density decreased, enzyme penetration was enhanced and more substrates were available for hydrolysis.

7.5.4 Chitosan

The effect of *N*-acylation on the hydrolysis of chitosan by lysozyme and chitinase was studied by Hirano et al. [153]. Chitosan was *N*-acylated with a mixture of acetic anhydride and D-glucosamine. The carbon/nitrogen ratio obtained from elemental analysis was used to determine the degree of substitution. The extent of *N*-acylation was controlled by varying the content of acetic anhydride in the reaction mixture. The hydrolytic activity of lysozyme was dependent on the degree of chitosan modification. There was little or no hydrolysis of the modified chitosan if the degree of substitution was below 0.2. The extent of hydrolysis, however, increased as the degree of substitution increased from 0.4 to 0.8. The maximum rate of hydrolysis was observed when the degree of substitution was 0.8. This was nearly four times higher than the rate of hydrolysis observed when the degree of substitution was 1.0. The same type of relationship was observed with chitinase-catalyzed hydrolysis. The maximum rate of hydrolysis was observed, however, when the degree of substitution ranged from 0.4 to 0.8. The data demonstrate how chemical modification can influence the rate of enzyme-catalyzed hydrolysis by restricting or enhancing subsite-substrate interactions. A similar effect on subsite-substrate interactions between lysozyme and partially deacetylated chitin has also been reported [154].

7.5.5 Chondroitin-6-Sulfate

Chondroitin-6-sulfate is a mucopolysaccharide found in animal connective tissues, especially in cartilage. It consists of D-glucuronic acid linked to *N*-acetyl-D-galactosamine which is sulfated at C-6. Modified chondroitin sulfate has been suggested as an enzyme-degradable drug delivery system [155,156]. It can be degraded by anaerobic bacteria found in the large intestine of man. Rubinstein et al. [155] prepared crosslinked chondroitin-6-sulfate as a suitable matrix for indomethacin release in the colon. Chondroitin-6-sulfate was crosslinked with diaminododecane via dicyclohexylcarbodiimide activation. The release of indomethacin from the crosslinked chondroitin-6-sulfate tablets in the presence of rat cecal contents was degradation-dependent. As the degree of crosslinking increased, the amount of indomethacin released for 28 hours decreased.

7.6 REFERENCES

1. Stryer, L. 1975. *Biochemistry*. San Francisco, CA: W. H. Freeman and Company, Chapter 6.
2. Pauling, L. 1946. "Molecular Architecture and Biological Reactions," *Chem. Eng. News*, 24:1375–1377.
3. Spratt, T. E. and E. T. Kaiser. 1984. "Catalytic Versatility of Angiotensin Converting Enzyme: Catalysis of an a,b-Elimination Reaction," *J. Am. Chem. Soc.*, 106: 6440–6442.
4. White, H. and W. P. Jencks. 1976. "Mechanism and Specificity of Succinyl-CoA: 3-Ketoacid Coenzyme A Transferase," *J. Biol. Chem.*, 251:1688–1699.
5. White, H., F. Solomon and W. P. Jencks. 1976. "Utilization of the Inactivation Rate of Coenzyme A Transferase by Thiol Reagents to Determine Properties of the Enzyme-CoA Intermediate," *J. Biol. Chem.*, 251:1700–1707.
6. Webb, E. C., ed. 1984. *Enzyme Nomenclature, Recommendations of the Nomenclature Committee of the International Union of Biochemistry on the Nomenclature and Classification of Enzyme-Catalyzed Reactions*. Orlando, FL: Academic Press, Inc.
7. Sinnot, M. L. 1984. "Glycosyl Transfer," *New Compr. Biochem.*, 6:389–431.
8. Sinnot, M. L. 1987. "Glycosyl Group Transfer," in *Enzyme Mechanisms*, M. I. Page and A. Williams, eds., Northern Ireland: The Royal Society of Chemistry, pp. 259–297.
9. Schechter, I. and A. Berger. 1967. "The Size of the Active Site in Proteases. I. Papain," *Biochem. Biophys. Res. Commun.*, 27(2):157–162.
10. Schechter, I. and A. Berger. 1968. "The Active Site of Proteases. III. Mapping the Active Site of Papain. Specific Peptide Inhibitors of Papain," *Biochem. Biophys. Res. Commun.*, 32:898–902.
11. Fruton, J. S. 1977. "Some Aspects of Biochemical Catalysis," *Proc. Am. Philos. Soc.*, 121:309–315.
12. Steinbrink, D. R., M. D. Bond and H. E. VanWart. 1985. "Substrate Specificity of β-Collagenase from *Clostridium histolyticum*," *J. Biol. Chem.*, 260(5):2771–2776.
13. VanWart, H. E. and D. R. Steinbrink. 1985. "Complementary Substrate Specificities of Class I and Class II Collagenases from *Clostridium histolyticum*," *Biochemistry*, 24(23):6520–6526.
14. Ganong, W. F. 1991. *Review of Medical Physiology*. East Norwalk, CT: Appleton & Lange, pp. 437–447.
15. Gray, G. M. 1981. "Carbohydrate Absorption and Malabsorption," in *Physiology of the Gastrointestinal Tract, Vol. 2,* L. R. Johnson, J. M. Christensen, M. I. Grossman, E. D. Jacobson and S. G. Schultz, eds., New York, NY: Raven Press, pp. 1063–1072.
16. Guyton, A. C. 1986. *Textbook of Medical Physiology, Seventh Edition*. Philadelphia, PA: W. B. Saunders Company, pp. 787–797.
17. Patton, J. S. 1981. "Gastrointestinal Lipid Digestion," in *Physiology of the Gastrointestinal Tract, Vol. 2,* L. R. Johnson, J. M. Christensen, M. I. Grossman, E. D. Jacobson and S. G. Schultz, eds., New York, NY: Raven Press, pp. 1123–1146.
18. Gray, G. M. 1975. "Carbohydrate Digestion and Absorption. Role of the Small Intestine," *N. Engl. J. Med.*, 292(23):1225–1230.
19. Shoaf, C., R. M. Berko and W. D. Heizer. 1976. "Isolation and Characterization of

Four Peptide Hydrolases from the Brush Border of Rat Intestinal Mucosa,'' *Biochim. Biophys. Acta*, 445:694–719.

20. Heizer, W. D., R. L. Kerley and K. J. Isselbacher. 1972. ''Intestinal Peptide Hydrolases Differences between Brush Border and Cytoplasmic Enzymes,'' *Biochim. Biophys. Acta*, 264:450–461.
21. Josefsson, J. and H. Sjostrom. 1966. ''Intestinal Dipeptidases. IV. Studies on the Release and Subcellular Distribution of Intestinal Dipeptidases of the Mucosal Cells of the Pig,'' *Acta Physiol. Scand.*, 67:27–33.
22. Kim, Y. S., W. Birtwhistle and Y. W. Kim. 1972. ''Peptide Hydrolases in the Brush Border and Soluble Fractions of Small Intestinal Mucosa of Rat and Man,'' *J Clin. Invest.*, 51:1419–1430.
23. Kim, Y. S., Y. W. Kim and M. H. Sleisenger. 1974. ''Specificities of Peptide Hydrolases in Brush Border and Cytosol Fractions of Rat Small Intestine,'' *Biochim. Biophys. Acta*, 370:283–296.
24. Peters, T. J. 1970. ''The Subcellular Localization of Di- and Tripeptide Hydrolase Activity in Guinea Pig Small Intestine,'' *Biochem. J.*, 120:195–207.
25. Simon, G. L. and S. L. Gorbach. 1983. ''Bacteriology of the Colon,'' in *Colon, Structure and Function*, L. Bustos-Fernandez, ed., New York, NY: Plenum Publishing Co., pp. 103–119.
26. Cregan, J. and N. J. Hayward. 1953. ''The Bacterial Content of the Healthy Human Small Intestine,'' *Br. Med. J.*, 1:1356–1359.
27. Donaldson, R. M. 1978. ''The Relation of Enteric Bacterial Population to Gastrointestinal Function and Disease,'' in *Gastrointestinal Disease*, M. H. Sleisenger and J. S. Fordtran, eds., Philadelphia, PA: W. B. Saunders Co., pp. 79–92.
28. Drasar, B. S. and M. Shiner. 1969. ''Studies on the Intestinal Flora. II. Bacterial Flora of the Small Intestine in Patients with Gastrointestinal Disorders,'' *Gut*, 10:812–819.
29. Drasar, B. S., M. Shiner and G. M. McLeod. 1969. ''Studies on the Intestinal Flora. I. The Bacterial Flora of the Gastrointestinal Tract in Healthy and Achlorhydric Persons,'' *Gastroenterology*, 56:71–79.
30. Gorbach, S. L. 1971. ''Intestinal Microflora,'' *Gastroenterology*, 60:1110–1129.
31. Gorbach, S. L. and R. Levitan. 1970. ''Intestinal Flora in Health and in Gastrointestinal Diseases,'' in *Progress in Gastroenterology, Vol. 2*, G. B. L. Glass, ed., New York, NY: Grune & Stratton, pp. 252–275.
32. Gorbach, S. L., A. G. Plaut, L. Nahas and L. Weinstein. 1967. ''Studies of Intestinal Microflora. II. Microorganisms of the Small Intestine and Their Relations to Oral and Fecal Flora,'' *Gastroenterology*, 53:856–867.
33. Kalser, M. H., R. Cohen, I. Arteaga, E. Yawn, L. Mayoral, W. R. Hoffert and D. Frazier. 1966. ''Normal Viral and Bacterial Flora of the Human Small and Large Intestine,'' *N. Eng. J. Med.*, 274:500–505.
34. Cummings, J. H. and H. N. Englyst. 1987. ''Fermentation in the Human Large Intestine and the Available Substrates,'' *Am. J. Clin. Nutr.*, 45:1243–1255.
35. Levitt, M. D., P. Hirsh, C. A. Fetzer, M. Sherman and A. S. Levine. 1987. ''Hydrogen Excretion after Ingestion of Complex Carbohydrates,'' *Gastroenterology*, 92:383–389.
36. Steggerda, F. R. 1968. ''Gastrointestinal Gas Following Food Consumption,'' *Ann. N.Y. Acad. Sci.*, 150:57–66.

37. Calloway, D. H. and E. L. Murphy. 1968. "The Use of Expired Air to Measure Intestinal Gas Formation," *Ann. N.Y. Acad. Sci.*, 150:82–95.

38. Van Soest, P. J. 1978. "Dietary Fibers: Their Definition and Nutritional Properties," *Am. J. Clin. Nutr.*, 31:S12–S20.

39. Reddy, N. R., J. K. Palmer, M. D. Pierson and R. J. Bothast. 1984. "Intracellular Glycosidases of Human Colon Bacteroides *ovatus*," *Appl. Environ. Microbiol.*, 48:890–892.

40. Drassar, B. S. and M. J. Hill. 1974. *Human Intestinal Flora*. New York, NY: Academic Press Inc., pp. 54–71.

41. Sugarbaker, S. P., A. Revhaug and D. W. Wilmore. 1987. "The Role of the Small Intestine in Ammonia Production after Gastric Blood Administration," *Ann. Surg.*, 206:5–17.

42. Macfarlane, G. T., J. H. Cummings, S. Macfarlane and G. R. Gibson. 1989. "Influence of Retention Time on Degradation of Pancreatic Enzymes by Human Colonic Bacteria Grown in a 3-Stage Continuous Culture System," *J. Appl. Bacteriol.*, 67(5):520–527.

43. Scheline, R. R. 1968. "Drug Metabolism by Intestinal Microorganisms," *J. Pharm. Sci.*, 57(12):2021–2037.

44. Rubinstein, A. 1990. "Microbially Controlled Drug Delivery to the Colon," *Biopharm. Drug Dispos.*, 11:465–475.

45. Guyton, A. C. 1986. *Textbook of Medical Physiology, Seventh Edition*. Philadelphia, PA: W. B. Saunders Company, pp. 51–59.

46. de Duve, C. 1969. "The Lysosome in Retrospect," in *Lysosomes in Biology and Pathology*, J. T. Dingle and H. B. Fell, eds., Amsterdam: North-Holland, pp. 1–40.

47. Coffey, J. W. and C. de Duve. 1968. "Digestive Activity of Lysosomes, I. The Digestion of Proteins by Extracts of Rat Liver Lysosomes," *J. Biol. Chem.*, 243: 3255–3263.

48. Aronson, N. N. and C. de Duve. 1968. "Digestive Activity of Lysosomes, II. The Digestion of Macromolecular Carbohydrates by Extracts of Rat Liver Lysosomes," *J. Biol. Chem.*, 243:4564–4573.

49. Fowler, S. and C. de Duve. 1969. "Digestive Activity of Lysosomes, III. The Digestion of Lipids by Extracts of Rat Liver Lysosomes," *J. Biol. Chem.*, 244:471–481.

50. Arsenis, C., J. S. Gordon and O. Touster. 1970. "Degradation of Nucleic Acids by Lysosomal Extracts of Rat Liver and Ehrlich Ascites Tumor Cells," *J. Biol. Chem.*, 245:205–211.

51. Holtzman, E. 1989. *Lysosomes*. New York, NY: Plenum Press, pp. 1–24.

52. Davies, P., R. C. Page and A. C. Allison. 1974. "Changes in Cellular Enzyme Levels and Extracellular Release of Lysosomal Acid Hydrolases in Macrophages Exposed to Group A Streptococcal Cell Wall Substances," *J. Exp. Med.*, 139:1262–1282.

53. Pantalone, R. M. and R. C. Page. 1975. "Lymphokine-Induced Production and Release of Lysosomal Enzymes by Macrophages," *Proc. Natl. Acad. Sci.*, 72:2091–2094.

54. Pantalone, R. M. and R. C. Page. 1977. "Enzyme Production and Secretion by Lymphokine-Activated Macrophages," *J. Reticuloendothel. Soc.*, 21:343–357.

55. Schorlemmer, H. U., P. Davies and A. C. Allison. 1976. "Ability of Activated

Complement Components to Induce Lysosomal Enzyme Release from Macrophages," *Nature*, 261:48–49.

56. Schorlemmer, H. U. and A. C. Allison. 1976. "Effects of Activated Complement Components on Enzyme Secretion by Macrophages," *Immunology*, 31:781–788.
57. Dessaint, J. P., G. Torpier, M. Capron, H. Bazin and A. Capron. 1979. "Cytophilic Binding of IgE to the Macrophage. I. Binding Characteristics of IgE on the Surface of Macrophages in the Rat," *Cell Immunol.*, 46:12–23.
58. Dessaint, J. P., A. Capron, M. Joseph and H. Bazin. 1979. "Cytophilic Binding of IgE to the Macrophage. II. Immunologic Release of Lysosomal Enzyme from Macrophages by IgE and anti-IgE in the Rat: A New Mechanism of Macrophage Activation," *Cell Immunol.*, 46:24–34.
59. Riches, D. W. H. and D. R. Stanworth. 1980. "Primary Amines Induce Selective Release of Lysosomal Enzymes from Mouse Macrophages," *Biochem. J.*, 188:933–936.
60. Riches, D. W. H., C. J. Morris and D. R. Stanworth. 1981. "Induction of Selective Acid Hydrolase Release from Mouse Macrophages during Exposure to Chloroquine and Quinine," *Biochem. Pharmacol.*, 30:629–634.
61. Jessup, W., P. Leoni, J. L. Bodmer and R. T. Dean. 1982. "The Effect of Weak Bases on Lysosomal Enzyme Secretion by Mononuclear Phagocytes," *Biochem. Pharmacol.*, 31:2657–2662.
62. Gordon, S. and R. A. Ezekowitz. 1985. "Macrophage Neutral Proteinases: Nature, Regulation, and Role," in *The Reticuloendothelial System: A Comprehensive Treatise, Vol. 7B, Physiology*, S. M. Reichard and J. P. Filkins, eds., New York, NY: Plenum Press, pp. 95–141.
63. Davies, P. and A. C. Allison. 1976. "Secretion of Macrophage Enzymes in Relation to the Pathogenesis of Chronic Inflammation," in *Immunobiology of the Macrophage*, D. S. Nelson, ed., New York: NY: Academic Press, pp. 427–461.
64. Axline, S. G. and Z. A. Cohn. 1970. "*In vitro* Induction of Lysosomal Enzymes by Phagocytosis," *J. Exp. Med.*, 131:1239–1260.
65. Gordon, S., J. Todd and Z. Cohn. 1974. " *In vitro* Synthesis and Secretion of Lysozyme by Mononuclear Phagocytes," *J. Exp. Med.*, 139:1228–1248.
66. Bauvois, B., J. Sanceau and J. Wietzerbin. 1992. "Human U937 Cell Surface Peptidase Activities: Characterization and Degradative Effect on Tumor Necrosis Factor-Alpha," *Eur. J. Immunol.*, 22(4):923–930.
67. Perez, H. D. 1992. "Acute Inflammation," in *Textbook of Internal Medicine, Second Edition*, W. N. Kelley, ed., Philadelphia, PA: J. B. Lippincott Co., pp. 901–903.
68. Zimmerman, H. 1979. "Role of Proteinases from Leukocytes in Inflammation," in *Biological Function of Proteinases*, H. Holzer and H. Tschesche, eds., New York, NY: Springer-Verlag, pp. 186–195.
69. Dayer, J. M. 1992. "Cells and Mediators in Immediate Hypersensitivity," in *Textbook of Internal Medicine, Second Edition*, W. N. Kelley, ed., Philadelphia, PA: J. B. Lippincott Co., pp. 895–901.
70. Caughey, G. H. 1991. "The Structure and Airway Biology of Mast Cell Proteinases," *Am. J. Resp. Cell. Mol. Biol.*, 4(5):387–394.
71. Butrus, S. I., K. I. Ochsner, M. B. Abelson and L. B. Schwartz. 1990. "The Level of Tryptase in Human Tears. An Indicator of Activation of Conjunctival Mast Cells," *Ophthalmology*, 97(12):1678–1683.

72. Schwartz, L. B. 1990. "Tryptase from Human Mast Cells: Biochemistry, Biology, and Clinical Utility," *Monogr. Allergy*, 27:90–113.

73. Horpacsy, G. 1985. "The RES and the Turnover of Circulating Lysosomal Enzymes in Shock," in *The Reticuloendothelial System: A Comprehensive Treatise, Vol. 7B, Physiology*, S. M. Reichard and J. P. Filkins, eds., New York, NY: Plenum Press, pp. 499–519.

74. Kinoshita, M., T. Hamakubo, I. Fukui, T. Murachi and H. Toyohara. 1990. "Significant Amount of Multicatalytic Proteinases Identified on Membrane from Human Erythrocyte," *J. Biochem.*, 107(3):440–444.

75. Kopeček, J. and P. Rejmanová. 1983. "Enzymatically Degradable Bonds in Synthetic Polymers," *Controlled Drug Delivery, Vol. 1*, S. D. Bruck, ed., Boca Raton, FL: CRC Press, pp. 81–124.

76. Taylor, J. W., R. J. Miller and E. T. Kaiser. 1982. "Structural Characterization of β-Endorphin through the Design, Synthesis, and Study of Model Peptides," *Mol. Pharmacol.*, 22:657–666.

77. Pytela, J., V. Saudek, J. Drobnik and F. Rypacek. 1989. "Poly(N^5-hydroxyalkylglutamines). IV. Enzymatic Degradation of N^5-(2-Hydroxyethyl)-L-Glutamine Homopolymers and Copolymers," *J. Controlled Release*, 10:17–25.

78. Hayashi, T., Y. Tabata and A. Nakajima. 1984. "Biodegradation of Poly(α-amino Acid) *in Vitro*. III. Biodegradation of Poly(*N*-hydroxyethyl)-D,L-glutamine by Papain," *Rep. Prog. Pol. Phys. Jpn.*, 27:617–620.

79. Chandy, T. and C. P. Sharma. 1991. "Effect of Plasma Glow, Glutaraldehyde and Carbodiimide Treatments on the Enzymatic Degradation of Poly(L-lactic Acid) and Poly(g-benzyl-L-glutamate) Films," *Biomaterials*, 12:677–682.

80. Hayashi, T. and Y. Ikada. 1990. "Enzymatic Hydrolysis of Copoly(*N*-hydroxyalkyl L-Glutamine/γ-Methyl L-Glutamate) Fibres," *Biomaterials*, 11:409–413.

81. Kopeček, J. 1984. "Controlled Biodegradability of Polymers–A Key to Drug Delivery Systems," *Biomaterials*, 5:19–25.

82. Rejmanová, P. and J. Kopeček. 1983. "Polymers Containing Enzymatically Degradable Bonds, 8, Degradation of Oligopeptide Sequences of *N*-(2-Hydroxypropyl) Methacrylamide Copolymers by Bovine Spleen Cathepsin B," *Makromol. Chem.*, 184:2009–2020.

83. Duncan, R., H. Cable, J. B. Lloyd, P. Rejmanová and J. Kopeček. 1983. "Polymers Containing Enzymatically Degradable Bonds, 7, Design of Oligopeptide Side Chains in Poly[*N*-(2-hydroxypropyl) Methacrylamide] Copolymers to Promote Efficient Degradation by Lysosomal Enzymes," *Makromol. Chem.*, 184:1997–2008.

84. Rejmanová, P., J. Kopeček, R. Duncan and J.B. Lloyd. 1985. "Stability in Rat Plasma and Serum of Lysosomally Degradable Oligopeptide Sequences in *N*-(2-Hydroxypropyl) Methacrylamide Copolymers," *Biomaterials*, 6:45–48.

85. Ringsdorf, H., B. Schmidt and K. Ulbrich. 1987. "Bis(2-Chlorethyl) Amine Bond to Copolymers of *N*-(2-Hydroxypropyl) Methacrylamide and Methacryloylated Oligopeptides via Biodegradable Bonds," *Makromol. Chem.*, 188:257–264.

86. Říhová, B., V. Vetuicka, J. Strohalm, K. Ulbrich and J. Kopeček. 1989. "Action of Polymeric Prodrugs Based on *N*-(2-Hydroxypropyl) Methacrylamide Copolymers. I. Suppression of Antibody Response and Proliferation of Mouse Splenocytes *in Vitro*," *J. Controlled Release*, 9:21–32.

87. Říhová, B. and J. Kopeček. 1985. "Biological Properties of Targetable Poly[*N*-(2-

hydroxypropyl) Methacrylamide]-Antibody Conjugates,'' *J. Controlled Release*, 2:289–310.

88. Park, K. 1988. ''Enzyme-Digestible Swelling Hydrogels as Platforms for Long-Term Oral Drug Delivery: Synthesis and Characterization,'' *Biomaterials*, 9:435–441.
89. Shalaby, W. S. W. and K. Park. 1990. ''Biochemical and Mechanical Characterization of Enzyme-Digestible Hydrogels,'' *Pharm. Res.*, 7(8):816–823.
90. Dayhoff, M. O., ed. 1978. *Atlas of Protein Sequence and Structure, Vol. 5, Suppl. 3.* Silver Springs, MD: The National Biomedical Research Foundation, p 306.
91. Laurent, T. C. 1971. ''Enzyme Reactions in Polymer Media,'' *Eur. J. Biochem.*, 21:498–506.
92. Giddings, J. C. 1970. ''Effect of Membranes and Other Porous Networks on the Equilibrium and Rate Constants of Macromolecular Reactions,'' *J. Phys. Chem.*, 74(6):1368–1370.
93. Giddings, J. C., E. Kucera, C. P. Russel and M. N. Myers. 1968. ''Statistical Theory for the Equilibrium Distribution of Rigid Molecules in Inert Porous Networks. Exclusion Chromatography,'' *J. Phys. Chem.*, 72(13):4397–4408.
94. Kopeček, J., P. Rejmanová and V. Chytrý. 1981. ''Polymers Containing Enzymatically Degradable Bonds. I. Chymotrypsin Catalyzed Hydrolysis of *P*-Nitroanilides of Phenylalanine and Tyrosine Attached to Side Chains of Copolymers of *N*-(2-Hydroxypropyl) Methacrylamide,'' *Makromol. Chem.*, 182:799–809.
95. Ulbrich, K., J. Strohalm and J. Kopeček. 1981. ''Polymers Containing Enzymatically Degradable Bonds. III. Poly[*N*-(2-hydroxypropyl) Methacrylamide] Chains Connected by Oligopeptide Sequences Cleavable by Trypsin,'' *Makromol. Chem.*, 182:1917–1928.
96. Anatha Samy, T. S. 1967. ''Chemical Modification of Sheep Plasma Glycoprotein,'' *Arch. Biochem. Biophys.*, 121:703–710.
97. Sri Ram, J. and P. Maurer. 1959. ''Modified Bovine Serum Albumin. VIII. Estimation and Some Physicochemical Studies of the Methylated Derivative,'' *Arch. Biochem. Biophys.*, 85:512–520.
98. Shalaby, W. S. W., M. Chen and K. Park. 1992. ''A Mechanistic Assessment of Enzyme-Induced Degradation of Albumin-Crosslinked Hydrogels,'' *J. Bioact. Compatible Polym.*, 7(3):257–274.
99. Morimoto, Y. and S. Fujimoto. 1985. ''Albumin Microspheres as Drug Carriers,'' *Crit. Rev. Ther. Drug Carrier Syst.*, 2(1):19–63.
100. Willmott, N., Y. Chen, J. Goldberg, C. Meardle and A. T. Florence. 1989. ''Biodegradation Rate of Embolized Protein Microspheres in Lung, Liver, Kidney of Rats,'' *J. Pharm. Pharmacol.*, 41:433–438.
101. Mahato, R. I., N. Willmott and W. R. Vezin. 1991. ''Preparative Techniques for Albumin Microspheres,'' *Proceed. Intern. Symp. Control. Rel. Bioact. Mater.*, 18:375–376.
102. Lee, T. K., T. D. Sokoloski and G. P. Royer. 1981. ''Serum Albumin Beads: An Injectable, Biodegradable System for the Sustained Release of Drugs,'' *Science*, 213:233–235.
103. Willmott, N., J. Cummings, J. F. B. Stuart and A. T. Florence. 1985. ''Adriamycin-Loaded Albumin Microspheres: Preparation, *in vivo* Distribution and Release in the Rat,'' *Biopharm. Drug Dispos.*, 6:91–104.
104. Jones, C., M. A. Burton and B. N. Gray. 1989. ''Albumin Microspheres as Vehicles

for the Sustained and Controlled Release of Doxorubicin," *J. Pharm. Pharmacol.*, 41:813–816.

105. Kamath, K. R. and K. Park. 1992. "Use of Gamma Irradiation for the Preparation of Hydrogels from Natural Polymers," *Proceed. Intern. Symp. Control. Rel. Bioact. Mater.*, 19:42–43.
106. Tabata, Y. and Y. Ikada. 1989. "Synthesis of Gelatin Microspheres Containing Interferon," *Pharm. Res.*, 6(5):422–427.
107. McGrath, R. 1972. "Protein Measurement by Ninhydrin Determination of Amino Acids Released by Alkaline Hydrolysis," *Anal. Biochem.*, 49:95–102.
108. Yan, C., X. Li, X. Chen, D. Wang, D. Zhang, T. Tan and H. Kitano. 1991. "Anticancer Gelatin Microspheres with Multiple Functions," *Biomaterials*, 12:640–644.
109. Nucci, M. L., R. Shorr and A. Abuchowski. 1991. "The Therapeutic Value of Poly(ethylene Glycol)-Modified Proteins," *Adv. Drug Deliv. Rev.*, 6:133–151.
110. Pizzo, S. V. 1991. "Preparation, *in vivo* Properties and Proposed Clinical Use of Polyethylene-Modified Tissue Plasminogen Activator and Streptokinase," *Adv. Drug Deliv. Rev.*, 6:153–166.
111. Maeda, H. 1991. "SMANCS and Polymer-Conjugated Macromolecular Drugs: Advantages in Cancer Chemotherapy," *Adv. Drug Deliv. Rev.*, 6:181–202.
112. Sehon, A. H. 1991. "Suppression of Antibody Responses by Conjugates of Antigens and Monomethoxypoly(ethylene Glycol)," *Adv. Drug Deliv. Rev.*, 6:203–217.
113. Geiger, B., B. Von Specht and R. Arnon. 1977. "Stabilization of Human β-D-*N*-Acetylhexosaminidase A towards Proteolytic Inactivation by Coupling it to Poly(*N*-vinylpyrrolidone)," *Eur. J. Biochem.*, 73:141–147.
114. Von Specht, B. and W. Brendel. 1977. "Preparation and Properties of Trypsin and Chymotrypsin Coupled Covalently to Poly(*N*-vinylpyrrolidone)," *Biochim. Biophys. Acta*, 484:109–114.
115. Veronese, F. M., L. Sartore, P. Caliceti and O. Schiavon. 1990. "Hydroxy-Terminated Polyvinylpyrrolidone for the Modification of Polypeptides," *J. Bioact. Compatible Polym.*, 5:167–178.
116. Brucato, F. H. and S. V. Pizzo. 1990. "Catabolism of Streptokinase and Polyethylene Glycol-Streptokinase: Evidence for Transport of Intact Forms through the Biliary System in Mouse," *Blood*, 76(1):73–79.
117. Berger, H. and S. V. Pizzo. 1988. "Preparation of Polyethylene Glycol-Tissue Plasminogen Activator Adducts That Retain Functional Activity: Characteristics and Behavior in Three Animal Species," *Blood*, 71(6):1641–1647.
118. Abuchowski, A., J. R. McCoy, N. C. Palczuk, T. Van Es and F. F. Davis. 1977. "Effect of Covalent Attachment of Polyethylene Glycol on Immunogenicity and Circulating Life of Bovine Liver Catalase," *J. Biol. Chem.*, 252(11):3582–3586.
119. Abuchowski, A. and F. F. Davis. 1979. "Preparation and Properties of Polyethylene Glycol-Trypsin Adducts," *Biochim. Biophys. Acta*, 578:41–46.
120. Wieder, K. J., N. C. Palczuk, T. Van Es and F. F. Davis. 1979. "Some Properties of Polyethylene Glycol:Phenylalanine Ammonia-lyase," *J. Biol. Chem.*, 254(24): 12579–12587.
121. Matsushima, A., H. Nishimura, Y. Ashihara, Y. Yokota and Y. Inada. 1980. "Modification of *E. coli* Asparaginase with 2,4-bis(*O*-Methoxypolyethylene Glycol)-6-chloro-*S*-triazine (Activated PEG 2); Disappearance of Binding Ability towards Anti-Serum and Retention of Enzymic Activity," *Chem. Lett.*, 7:773–776.

122. Kamisaki, Y., H. Wada, T. Yagura, A. Matsushima and Y. Inada. 1981. ''Reduction in Immunogenicity and Clearance Rate of *Escherichia coli* L-Asparaginase by Modification with Monomethoxypolyethylene Glycol,'' *J. Pharmacol. Exp. Ther.*, 216:410–414.

123. Boccu, E., G. P. Velo and F. M. Veronese. 1982. ''Pharmacokinetic Properties of Polyethylene Glycol Derivatized Superoxide Dismutase,'' *Pharmacol. Res. Commun.*, 14(2):113–120.

124. Tomiya, N., K. Watanabe, J. Awaya, M. Kurono and S. Fujii. 1985. ''Modification of Acyl-Plasmin-Streptokinase Complex with Polyethylene Glycol,'' *FEBS Lett.*, 193(1):44–48.

125. Veronese, F. M., P. Caliceti, A. Pastorino, O. Schiavon and L. Sartore. 1989. ''Preparation, Physico-Chemical and Pharmacokinetic Characterization of Monomethoxypoly(ethylene Glycol)-Derivatized Superoxide Dismutase,'' *J. Controlled Release*, 10:145–154.

126. Katre, N. V., M. J. Knauf and W. J. Laird. 1987. ''Chemical Modification of Recombinant Interleukin-2 by Polyethylene Glycol Increases Its Potency in the Murine Meth A Sarcoma Model,'' *Proc. Natl. Acad. Sci.*, 84:1487–1491.

127. Vegarud, G. and T. B. Christensen. 1975. ''Glycosylation of Proteins: A New Method of Enzyme Stabilization,'' *Biotechnol. Bioeng.*, 17:1391–1397.

128. Marshall, J. J., J. D. Humphreys and S. L. Abramson. 1977. ''Attachment of Carbohydrate to Enzymes Increases Their Circulatory Lifetimes,'' *FEBS Lett.*, 83(2): 249–252.

129. Marshall, J. J. 1978. ''Manipulation of the Properties of Enzymes by Covalent Attachment of Carbohydrate,'' *Trends Biochem. Sci.*, 3:79–83.

130. Marshall, J. J. and M. L. Rabinowitz. 1976. ''Stabilization of Catalase by Covalent Attachment to Dextran,'' *Biotechnol. Bioeng.*, 18:1325–1329.

131. Melton, R. G., C. N. Wiblin, R. L. Foster and R. F. Sherwood. 1987. ''Covalent Linkage of Carboxypeptidase G_2 to Soluble Dextrans–I, Properties of Conjugates and Effects on Plasma Persistence in Mice,'' *Biochem. Pharmacol.*, 36(1):105–112.

132. Melton, R. G., C. N. Wiblin, A. Baskerville, R. L. Foster and R. F. Sherwood. 1987. ''Covalent Linkage of Carboxypeptidase G_2 to Soluble Dextrans–II, *in vivo* Distribution and Fate of Conjugates,'' *Biochem. Pharmacol.*, 36(1):113–121.

133. Wileman, T. E., M. Bennett and J. Lilleymann. 1983. ''Potential Use of an Asparaginase-Dextran Conjugate in Acute Lymphoblastic Leukemia,'' *J. Pharm. Pharmacol.*, 35:762–765.

134. Benbough, J. E., C. N. Wiblin, T. N. A. Rafter and J. Lee. 1979. ''The Effect of Chemical Modification of L-Asparaginase on Its Persistence in Circulating Blood of Animals,'' *Biochem. Pharmacol.*, 28:833–839.

135. Wileman, T. E., R. L. Foster and P. N. C. Elliot. 1986. ''Soluble Asparaginase-Dextran Conjugates Show Increase Circulatory Persistence and Lowered Antigen Reactivity,'' *J. Pharm. Pharmacol.*, 38:264–271.

136. Wileman, T. E. 1991. ''Properties of Asparaginase-Dextran Conjugates,'' *Adv. Drug Deliv. Rev.*, 6:167–180.

137. Schacht, E., F. Vandoorne, J. Vermeersch and R. Duncan. 1987. ''Polysaccharides as Drug Carriers, Activation Procedures and Biodegradation Studies,'' in *Controlled-Release Technology, Pharmaceutical Applications, ACS Symposium Series No. 348*, P. I. Lee and W. R. Good, eds., Washington, DC: American Chemical Society, pp. 188–200.

138. Schacht, E., R. Vercauteren and S. Vansteenkiste. 1988. "Some Aspects of the Application of Dextran in Prodrug Design," *J. Bioact. Compatible Polym.*, 3:72–80.

139. Vercauteren, R., D. Brunnel and E. Schacht. 1990. "Effect of the Chemical Modification of Dextran on the Degradation by Dextranase," *J. Bioact. Compatible Polym.*, 5:4–15.

140. Chaves, M. S. and F. Arranz. 1985. "Water-Insoluble Dextrans by Grafting, 2 Reactions of Dextrans with *N*-Alkyl Chloroformates. Chemical and Enzymatic Hydrolysis," *Makromol. Chem.*, 186:17–29.

141. Crepon, B., J. Jozefonvicz, V. Chytrý, B. Říhová and J. Kopeček. 1991. "Enzymatic Degradation and Immunogenic Properties of Derivatized Dextrans," *Biomaterials*, 12:550–554.

142. Edman, P. and I. Sjöholm. 1983. "Acrylic Microspheres *in vitro* VIII: Distribution and Elimination of Polyacryldextran Particles in Mice," *J. Pharm. Sci.*, 72(7): 796–799.

143. Edman, P., I. Sjöholm and U. Brunk. 1983. "Acrylic Microspheres *in vivo* VII: Morphological Studies on Mice and Cultured Macrophages," *J. Pharm. Sci.*, 72(6): 658–665.

144. Edman, P. and I. Sjöholm. 1982. "Treatment of Artificially Induced Storage Disease with Lysosomotropic Microparticles," *Life Sci.*, 30:327–330.

145. Heller, J., S. H. Pangburn and K. V. Roskos. 1990. "Development of Enzymatically Degradable Protective Coatings for Use in Triggered Drug Delivery Systems: Derivatized Starch Hydrogels," *Biomaterials*, 11:345–350.

146. Artursson, P., P. Edman, T. Laakso and I. Sjöholm. 1984. "Characterization of Polyacryl Starch Microparticles as Carriers for Proteins and Drugs," *J. Pharm. Sci.*, 73(11):1507–1513.

147. Laakso, T. and I. Sjöholm. 1987. "Biodegradable Microspheres X: Some Properties of Polyacryl Starch Microparticles Prepared from Acrylic Acid-Esterified Starch," *J. Pharm. Sci.*, 76(12):935–939.

148. Laakso, T., P. Artursson and I. Sjöholm. 1986. "Biodegradable Microspheres IV: Factors Affecting the Distribution and Degradation of Polyacryl Starch Microparticles," *J. Pharm. Sci.*, 75(10):962–967.

149. Artursson, P., P. Edman and I. Sjöholm. 1984. "Biodegradable Microspheres I: Duration of Action of Dextranase Entrapped in Polyacryl Starch Microparticles *in Vivo*," *J. Pharm. Exp. Ther.*, 231(3):705–712.

150. Laakso, T., P. Stjarnkvist and I. Sjöholm. 1987. "Biodegradable Microspheres. VI: Lysosomal Release of Covalently Bound Antiparasitic Drugs from Starch Microparticles," *J. Pharm. Sci.*, 76(2):134–140.

151. Stjarnkvist, P., L. Degling and I. Sjöholm. 1991. "Biodegradable Microspheres. XIII. Immune Response to the DNP Hapten Conjugated to Polyacryl Starch Microparticles," *J. Pharm. Sci.*, 80(5):436–440.

152. Mateescu, M. A. and H. D. Schell. 1983. "A New Amyloclastic Method for the Selective Determination of a-Amylase Using Cross-Linked Amylose as an Insoluble Substrate," *Carbohydr. Res.*, 124:319–323.

153. Hirano, S., H. Tsuchida and N. Nagao. 1989. "*N*-Acetylation in Chitosan and the Rate of Its Enzymic Hydrolysis," *Biomaterials*, 10:574–576.

154. Pangburn, S. H., P. V. Trescony and J. Heller. 1982. "Lysozyme Degradation of Partially Deacetylated Chitin, Its Film and Hydrogels," *Biomaterials*, 3:105–108.

155. Rubinstein, A., D. Nakar and A. Sintov. 1992. "Colonic Drug Delivery: Enhanced Release of Indomethacin from Cross-Linked Chondroitin Matrix in Rat Cecal Content," *Pharm. Res.*, 9(2):276–278.

156. Sintov, A., D. Nakar and A. Rubinstein. 1991. "Colonic Administration of Indomethacin Using Modified Chondroitin in a Cannulated Dog Model," *Proceed. Intern. Symp. Control. Rel. Bioact. Mater.*, 18:381–382.

CHAPTER 8

Biodegradable Drug Delivery Systems

Drug release can be controlled by several means, such as diffusion through a rate-controlling membrane or a matrix, osmosis, ion exchange, or degradation of a matrix or a part of a matrix. The biggest advantage of using biodegradation for controlling drug delivery is that the drug delivery device may not have to be removed from the site of action after drug delivery is completed. For systemic applications, bioresorbable polymers, which degrade to low molecular weight fragments and are eliminated from the body, are preferred. In topical applications, however, it may be desirable that polymers retain the high molecular weight of the degradation products to inhibit systemic absorption [1].

The degradation mode of the delivery systems is one of the factors controlling drug release. There are two different modes of biodegradation: surface and bulk degradations. The surface and bulk degradations are also called heterogeneous and homogeneous degradations, respectively. These two degradation modes are based on the location of polymer breakdown. In a bulk-degrading polymer, degradation occurs homogeneously throughout the bulk of the material. In a surface-degrading polymer, however, degradation is confined to the outer surface of a device [1,2]. The predominant mechanism of drug release from a bulk-degrading polymer is the simple diffusion of a drug occurring prior to or concurrent with the degradation of the polymer matrix [3]. On the other hand, drug release from a surface-degrading polymer is determined by the relative contribution between the drug diffusion and the degradation of the matrix [4]. In cases where drug molecules are covalently bound to the polymer chains, drug release is determined mainly by the degradation of labile bonds between the drug and the polymer backbone regardless of the degradation mode of the polymer matrix.

In biodegradable hydrogels, drugs are usually in contact with water and thus the drug solubility is an important factor in drug release. The release of drugs with appreciable water solubility will be rapid and independent of the matrix degradation rate. Thus, in general, hydrogels may not be suitable for the controlled release of most low molecular weight, water-soluble drugs. The biodegradable hydrogel systems, however, are useful for the delivery of macromolecular drugs, such as peptides and proteins, which are entrapped in the gel network until the gel is degraded.

The mathematical expressions for drug release from biodegradable hydrogels are not well established yet due to the high complexity involved in the modeling. To gain insight into the mechanisms of drug release from biodegradable hydrogels, however, we will briefly review mathematical expressions derived for diffusion-controlled and swelling-controlled drug release, which are well documented in the literature [5–8]. Certainly, these mathematical treatments of nondegrading devices can be used for the description of drug release from biodegradable hydrogel systems, if the duration of drug release is considerably shorter than the lifetime of the biodegradable polymer [3,7]. In addition, models describing drug release from nonswelling, biodegradable devices are presented. These expressions may be used for describing drug release from hydrogel systems which rapidly reach equilibrium swelling in an aqueous medium.

8.1 MECHANISMS OF DRUG RELEASE

8.1.1 Diffusion-Controlled Systems

Diffusion-controlled systems can be divided into reservoir devices and monolithic devices [9,10]. These are systems which do not change dimensions or physicochemical properties during drug release. In reservoir systems, the drug core is encapsulated by an inert membrane. Reservoir devices have the advantage of providing a constant rate of release over a substantial portion of their lifetime. The release rate remains constant as long as drug solubility is maintained by the presence of excess drug. Mathematical modeling of drug diffusion through a rate-limiting membrane has been well documented [5,7,11–14]. In the monolithic device, the drug is dispersed or dissolved in an inert polymer. Slab, cylindrical, and spherical configurations are commonly used for controlled release devices and the kinetics of diffusional release from these devices can be correlated with theoretical models [15–18]. As in reservoir systems, drug diffusion through the polymer matrix is the rate-limiting step. The release rates are determined by the choice of polymer and the diffusion and partition coefficients of the drug in the polymer [7,9].

In a monolithic solution where the drug is dissolved in the polymer medium, the release of the drug follows Fick's law. In the case of slab geometry, drug release is described by the following equation [7,19,20]:

$$M_t/M_o = 4(Dt/\pi h^2)^{1/2} \tag{8.1}$$

where M_t is the amount of drug released in time t, M_o is the total mass of the drug incorporated into the device, D is the diffusion coefficient of a drug, π is 3.14, and h is the thickness of the device. Sometimes M_o is replaced by M_∞, the cumulative amount of drug released as time approaches infinity. Equation (8.1) is an early time approximation which holds for the release of the first 60 percent of the total drug (i.e., $0 \leq M_t/M_o \leq 0.6$). The late time approximation, which holds for the final portion of the drug release, is described by following equation:

$$M_t/M_o = 1 - (8/\pi^2)\exp[(-\pi^2 Dt)/h^2] \tag{8.2}$$

for $0.4 \leq M_t/M_o \leq 1.0$

Mathematical solutions for the release of a drug from the devices of other geometries can be found in the literature [6,7,15–18].

In monolithic dispersions where excess drug is dispersed, the drug release profile varies not only with the geometry of the device but also with the loading dose [6,15,16,21,22]. When the drug concentration is less than 5 vol percent, the drug release from the slab device can be described by the Higuchi model [17,21,22]

$$\begin{aligned} M_t &= A[DC_s(2C_o - C_s)t]^{1/2} \\ &= A(2DC_sC_o\,t)^{1/2} \end{aligned} \tag{8.3}$$

for $C_o >> C_s$

where M_t is the amount of drug released in time t, A is the total area of the slab, D is the diffusion coefficient of a drug in the polymer, C_s is the drug solubility in the polymer matrix, and C_o is the initial total drug concentration which includes both dissolved and dispersed drugs.

At higher drug loadings where C_o is in the range of 5–15 vol percent, the drug release profile deviates from the expected one due to the formation of fluid-filling cavities created by dissolution of drug particles near the surface. The presence of cavities leads to the increase in the system's apparent permeability to the drug. For these systems the drug release is described by the modified Higuchi equation [7]

$$M_t = A[2DC_sC_o\,t(1 + 2C_o/\varrho)/(1 - C_o/\varrho)]^{1/2} \tag{8.4}$$

where ϱ is the density of the permeant, i.e., water.

At drug loadings of approximately 15–20 vol percent, all the drug particles dispersed in the polymer matrix are in contact with one another. In this case, the drug is released through the pores formed by dissolution of the drug. Thus, the partition coefficient is replaced by porosity (ϵ) which is the volume fraction of the pores in the membrane filled with fluid. D is replaced by D_w, which is the diffusion coefficient of the drug in the aqueous pores. Since the formed pores are not straight, drug molecules have to diffuse a longer distance than the thickness of the membrane. This means that D_w is reduced by a factor of τ, the tortuosity of the membrane. Thus, the drug release from the slab is described by [7,15,23]

$$M_t = A[2(\epsilon/\tau)D_w C_w C_o t]^{1/2} \tag{8.5}$$

where C_w is the solubility of the drug in water. The tortuosity, τ, reflects the average entire distance that the drug must diffuse to be released from the device.

8.1.2 Swelling-Controlled Systems

In the reservoir and monolithic systems described above, it was assumed that their dimensions and physical properties did not change during the release process. In hydrogel systems, however, absorption of water from the environment changes the dimensions and/or physical properties of the system and thus the drug release kinetics. A model based on the work of Alfrey et al. [24] describes the swelling membrane which consists of three zones. Adjacent to the bulk water is a layer of completely swollen gel. Then there is a fairly thin layer in which the polymer chains are slowly hydrating and relaxing. The third zone is a matrix of unswollen, completely dried, rigid polymer. The diffusion of permeant (i.e., water in hydrogels) was classified into three different types based on the relative rates of diffusion and polymer relaxation [24]. They are:

(1) Case I or Fickian diffusion occurs when the rate of diffusion is much less than that of relaxation.
(2) Case II diffusion (relaxation-controlled transport) occurs when diffusion is very rapid compared with the relaxation process.
(3) Non-Fickian or anomalous diffusion occurs when the diffusion and relaxation rates are comparable.

The significance of this classification is that both Case I and Case II diffusion can be described in terms of a single parameter. The diffusion in Case I systems is described by the diffusion coefficient. Case II diffusion is described by the constant velocity of an advancing water front which marks the boundary between the swollen hydrogel and the glassy core. Non-Fickian or anomalous systems need two or more parameters to describe the release resulting from both diffusion and polymer relaxation.

The above classification of the diffusion of permeant can also be used to classify the drug release profiles from the swelling polymer [25]. The diffusion coefficient of a drug in the dried hydrogel is very low initially, but it increases significantly as the gel absorbs water. Thus, the drug release from the device is a function of the rate of water uptake from the environment as well as the drug diffusion. For such a device, it would be difficult to predict the effect of the water uptake rate on the rate of drug release. The drug release rate is governed by the dependence of the diffusion coefficient of the particular drug on the water content in the polymer.

8.1.2.1 CASE I OR SIMPLE FICKIAN DIFFUSION

When the drug is loaded into the hydrogel by equilibrium swelling in the drug solution, drug release from the swollen gel follows Fick's law. Thus, the rate of drug release from the equilibrated slab device can be described by Equation (8.1) [7,20,26]. It is noted again that drug release from Case I systems is dependent on $t^{1/2}$.

8.1.2.2 CASE II AND ANOMALOUS DIFFUSION

In Case II systems, diffusion of water through the previously swollen shell is rapid compared with the swelling-induced relaxations of polymer chains. Thus, the rate of water penetration is controlled by the polymer relaxation. For film specimens, the swelling zone moves into the membrane at a uniform rate and the weight gain increases in direct proportion to time [24,25]. If the hydrogel contains a water-soluble drug, the drug is essentially immobile in a glassy polymer, but begins to diffuse out as the polymer swells by absorbing water. Thus, drug release depends on two simultaneous rate processes, water migration into the device and drug diffusion through the swollen gel. The mathematical modeling of the water absorption and drug diffusion through continuously swelling hydrogels is highly complicated, especially in the case of non-Fickian transport. Although there are a number of reports dealing with the mathematical modeling of drug release from swellable polymeric systems [27–32], no single model successfully predicts all the experimental observations.

Since most complex models do not yield a convenient formula and require numerical solution techniques, the generalized empirical equation has been widely used to describe both the water uptake through the swellable glassy polymers and the drug release from these devices. In the case of water uptake, the weight gain, M_s, is described by the following empirical equation [24,33]:

$$M_s = kt^n \tag{8.6}$$

where k and n are constant. Normal Fickian diffusion is characterized by

$n = 0.5$, while Case II diffusion by $n = 1.0$. A value of n between 0.5 and 1.0 indicates a mixture of Fickian and Case II diffusion, which is usually called non-Fickian or anomalous diffusion [24].

Hopfenberg and Hsu [34] observed that a glassy polymer which absorbed liquid at a constant rate (Case II transport) also released a drug at a constant rate. They proposed that the mechanism was analogous to that of surface eroding delivery systems in which drug delivery is controlled by the chemical relaxation such as hydrolysis or dissolution of the polymer. In Case II transport systems, drug release was controlled by the rate of polymer chain relaxation. The polymer chain relaxation was described as a "phase erosion" compared to a "mass erosion" in the surface-eroding systems [34]. Ritger and Peppas showed that the above power law expression can be used for the evaluation of drug release from swellable systems [20,26]. In this case, M_t/M_∞ replaces M_s in Equation (8.6) to give

$$M_t/M_\infty = kt^n \tag{8.7}$$

where M_t/M_∞ is the fractional release of the drug in time t, k is a constant characteristic of the drug-polymer system, and n is the diffusional exponent characteristic of the release mechanism. A more detailed description of solvent swelling and Case II diffusion can be found in the reviews by Hopfenberg et al. [34,35] and Windle [36].

For Fickian release from a slab, fractional release can be characterized by some constant multiplied by the square root of time. This resembles Equations (8.1) and (8.3). For Case II transport, the fractional release is a linear function of time until the two penetration fronts meet at the center of the slab. Many diffusion processes from swellable polymers show n values changing sigmoidally between 0.5 and 1.0. Thus, the following semi-empirical expression was suggested by Alfrey et al. [24]

$$M_s = k_1 t + k_2 t^{1/2} \tag{8.8}$$

which includes expressions for both Fickian and Case II transport mechanisms. Equation (8.8) was adapted to describe some drug release processes from swellable polymers [20].

8.1.3 Biodegradable Systems

During the past several years many biodegradable drug delivery systems have been developed. Biodegradable systems can be divided into surface-degrading and bulk-degrading systems. Except for a few surface-degrading systems such as polyorthoesters [2,36,37] and polyanhydrides [39,40,41], most systems degrade by a combination of the two mechanisms.

8.1.3.1 DEGRADATION-CONTROLLED MONOLITHIC SYSTEMS

In degradation-controlled monolithic systems, the drug is dispersed uniformly throughout the polymer matrix, and diffusion of the drug is slow compared with polymer degradation. For surface-degrading systems, drug release is affected by the surface-to-volume ratio and the geometry of the device. The effect of the shape of the device on the dissolution of pharmaceutical tablets is well known [42–45]. Here we will briefly describe an example of drug release from surface-degrading devices following the analysis of Hopfenberg [46].

If M_t is the amount of drug released from a sphere of radius r in time t, then the kinetic expression describing release would be given by:

$$dM_t/dt = k_o 4\pi r^2 \tag{8.9}$$

where k_o is a rate constant. Since the rate-determining erosion process occurring at position r is carried out at an ever decreasing radius, the drug release rate, dM_t/dt decreases continuously. From the mass balance, M_t is also given by:

$$M_t = (4\pi/3)C_o[a^3 - r^3] \tag{8.10}$$

where C_o is the uniform initial concentration of drug dispersed throughout the device and a is the initial radius at time $t = 0$. Substitution of Equation (8.10) into Equation (8.9) and differentiation gives the following relationships:

$$r = a - (k_o/C_o)t \tag{8.11}$$

Substituting Equation (8.11) into Equation (8.10) yields

$$M_t = (4\pi/3)C_o\{a^3 - [a - (k_o/C_o)t]^3\} \tag{8.12}$$

and since $(4\pi/3)C_oa^3$ is equal to M_∞, which is the cumulative amount of drug released as time approaches infinity, we have

$$M_t/M_\infty = 1 - [1 - (k_o t)/(C_o a)]^3 \tag{8.13}$$

Similar derivation for a cylinder of radius a and length L provides the following equation:

$$M_t/M_\infty = 1 - [1 - (k_o t)/(C_o a)]^2 \tag{8.14a}$$

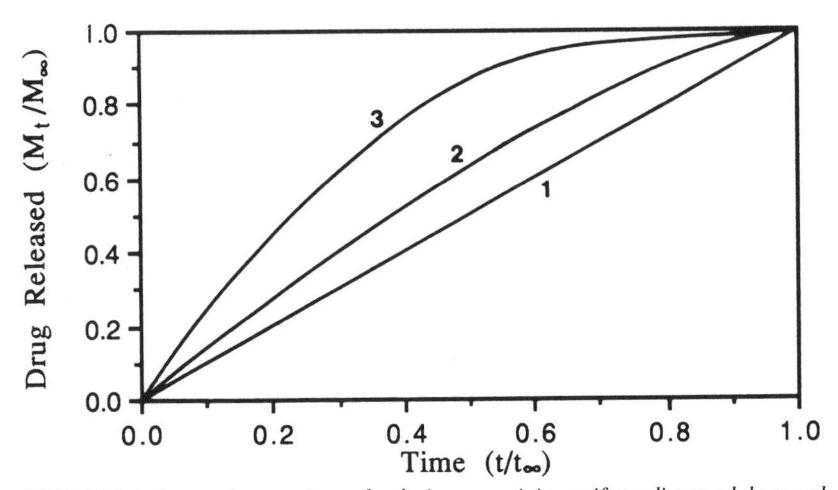

FIGURE 8.1. Drug release patterns for devices containing uniform dispersed drug and degrading by a surface degradation mechanism; slab (1), cylinder (2), and sphere (3) (from Reference [47]).

and the equivalent expression for a slab with a thickness of $2a$ would be

$$M_t/M_\infty = 1 - [1 - (k_o\,t)/(C_o\,a)]^1$$

$$= k_o\,t/(C_o\,a) \qquad (8.14b)$$

Plots of fraction of drug released from sphere, cylinder, and slab against time are given in Figure 8.1. It is apparent that only slab provides a zero-order release.

8.1.3.2 DIFFUSION-CONTROLLED MONOLITHIC SYSTEMS

A detailed understanding of drug release from a matrix undergoing bulk degradation (without swelling) is difficult because drug release occurs by a combination of diffusion and degradation. If the drug diffuses rapidly from the device relative to the degradation of the polymer, the drug release is governed by the equations derived for simple diffusion-controlled systems. The situation becomes more complex if the drug is released during bulk degradation, since the matrix is gradually loosened by degradation. In this case, the drug permeability through the polymer will increase with time, offsetting the decrease in concentration gradient. The following is a model used by Baker for estimation of the permeability changing with time [47].

The release of dispersed drug from nondegradable devices is described by the following Higuchi equation:

$$M_t = A(2DC_sC_o\, t)^{1/2} \qquad (8.3)$$

for $C_o >> C_s$

$$= A(2PC_o\, t)^{1/2} \qquad (8.15)$$

where P is the drug permeability ($P = DC_s$). If the matrix is gradually loosened by degradation, the drug permeability will increase with time and such an increase will offset the decline of drug release. If it is assumed that the fractional increase in permeability is inversely proportional to the number of intact (undegraded) bonds in the polymer matrix, then

$$P/P_o = (\text{initial number of bonds})/(\text{number of bonds remaining})$$

$$= N/(N - Z) \qquad (8.16)$$

where P_o is the original permeability and P is the permeability after Z cleavages of polymer chains. If we further assume that bond cleavage follows simple first-order kinetics, then

$$dZ/dt = k(N - Z) \qquad (8.17)$$

where k is the first-order rate constant. This leads to

$$\ln\,[N/(N - Z)] = kt \qquad (8.18)$$

Substitution of Equations (8.16) and (8.18) into Equation (8.15) yields

$$M_t = A(2P_o\, e^{kt} C_o\, t)^{1/2} \qquad (8.19)$$

Equation (8.19) is plotted in Figure 8.2. The polymer degradation causes the drug permeability to increase. Figure 8.2 also shows the diffusional release of drug in the absence of erosion, which slowly declines and follows the $t^{1/2}$ kinetics. Figure 8.3 shows an example of hydrocortisone release from a degradable polyester device. It is noted that drug release was accelerated as a result of the increase in permeability above the initial value by matrix degradation. If the matrix disintegrates before drug depletion, a large burst in drug release will be observed [48].

8.1.3.3 BIODEGRADABLE HYDROGEL SYSTEMS

Information on the degradation kinetics of biodegradable hydrogels and mathematical modeling of drug release from degrading hydrogels is rather limited at this time. A detailed understanding of drug release from a

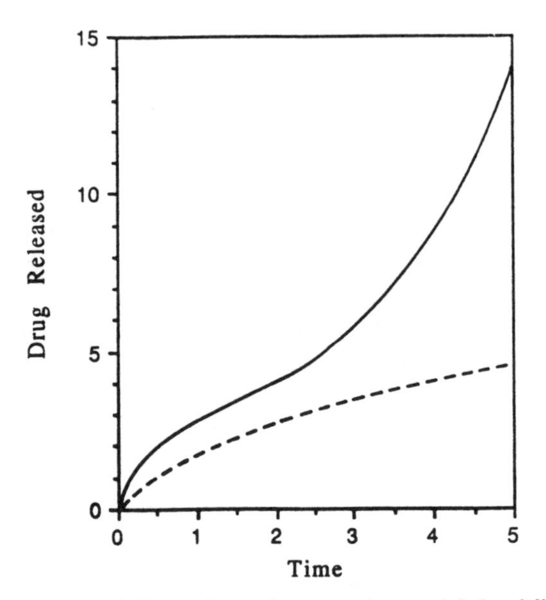

FIGURE 8.2. Theoretical drug release from a polymer slab by diffusion, and by combined erosion and diffusion; erosion and diffusion (———), and diffusion (- - - - -) (from Reference [49]).

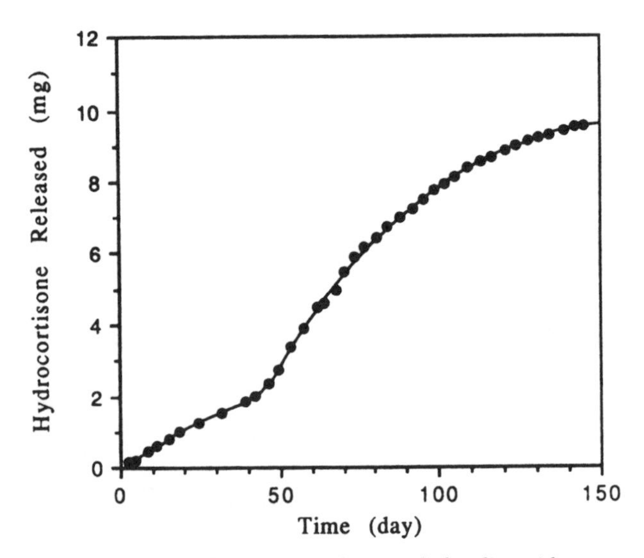

FIGURE 8.3. Release of hydrocortisone from a diglycolic acid trans-cyclohexane dimethanol polyester (from Reference [49]).

hydrogel undergoing a homogeneous degradation process is difficult. This is because the degradation and release characteristics of most degradable hydrogel systems depend on a complex, interrelated set of events. They include, but are not limited to, chemical reaction, multicomponent diffusion, and physical changes in the system such as tortuosity and porosity. During the degradation process, the drug permeability continuously changes as a result of the combination of all these interrelated events and thus, it is often difficult to assess the overall performance [50]. A generalized mathematical model which incorporates all the important events will be useful in the design of biodegradable hydrogel drug delivery systems. Recent rapid growing interest in the delivery of peptides and proteins will undoubtedly attract more attention to biodegradable hydrogel drug delivery systems in the near future.

Many studies have shown that drug release is influenced by various formulation variables and/or physicochemical properties of the components in biodegradable drug delivery systems. In addition to the rate of polymer degradation, drug release is affected by the physical parameters of the polymer such as the water content, degree of crosslinking, crystallinity, and phase separation [51]. The physicochemical properties of the drug, particularly its solubility in the polymer and in the aqueous medium, and the amount of drug loading are also expected to have profound effects on the release characteristics of the drug-polymer composite. For these biodegradable systems the release rate can be achieved by adjusting these parameters.

8.2 DRUG RELEASE FROM HYDROGELS WITH BIODEGRADABLE BACKBONES

8.2.1 Hydrogels Based on Protein/Polypeptides

8.2.1.1 ALBUMIN BEADS

Albumin microspheres have been used as drug carriers for studying passive as well as active drug targeting [52–54]. The factors influencing the release of drugs from microspherical carriers have been discussed in detail by Tomlinson et al. [55] and more recently by Gupta and Hung [56]. They are listed in Table 8.1. Effects of those factors on the release of drugs from albumin microspheres have been examined by various investigators [53,54,57–62].

Ever since the use of albumin microspheres as drug delivery devices by Kramer [60], many different therapeutic and/or diagnostic agents have been incorporated into albumin microspheres. The characteristics of various albumin microspheres investigated so far have been described in detail by Gupta and Hung [56].

TABLE 8.1. Factors Affecting the Drug Release from Microspheres [55,56].

1.	Method of drug incorporation (affects the amount and the position of the drug in the microsphere)
2.	Size and density of the microsphere
3.	Extent and nature of crosslinking
4.	Physicochemical properties of the drug
5.	Presence of adjuvants in the microsphere
6.	Interaction between the drug and the matrix material
7.	Concentration of the matrix material
8.	Release environment, such as the presence of enzymes

8.2.1.1.1 Drug Release Kinetics

Some studies have shown that the release of drugs from albumin particles follows first-order kinetics [63–65]. The release of triamcinolone diacetate from crosslinked human serum albumin nanoparticles was described by the following first-order equation [63]:

$$\ln M = \ln M_o - kt \tag{8.20}$$

where M_o is the original amount of drug present in the microspheres, M is the amount of drug remaining in the microspheres, k is the first-order release rate constant, and t is the time during which drug release occurs. Many other studies, however, have shown that the *in vitro* release of low molecular weight drugs from monolithic microspheres is biphasic with an initial burst release followed by a much slower release [55,66–68]. Thus, drug release was described using the following biexponential equation [55]:

$$M = Ae^{\alpha t} + Be^{\beta t} \tag{8.21}$$

where M is determined by the two constants A and B, which are zero-time intercepts for the initial and terminal phases, and α and β are the apparent first-order initial and terminal release rate constants, respectively. Figure 8.4 shows the biphasic release of sodium cromoglycate (SCG) from albumin microspheres. The mass balance calculations showed that 88 percent of the drug originally incorporated in the microspheres was released within 10 minutes. After this burst, release continued to be first order but at a much slower rate. The burst effect can be reduced by repeated washings or ultrasonication of the spheres in isotonic saline solution for a few minutes, or by increasing the crosslinking density of the albumin microspheres [55]. Drug release at the terminal phase can be increased by using microspheres of a smaller size due to the larger surface area available for release and the shorter path length for diffusion [55].

The Higuchi equation [Equation (8.3)] was also used to fit the release of drugs from albumin microspheres. The Higuchi equation was able to fit the release data of 1-norgestrel [69] and progesterone [70] adequately. In most cases, however, it was useful only for the initial or partial phase of drug release [65,71,72].

Gupta et al. [68] showed that the *in vitro* release of adriamycin from albumin microspheres could be described by the biphasic zero-order model as shown in Figure 8.5. In their study, the drug-loaded albumin microspheres were washed four times and this might have reduced the drug release in the burst phase. In the sink condition, the decrease in M, the amount of drug in the microspheres, can be written as

$$dM/dt = -ADC_m/h \tag{8.22}$$

where A is the exposed surface area, D is the drug diffusivity in the matrix, C_m is the drug concentration inside the microspheres, and h is the thickness of the diffusion layer. The increase in drug diffusivity and particle surface area resulting from particle hydration was thought to be compensated for by an increase in the diffusion layer thickness and depletion of the drug.

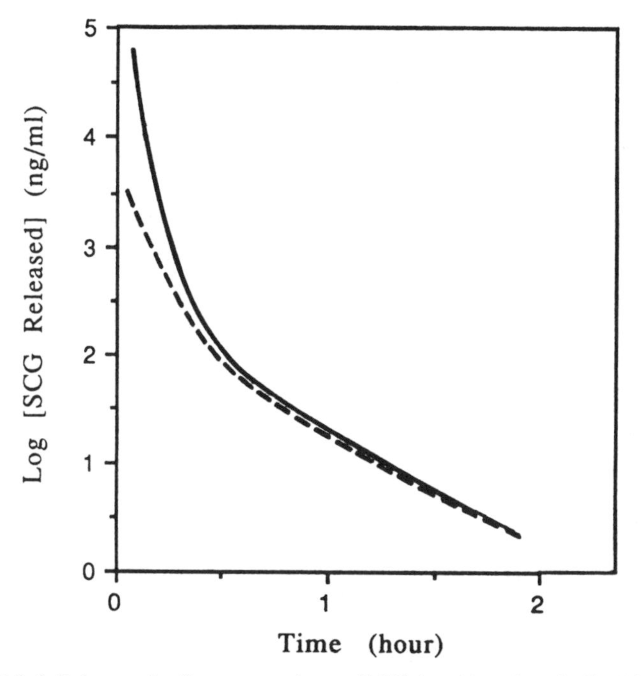

FIGURE 8.4. *Release of sodium cromoglycate (SCG) into phosphate buffer (pH 7) from HSA microspheres stabilized by crosslinking with 5 percent glutaraldehyde for 18 hours; normal treatment (———), and after ultrasonication (- - - - - -) (from Reference [55]).*

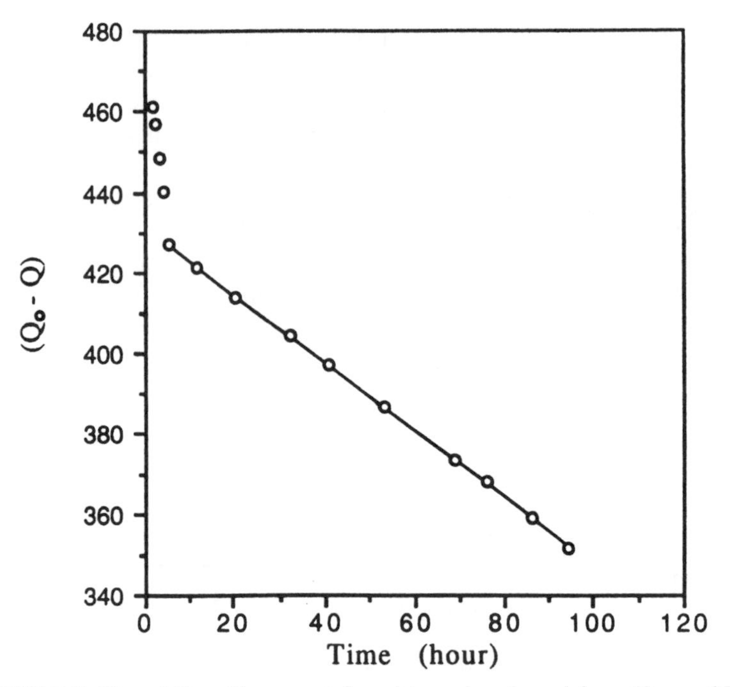

FIGURE 8.5. Plot of $(Q_0 - Q)$ *versus* t *for adriamycin released from 40 mg of BSA microspheres (from Reference [68]).*

This situation yielded a constant value for Equation (8.22) and thus a zero-order release.

In general, any drug release from albumin microspheres can be described by the bi- or triphasic model [68,73]. The initial fast release phase is due to the release of surface-associated drug and particle hydration, while the terminal slow release phase or phases are primarily due to the release of entrapped drug [67].

8.2.1.1.2 In vivo *Delivery of Low Molecular Weight Drugs*

Albumin beads were used to deliver various drugs such as antitumor agents [53,54,59,63–65,73–76], steroids [59,66,70], and proteins [69,77–79]. In most studies the serum drug levels after administration of microspheres exhibited an initial burst effect followed by a more protracted release phase. This initial rapid release appeared to be a common feature of albumin microsphere systems for both *in vitro* and *in vivo* drug release.

Lee et al. [70] administered progesterone-containing albumin beads crosslinked with glutaraldehyde into rabbits by intramuscular or subcutaneous injections. The serum concentration vs. time profile showed the initial burst release of the drug. Subsequently, the drug was released at a near constant rate for 30 days. This was probably due to degradation of the albumin matrix by intrinsic proteolytic enzymes as well as to diffusion of progesterone from the matrix. Similar *in vivo* drug release profiles were observed in other studies [66,76].

Burgess and Davis investigated the absorption of radiolabeled prednisolone or triamcinolone from heat-stabilized albumin microspheres following intraarticular and intramuscular administration [66]. As shown in Figure 8.6, biphasic first-order kinetics was observed. The initial absorption phase was dependent on the formulation, while the second sustained release phase was independent of the formulation. The drug absorption rate was dependent on both the site of injection and drug preparation. Triamcinolone, the lipophilic drug, was absorbed more slowly from the intramuscular site than from the intraarticular site.

8.2.1.1.3 Delivery of Macromolecular Drugs

A vast majority of drugs incorporated into albumin microspheres are low molecular weight drugs. There is a need to deliver high molecular weight drugs such as polypeptides and proteins which have very low bioavailability

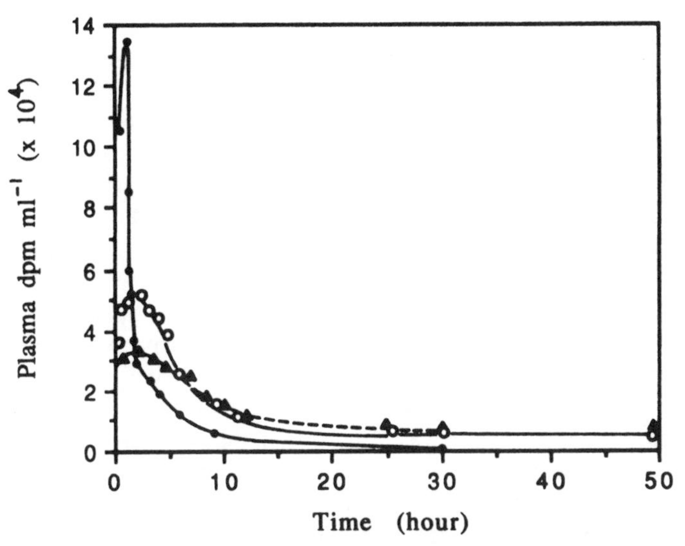

FIGURE 8.6. *Plasma radioactivity-time profiles following intraarticular injection of [³H]prednisolone into NZW rabbits; suspension (●), microsphere-stabilized for 10 hours at 150°C (○), and microsphere-stabilized for 22 hours at 150°C (▲) (from Reference [66]).*

after oral administration, and short plasma half-lives after injection. The large size and high hydrophilicity of proteins make diffusion through hydrophobic polymer membranes impractically slow. Thus, hydrogel systems and biodegradable polymers were explored for the delivery of proteins [69,77–79].

Goosen et al. prepared implantable, biodegradable albumin microbeads containing insulin [77]. Insulin was incorporated in the albumin microbeads by physical entrapment during crosslinking of the beads with glutaraldehyde. The efficacy of the insulin in the microbeads was measured in diabetic rats. The blood insulin in treated rats was sustained at levels between 10 and 67 μU over a 60-day period, while complete biodegradation of the microbeads required longer than five months. It appears that the insulin incorporated into the albumin beads during glutaraldehyde crosslinking was bioactive. It should be noted, however, that glutaraldehyde is a nonspecific crosslinking agent and thus the incorporated protein drugs may be crosslinked intra- and/or intermolecularly during the crosslinking process. Royer et al. [69] improved the system by protecting amine groups of insulin before glutaraldehyde crosslinking. Prior to encapsulation, amine groups of insulin were citraconylated at pH 8–10 to protect them from nonspecific glutaraldehyde crosslinking, and deprotection was accomplished by soaking the insulin-albumin beads in an acidic solution. It was shown that the insulin was released very rapidly *in vitro*. Almost 60 percent of the insulin was released in the first five minutes. A second release phase lasted for nine hours.

Recently, Kwon et al. [79] used glutaraldehyde-crosslinked albumin-heparin microspheres for the delivery of dextran (weight average molecular weight of 17,200), a model macromolecule. Dextran was loaded either during microsphere preparation or after microsphere preparation. In both loading conditions, *in vitro* release of dextran was characterized by an initial fast release followed by a much slower release. If dextran was loaded (up to 7 percent) during crosslinking, about one-third of the incorporated dextran was released in the first hour, and more than half of the dextran was retained in the microspheres even after 50 hours. In contrast, for microspheres where dextran was loaded after microsphere preparation through the use of the high swelling property of albumin-heparin microspheres at high pH and low ionic strength medium, a greater fraction of the dextran was released and the release was faster than in the other case. However, the total amount of dextran loaded using this method was low, only around 0.5 wt percent. Protein loading during microparticle preparation may result in the loss of a large fraction of the incorporated protein and the uncertainty of reservation of bioactivity during the multiple preparation process. Protein loading after microparticle preparation will significantly reduce the amount of incorporated protein and result in a faster release.

8.2.1.2 GELATIN MICROSPHERES

Gelatin is known to be a nontoxic, biodegradable, natural polymer with low antigenicity. Simple coacervation with gelatin has been known for many years and has been studied as a means of encapsulating various pharmaceuticals and chemicals. Madan et al. prepared a reservoir type device with gelatin by simple coacervation using sodium sulfate as a coacervating agent and formaldehyde as a hardening agent [80]. Droplets of clofibrate, a liquid hypocholesterolemic agent in the form of monodisperse spheres, were added to the 10 percent gelatin solution under constant stirring. Formed microcapsules were hardened in the chilled sodium sulfate solution and further crosslinked with formaldehyde. The *in vitro* drug release from these uniform microparticles was zero order between 10 percent and 90 percent of release, irrespective of the effect of hardening as shown in Figure 8.7. The release rates of the microcapsules were related directly to the hardening times of the microcapsules.

Gelatin microspheres are usually prepared by crosslinking with formaldehyde or glutaraldehyde [81,82]. Drug release from gelatin microparticles can be controlled by the relative amount of gelatin employed in making gelatin-drug microparticles and by the crosslinking density in the system [83 – 85]. The release of some drugs from gelatin particles followed first-order kinetics [81,86,87]. *In vitro* release of sulfonamides (sulfanilamide, sulfisomidine, sulfamethizole) from gelatin microparticles was first order regardless of drug content up to 67 wt percent [86].

Tabata et al. prepared glutaraldehyde-crosslinked gelatin microspheres for targeting interferon (IFN) to macrophages [84]. Interferon was incorporated into the gelatin microspheres by crosslinking gelatin in the presence of the drug. The rate of IFN release from the microspheres was found to be dependent on the extent of crosslinking. Incubation in phosphate buffered saline solution (PBS) without collagenase did not result in degradation of gelatin microspheres or in IFN release within two days of observation. This indicates that IFN was released only by degradation of the gelatin backbone. As the amount of glutaraldehyde was increased from 0.03 to 1.33 mg/mg gelatin, the percentage of degraded microspheres decreased from 98 percent to 60 percent, and IFN release from the microspheres into the collagenase-containing buffer in 24 hours correspondingly decreased from 100 percent to 20 percent (Figure 8.8). When microspheres containing ^{125}I-labeled IFN were added to the suspension of mouse peritoneal macrophages, the microspheres were phagocytized and degraded gradually in the interior of the macrophages. This resulted in the slower release of the incorporated IFN in the cells compared to the *in vitro* release. The study showed that the IFN release rates, both *in vitro* and *in vivo*, were related to the glutaraldehyde-crosslinking density, and thus the release of IFN could be controlled by

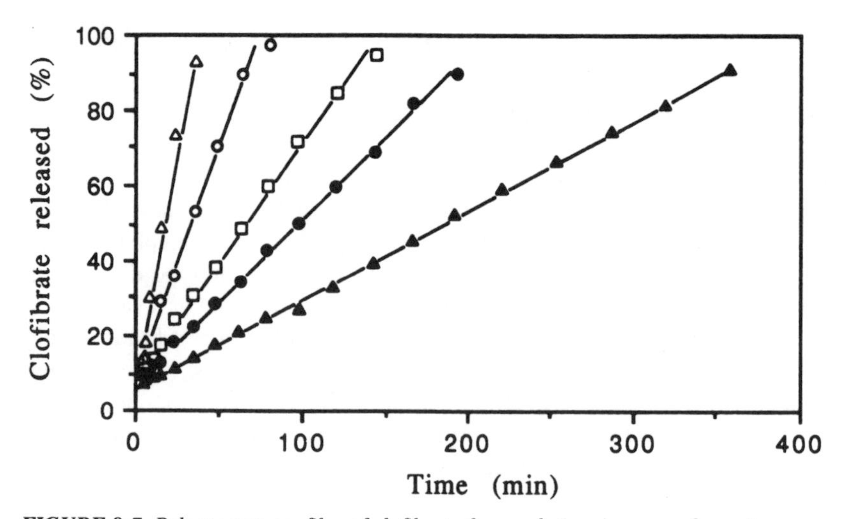

FIGURE 8.7. Release rate profiles of clofibrate from gelatin microcapsules; microcapsules hardened for 1 hour (○), microcapsules hardened for 2 hours (□), microcapsules hardened for 4 hours (●), microcapsules hardened for 8 hours (▲), and unhardened microcapsules (△) (from Reference [80]).

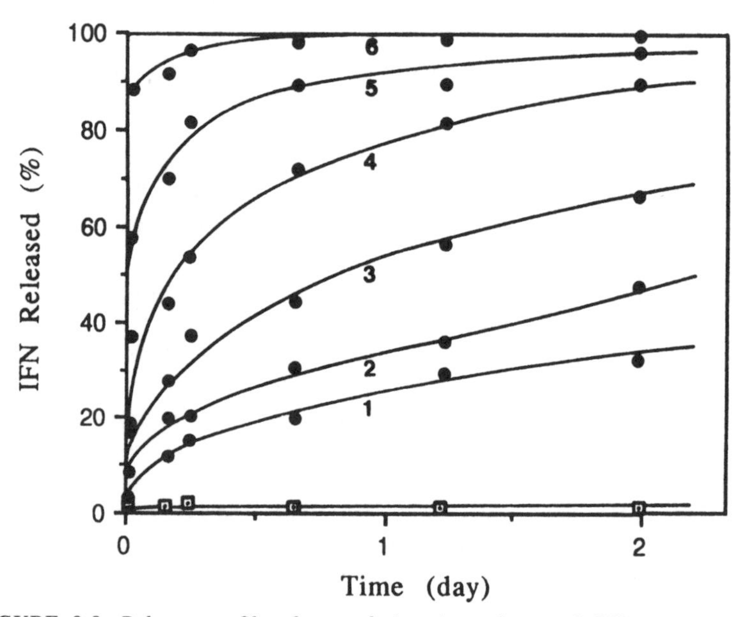

FIGURE 8.8. Release profiles from gelatin microspheres of different extents of crosslinking in the presence of collagenase; amount of glutaraldehyde added (mg/mg gelatin) was 1.33 (1), 0.71 (2), 0.28 (3), 0.14 (4), 0.05 (5), and 0.03 (6). Release was measured either in PBS (⊡) or in PBS-containing collagenase (●) (from Reference [84]).

changing crosslinking density with glutaraldehyde. It should be noted again, however, that IFN may be crosslinked intra- and intermolecularly during the crosslinking process. Thus, the bioactivity of the released IFN must be demonstrated.

8.2.2 Hydrogels Based on Polysaccharides

8.2.2.1 MODIFIED STARCH HYDROGELS

Of the many polysaccharides, starch has been used most widely as a drug delivery system. Starch is usually derivatized (e.g., with glycidyl acrylate) to introduce acryloyl groups which are necessary for polymerization into microspheres. The starch microspheres are biocompatible and readily degradable in biological fluids by amyloglycosidase or α-amylase, if the extent of derivatization is low [88–90]. When the particles are incubated in the lysosomal fraction, their degradation is dependent on the extent of derivatization. When 0.2 wt percent (final concentration) of glycidyl acrylate is used to derivatize starch, microspheres are rapidly broken down and about 30 percent of the initial radioactivity is found in the microspheres after six hours of incubation [90].

Sjöholm and his coworkers investigated the potential of injectable starch microspheres for drug targeting to the reticuloendothelial system (RES) [90–95,97]. Macromolecules, such as lysozyme [90], human serum albumin (HSA) [90,96], carbonic anhydrase [90], and immunoglobulin G [90], were entrapped in the starch microparticles and targeted to the lysosomes of the RES. When injected intravenously into mice, microspheres were found to be phagocytized by macrophages of the RES. The residence time of the spheres in central circulation was about several minutes [91]. The studies showed that the degradation rate of starch microspheres, and thus the release of the content, could be controlled by adjusting the degree of starch modification [88,93].

Starch has also been crosslinked with di- or multivalent cations, such as calcium ions, capable of reacting with the hydroxyl groups on starch molecules. The formation of calcium crosslinks tightens the starch matrix and therefore affects its degradation rate as well as the diffusion of molecules through the matrix [98]. Kost et al. examined the release of low and high molecular weight drugs from the calcium-crosslinked starch matrix in simulated gastric and intestinal solutions containing α-amylase [96]. The release of macromolecules, such as myoglobulin and bovine serum albumin, was affected by the amylase activity, while the release of salicylic acid was not (Figure 8.9). The release of albumin was degradation controlled, while that of salicylic acid was diffusion controlled.

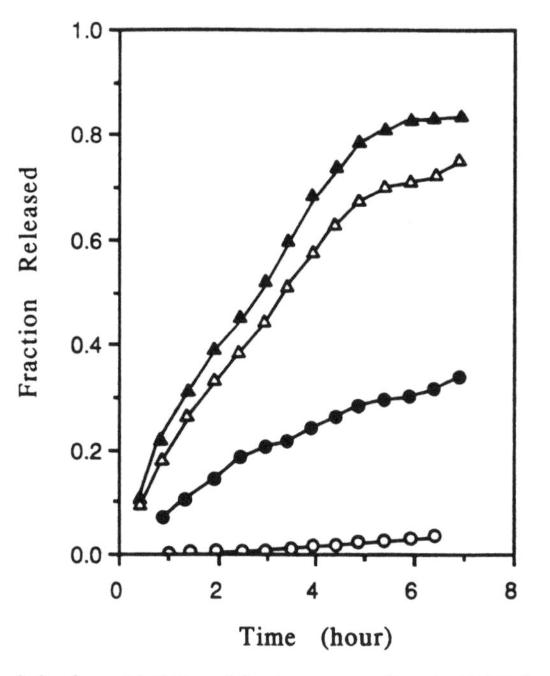

FIGURE 8.9. *Salicylic acid (SA) and bovine serum albumin (BSA) fraction released from calcium-modified corn starch, in simulated intestine solution (pH 7.4), in the presence (+E) and absence (−E) of 0.5 unit/ml of α-amylase; SA + E (▲), SA − E (△), BSA + E (●), and BSA − E (○) (from Reference [96]).*

Heller et al. used starch to prepare a self-regulating naltrexone delivery system [99]. Starch was derivatized by reacting with glycidyl methacrylate and further crosslinked with the addition of unsaturated acid monomers such as acrylic acid, methacrylic acid, maleic acid, or itaconic acid. Naltrexone was first dispersed in the bioerodible copolymer of the *n*-hexyl half ester of methyl vinyl ether and maleic anhydride. This drug core was coated with derivatized acidic starch and then with a layer of enzymes (α-amylase) which were inhibited by antibodies against morphine (Figure 8.10). The entire device was enclosed within a microporous membrane permeable only to morphine and naltrexone. Upon introduction of morphine, the enzyme-hapten-antibody complex dissociated and the enzymes became active. The acidic starch gel was then degraded by the enzymes and naltrexone was released when the pH sensitive polymer is exposed to the physiological pH.

Since the *n*-hexyl half ester of methyl vinyl ether and maleic anhydride in the drug core was insoluble at pH 6 or lower and became soluble above that pH, the release of naltrexone from the polymer matrix was very rapid only at pH 7.4 as shown in Figure 8.11. The acidic starch coating provided the necessary stability of the polymer matrix and prevented the premature release of the drug from the device before activation.

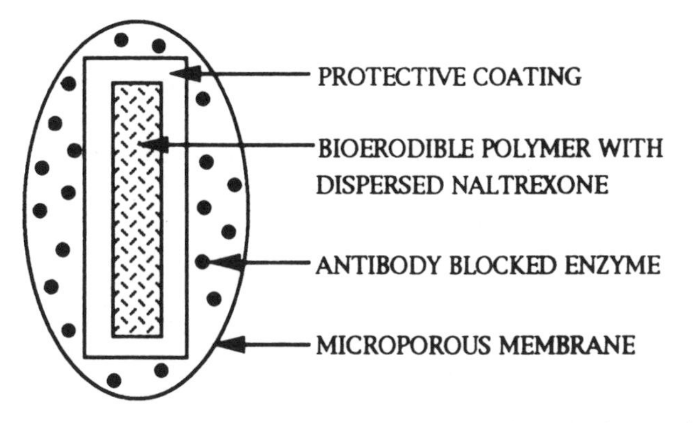

FIGURE 8.10. Proposed triggered delivery system with naltrexone dispersed in rate-controlled polymer (from Reference [99]).

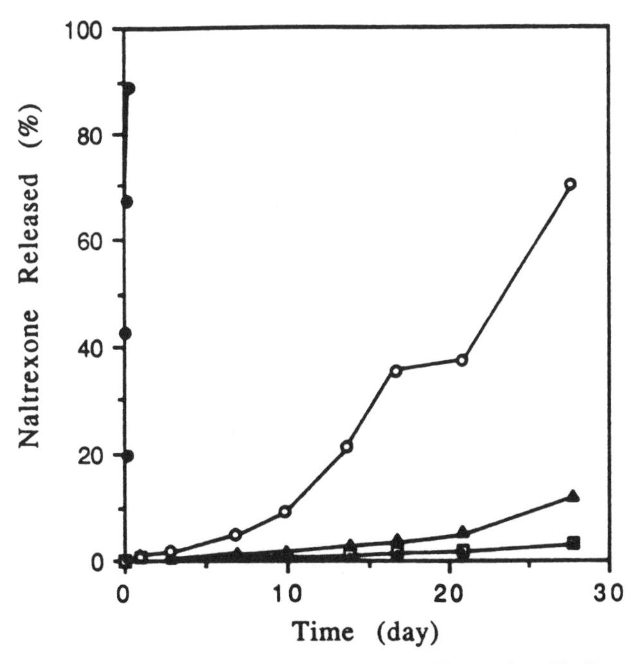

FIGURE 8.11. Cumulative percent naltrexone released from n*-hexyl half ester of methyl vinyl ether and maleic anhydride discs as a function of pH in 0.01 M phosphate buffer, with the addition of 0.9 percent NaCl; pH 3 (⊡), pH 4 (▲), pH 5 (○), and pH 7.4 (●) (from Reference [99]).*

8.2.2.2 DEXTRAN HYDROGELS

Dextran has also been widely used as a drug carrier or as a prodrug for macromolecular drugs. Dextrans and derivatized dextrans are degraded by dextranases which are present in the liver, intestinal mucosa, colon, spleen, and kidneys of humans [100–105].

Several proteins of various sizes including human serum albumin, immunoglobulin G, bovine carbonic anhydrase, and catalase, were physically entrapped in dextran microspheres [103]. Dextran was first modified with glycidyl acrylate to introduce acryloyl groups and the derivatized dextran was then crosslinked with *N,N'*-methylenebisacrylamide (BIS). Proteins were added during crosslinking polymerization of the derivatized dextran. When dextran microparticles containing immobilized proteins were suspended in PBS (pH 7.4), the proteins were slowly released as shown in Figure 8.12. The *in vitro* release results showed some relationship with the molecular weight of the proteins, although the release mechanism was not clearly understood. The amounts of proteins released from the dextran microspheres after 80 days were 65 percent for carbonic anhydrase (MW 31,000), 33 percent for human serum albumin (MW 66,500), 44 percent for immunoglobulin G (MW 150,000), and 30 percent for catalase (MW 240,000).

When the BIS-crosslinked dextran microspheres were injected in mice, 60–80 percent of the particles were taken up by the cells in the spleen and liver and stored in the lysosomal vacuole [106–108]. The study with

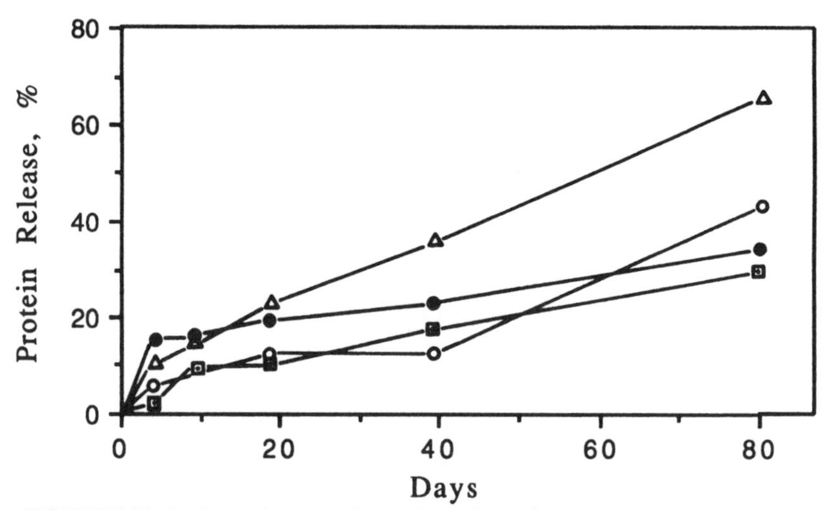

FIGURE 8.12. *Leakage of proteins from microspheres during storage at room temperature and pH 7.4; the immobilized proteins were carbonic anhydrase (△), human serum albumin (●), immunoglobulin G (○), and catalase (⊡) (from Reference [103]).*

radiolabeled BIS showed that the particles were eliminated by first-order kinetics with a half-life of about 12 to 30 weeks [108]. Injection of soluble dextranase did not affect the elimination rate of the dextran microspheres. This indicated that dextranase could not reach the lysosomal vacuole intact. The use of dextranase entrapped in microspheres, however, significantly accelerated the elimination rate of the dextran microspheres. This illustrated the use of self-destroying microparticles as lysosome-directed carriers of enzymes.

Due to the presence of a large number of hydroxyl groups, dextrans were frequently used to attach low molecular weight drugs as reviewed by Larsen [100]. Drug molecules were conjugated to dextran by direct esterification, periodate oxidation, or cyanogen bromide activation [109–111]. Conjugates of mitomycin C and dextran (MMC-D) were synthesized and their biological and pharmacological properties were investigated [112–114]. Dextran was first activated by cyanogen bromide and then coupled with a spacer, either ϵ-aminocaproic acid or 6-bromohexanoic acid, through which MMC was attached. Since not all spacer arms were used for MMC coupling, dextran became charged. *In vitro* release of MMC into PBS followed first-order kinetics with a half-life of 24 hours and the release rates were not accelerated by the addition of rat plasma or tissue homogenates [112]. The negatively charged MMC-D showed the strongest *in vivo* antitumor effect after intravenous injection. The effect was thought to result from the altered pharmacokinetic behavior of MMC-D.

8.2.2.3 CELLULOSE HYDROGELS

Cellulose ethers have been used in a variety of formulations including topical and ophthalmic preparations, enteric polymer film coats, microcapsules, and matrix systems [115]. Pure cellulose is not water-soluble because of its high crystallinity. Substitution of the hydroxyl groups decreases crystallinity by reducing the regularity of the polymer chains and interchain hydrogen bonding [115]. Cellulose derivatives commonly used in the pharmaceutical area are methylcellulose (MC), hydroxyethylmethylcellulose (HEMC), hydroxypropylmethylcellulose (HPMC), ethylhydroxyethylcellulose (EHEC), hydroxyethylcellulose (HEC), and hydroxypropylcellulose (HPC). Physical properties of cellulose compounds are governed by the type and amounts of the substituent groups [115].

HPMC is one of the most widely used cellulose products. The release of water-soluble drugs through uncrosslinked HPMC occurs by a combination of diffusion and dissolution of the matrix itself following hydration [116]. Diffusion, however, may not be significant to the release of hydrophobic drugs. In general, drug release can be modified by various formulation factors, such as type of polymer, polymer concentration, drug particle size, and the presence of additives [117–121]. Ford et al. investigated the influence of these formulation factors on drug release from HPMC matrix tablets [117,119]. A typical example of the release of water-soluble drugs from the HPMC tablet matrix is shown in Figure 8.13 [119]. Drug release of up to 90 percent of the total drug fits the Higuchi equation [Equation (8.3)]. This indicates that erosion of the HPMC does not contribute to the release of water-soluble drugs. Varying the polymer concentration is most efficient in controlling the drug release kinetics [119,121,122]. Increasing the polymer concentration reduces the drug release rates.

The release of a series of drugs, with different water-solubility, from HPMC matrices was examined using Equation (8.7) [117]. The n value for water-soluble drugs, such as promethazine hydrochloride, aminophylline, and propranolol, was 0.67. The n values for indomethacin and diazepam were 0.9 and 0.82, indicating near zero-order release. The release of the poorly soluble drug, tetracycline hydrochloride, showed sigmoidal dissolution profiles which indicate the complex nature of release (see Figure 8.14).

The addition of a hydrophobic lubricant, such as magnesium stearate, stearic acid, or cetyl alcohol, did not affect the release of highly water-soluble drugs [122]. The addition of ionic surfactants, however, reduced

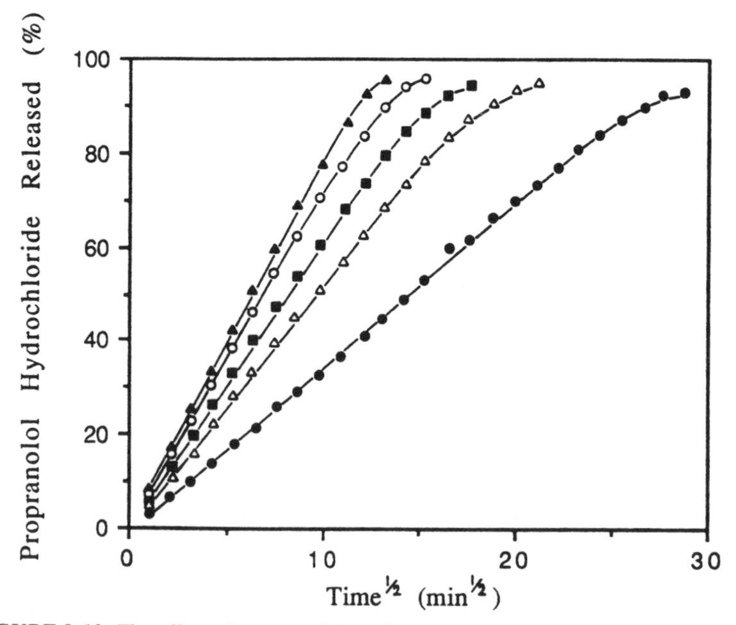

FIGURE 8.13. The effect of propranolol: hydroxypropylmethylcellulose K4M variation on the release of 160 mg propranolol hydrochloride into 1,000 ml water at 37°C from tablets containing 57 (▲), 71 (○), 95 (■), 140 (△), and 285 (●) mg of HPMC K4M (from Reference [119]).

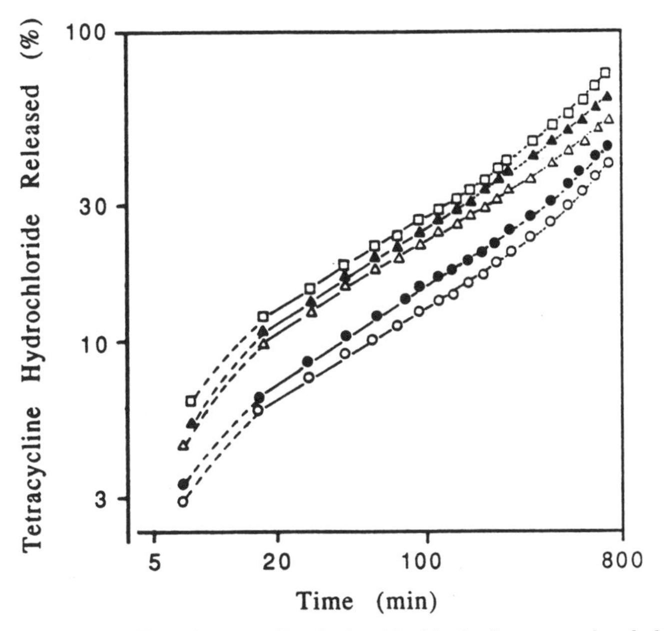

FIGURE 8.14. The effect of tetracycline hydrochloride: hydroxypropylmethylcellulose K15M variation on the release of 250 mg tetracycline hydrochloride in 1,000 ml water at 37°C from tablets containing 45 (□), 60 (▲), 90 (△), 180 (●), and 270 (○) mg of HPMC (from Reference [117]).

the release rate by forming drug/surfactant complexes [118]. When 5–20 percent of sodium dodecyl sulfate (SDS) was included in the propranolol tablet, the dissolution rate decreased due to the formation of propranolol dodecyl sulfate precipitates [123]. The precipitate was poorly soluble and formed a lyotropic liquid crystal on contact with water. The presence of HPMC appeared to facilitate the precipitation.

Ethylcellulose is useful in the preparation of membranes which are permeable to relatively polar, hydrophilic drugs. Drug release through the ethylcellulose membrane is dependent on the topography of the membrane which is changed by the porosity, the capillarity of the membrane, the cooling rate during temperature induced coacervation, and the rate of solvent removal [124,125]. The drug release pattern through ethylcellulose membranes is usually described by either the Higuchi model [126–129] or the first-order release kinetics [130,131].

8.2.3 Biodegradable Hydrogels Based on Polyesters

Heller prepared hydrogels in which degradable polyester prepolymers were crosslinked with polyvinylpyrrolidone (PVP) chains (see Chapter 3, Section 3.1.1.2). The release of physically entrapped bovine serum albumin

(BSA) from the hydrogels was regulated by the amount of PVP. Figure 8.15 shows the dependency of the release of BSA on the amount of *N*-vinylpyrrolidone, a monomer of PVP [132].

Since BSA could not diffuse out from the gel, the release of BSA was dependent solely on the degradation of the polyester chains. The hydrolysis rate of the polyesters at pH 7.4 and 37°C, however, was slow. Only less than 40 percent of the total BSA was released in 40 days. To enhance the hydrolytic degradation of the ester linkages, polyesters containing dicarboxylic acids, such as ketomalonic acid, diglycolic acid, and ketoglutaric acid, were prepared [132]. Due to an increased hydrolytic instability of the "activated" polyesters, the release rate of BSA was substantially increased as shown in Figure 8.16. The duration of the BSA release could be varied from 10 days to more than 12 weeks. The duration was controlled mainly by varying the crosslink density of the gel (i.e., the fraction of *N*-vinylpyrrolidone) and/or by the addition of the electron-withdrawing groups adjacent to the carboxyl groups.

Since the degradation of the PVP-crosslinked hydrogels produced the water-soluble, but nondegradable polymer, PVP, as a by-product, another type of hydrogel was prepared from water-soluble polyesters containing unsaturated pendant groups, such as itaconic and allylmalonic acid [132]. These were able to undergo crosslinking by free radical polymerization

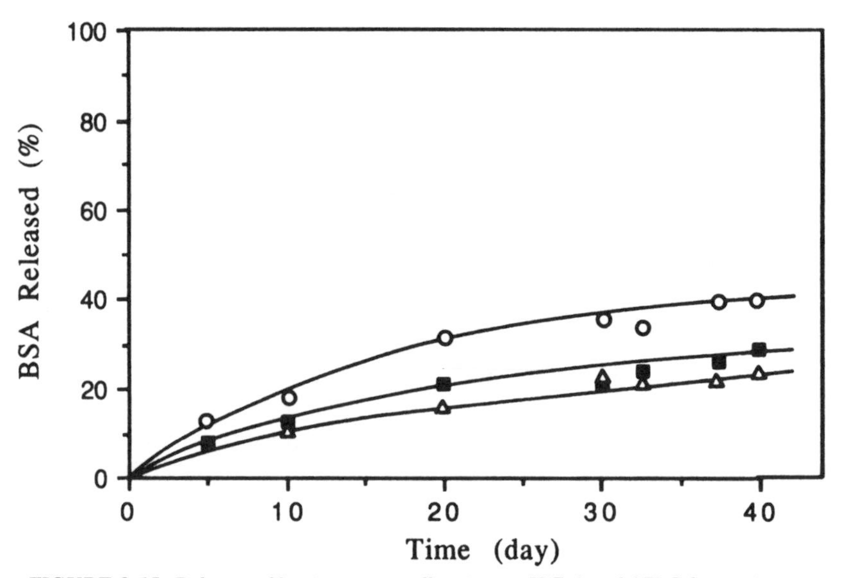

FIGURE 8.15. *Release of bovine serum albumin at pH 7.4 and 37°C from microparticles of a water-soluble fumaric acid polyester crosslinked with varying amounts of* N-*vinylpyrrolidone. Amount of* N-*vinylpyrrolidone was 20 percent (○), 40 percent (■), and 60 percent (△) (from Reference [132]).*

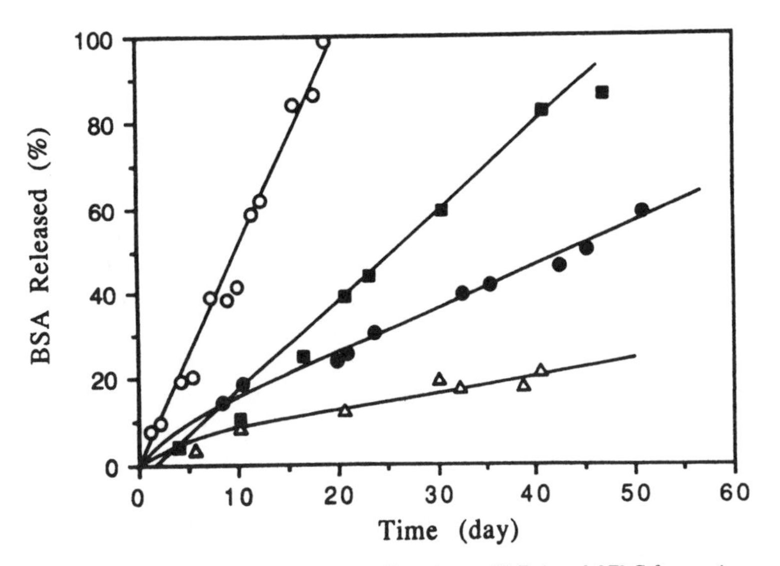

FIGURE 8.16. *Release of bovine serum albumin at pH 7.4 and 37°C from microparticles prepared from various water-soluble, unsaturated polyesters crosslinked with 60 wt percent* N-*vinylpyrrolidone; 4:1 ketomalonic/fumaric (○), 1:1 diglycolic/fumaric (●), 1:1 ketoglutaric/fumaric (■), and fumaric (△) (from Reference [132]).*

without the addition of vinyl comonomers. The rate of BSA release from these gels was governed by the chemical structure (itaconic vs. allylmalonic), and the concentration of the unsaturated ester in solution prior to crosslinking. The release of BSA from an itaconic hydrogel was slower than that from an allylmalonic hydrogel. The difference was considered to be related to the different efficiency of the crosslinking reaction which arose from the different reactivity of the two types of double bonds [132].

8.3 DRUG RELEASE FROM HYDROGELS WITH DEGRADABLE CROSSLINKING AGENTS

8.3.1 Hydrogels Crosslinked with Small Molecules

8.3.1.1 HYDROGELS CROSSLINKED WITH N,N′-METHYLENEBISACRYLAMIDE

Davis injected the insulin-containing hydrogels intraperitoneally to alloxan induced diabetic rats [133]. Hydrogels were prepared by polymerizing acrylamide in the presence of dissolved insulin and *N,N′*-methylenebisacrylamide (BIS), a crosslinking agent. The concentration of acrylamide was either 25 percent or 40 percent by weight, and the concentration of BIS was 2 percent. The diabetic rats implanted with a 40 percent polyacrylamide hydrogel sustained a normal growth rate for at least a few weeks as shown in Figure 8.17. On the other hand, the rats with a 25 percent polyacrylamide implant gained no weight due to the rapid release of insulin from the implant. Since the implanted hydrogels did not completely degrade at the BIS concentration used, the insulin depleted implants were surgically removed. Once the implants were removed, the symptoms of diabetes returned rapidly.

Torchilin et al. prepared PVP hydrogels crosslinked with BIS [134]. Chymotrypsin was incorporated into the gel by physical entrapment during polymerization. These hydrogels were degraded slowly by hydrolysis of the crosslinker. Degradation of the crosslinker was sensitive to the concentration of BIS used in the hydrogel preparation. Hydrogels crosslinked with 0.3 to 0.6 percent BIS degraded over a period of 2 – 10 days, while gels crosslinked with more than 1 percent BIS were essentially nondegradable. The gels with a very low crosslink density were unable to retain the entrapped chymotrypsin due to their highly porous structure. As a result, the drug was rapidly released by the diffusion mechanism rather than by the hydrogel degradation. The release of chymotrypsin from the gels was dependent on the crosslink density. When the concentrations of BIS were

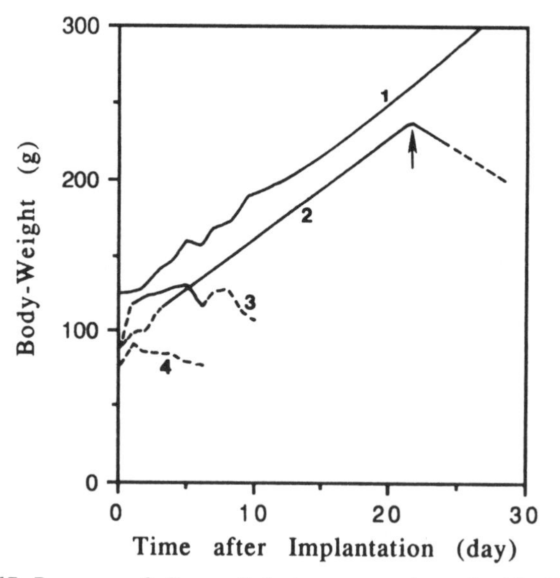

FIGURE 8.17. Response of alloxan diabetic rats to polyacrylamide (PAA) implants containing insulin. The arrow designates when implants were removed from rats in the 10 mg insulin, 40 percent PAA group; normal (———), and glycosuria (- - - - -), nondiabetic (1), 40 percent PAA/10 mg insulin (2), 25 percent PAA/10 mg insulin (3), and 25 percent PAA/1 mg insulin (4) (from Reference [133]).

0.3 percent, 0.6 percent, and 1.0 percent, the percentages of chymotrypsin released in 50 hours at 25 °C were 100, 81, and 64, respectively. The release rate of the enzyme from the hydrogels was much faster than the degradation rate of the hydrogels.

8.3.1.2 HYDROGELS CROSSLINKED WITH AZO REAGENTS

Saffran et al. [135] delivered two polypeptide drugs, vasopressin and insulin, by oral administration using biodegradable hydrogels. Vasopressin was loaded into gelatin capsules and insulin was fabricated into pellets. These drug cores were then coated with a polymer film. The film was made of poly(hydroxyethyl methacrylate-*co*-styrene) (at the ratio of 6:1) crosslinked with 1 or 2 percent divinylazobenzene. The polymer film protected the drugs from the harsh conditions in the stomach and from digestion by enzymes in the gastrointestinal tract. The polymer film, however, was degraded in the colon by the indigenous microflora which broke down the azo bonds of the the crosslinker. The degradation of the polymer film resulted in the release of polypeptide drugs in the lumen of the colon for local action or for local absorption. The polymer films were shown to lose strength and undergo degradation when incubated with human feces. When these drug delivery systems were deposited in the stomach of rats, significant biological responses, such as antidiuresis and hypoglycemia, were observed. Antidiuresis occurred 100 minutes after administration of capsules and lasted one hour, while it occurred after only a few minutes and lasted about 20 to 30 minutes following the administration of a vasopressin solution. When azoaromatic, polymer-coated insulin pellets were administered into the streptozotocin-induced diabetic rats, the blood glucose level was significantly reduced in the next nine hours as shown in Figure 8.18. In rats which were given control pellets, the blood glucose level varied from 83 to 99 percent of the initial value during the same period.

8.3.2 Hydrogels Crosslinked with Albumin

Polyvinylpyrrolidone (PVP) hydrogels crosslinked with functionalized albumin were used for long-term oral delivery of flavin mononucleotide (FMN) [136]. FMN was chosen as a model drug in the study because its absorption was restricted to the upper small intestine and its biological half-life was only about 70 minutes [137]. Thus, the effect of the long-term gastric retention of the hydrogels on the absorption of FMN could be easily examined. The albumin-crosslinked PVP hydrogels were loaded with FMN by swelling the gels in an FMN-saturated solution. The *in vitro* release of FMN from the hydrogels in simulated gastric fluid was examined. As shown in Figure 8.19, release occurred for more than 400 hours. After the initial

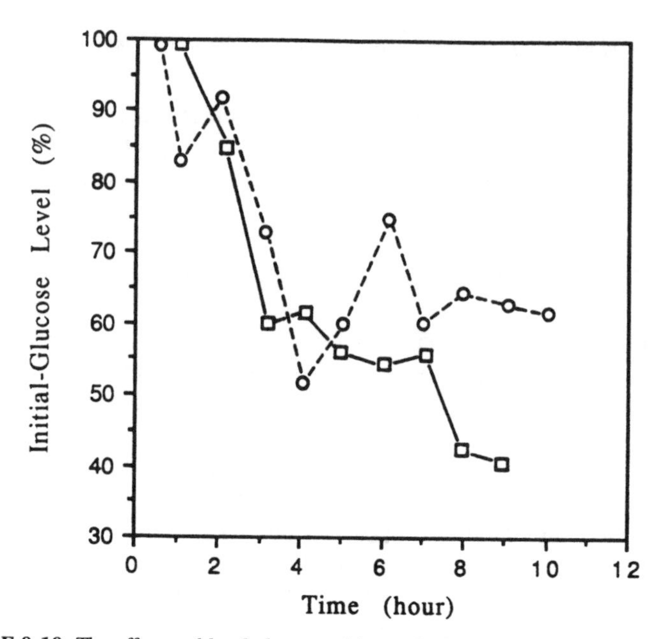

FIGURE 8.18. *The effect on blood glucose of the oral administration of an azoaromatic polymer-coated pellet containing 1 IU (about 28 nmol/kg body weight) of insulin on the blood glucose levels in two rats made diabetic with streptozotocin. The initial blood glucose levels were 400 mg per 100 ml (○), and 290 mg per 100 ml (□) (from Reference [135]).*

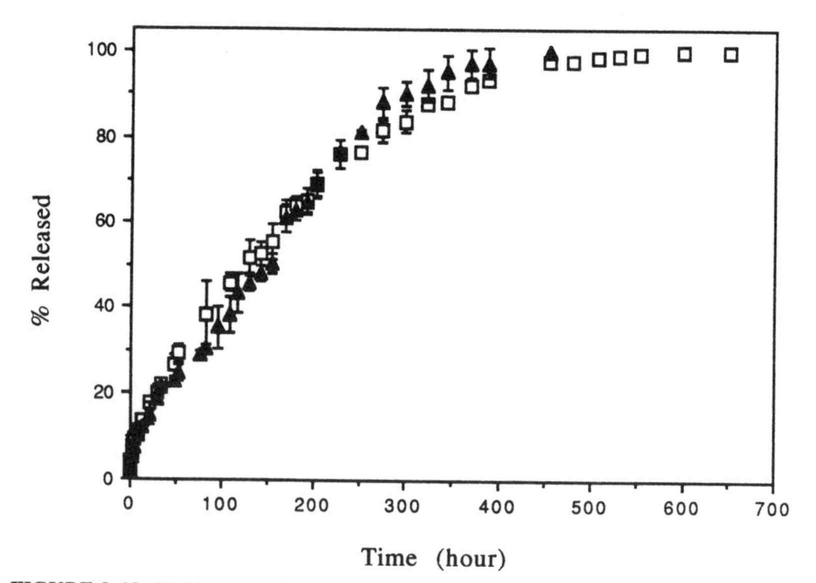

FIGURE 8.19. *FMN release from enzyme-digestible hydrogels in the simulated gastric fluid. The percentage of FMN released was observed in the presence of pepsin (250 units/ml) (▲), and in the absence of pepsin (□); average ± SD,* n = *3 (from Reference [136]).*

burst effect, the release of FMN was zero order up to 300 hours. The presence of pepsin in the simulated gastric fluid did not significantly alter the release profile.

The FMN-containing hydrogels were administered to dogs and the total riboflavin concentration in the blood was measured over time as shown in Figure 8.20. After the initial surge of the total riboflavin concentration resulting from the burst release of FMN, a relatively constant blood concentration was maintained up to 54 hours. Radiographic images made at each sampling confirmed the retention of the hydrogel in the stomach for up to 36 hours. Note that the dogs were under fast conditions for the first 24 hours. In control experiments where FMN was administered using gelatin capsules, the absorption of FMN reached a peak within one hour and then declined to an insignificant level within six hours. Bioavailability of FMN delivered by the hydrogel system was 3.7 times larger than that by the gelatin capsule. The study has shown that the albumin-crosslinked hydrogels were suitable as a once-a-day or once-every-other-day oral drug delivery system.

8.4 DRUG RELEASE FROM HYDROGELS WITH BIODEGRADABLE PENDANT CHAINS

Drug molecules chemically bound to the polymer backbone are released to the surrounding medium upon hydrolysis or enzymatic degradation of

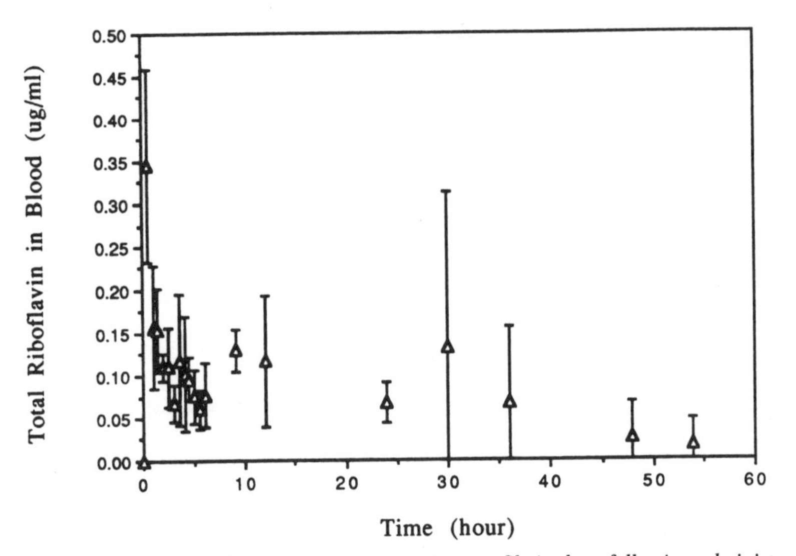

FIGURE 8.20. The FMN concentration versus time profile in dogs following administration of FMN-containing hydrogels. Dogs were maintained under fast conditions for the first 24 hours and were given meals once a day afterwards. FMN was expressed as total riboflavin in the blood; average ± SD, n = *4 (from Reference [136]).*

the bond between the drug and the polymer. The rate-limiting step of drug release from this type of system is usually hydrolysis or enzymatic degradation of the labile bond attaching the drug to the polymer [1,100]. Thus, one of the main factors controlling drug release is the nature of the labile bond [138–143,153].

Kopeček and his coworkers investigated the enzyme-induced release of *p*-nitroaniline (NAP) from the poly[*N*-(2-hydroxypropyl) methacrylamide] (PHPMA) backbone [138,144–148]. NAP was attached to PHPMA via the oligopeptide linkage which can be degraded by specific enzymes. The structure of oligopeptide linkages was varied and its effect on the release of NAP by bovine cathepsin B, an intracellular proteolytic enzyme, was examined [144]. The rate of hydrolysis of the oligopeptides was dependent on the oligopeptide structure. The degradation rates of Gly-Phe-Tyr-Ala and Gly-Phe-Leu-Gly tetrapeptides were the highest, while those of Gly-Phe-Gly-Phe tetrapeptide and Gly-Phe-Leu-Gly-Phe pentapeptide were the lowest (Figure 8.21).

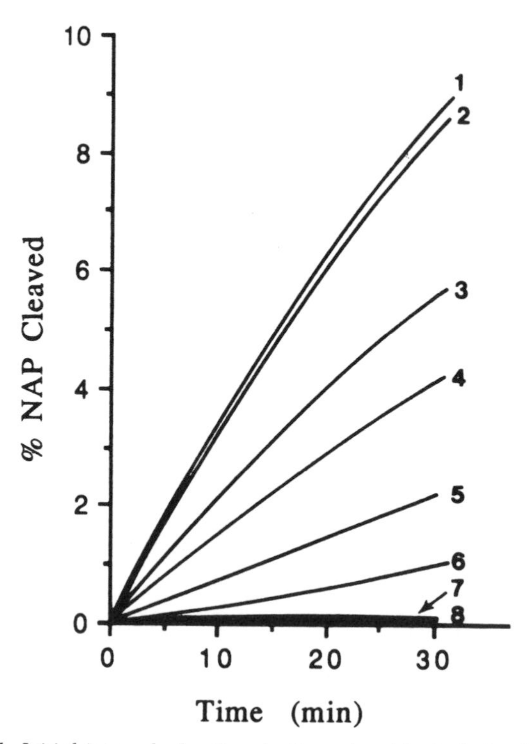

FIGURE 8.21. Initial interval of cathepsin B-catalyzed hydrolysis of polymeric substrates; P-Gly-Phe-Tyr-Ala-NAP (1), P-Gly-Phe-Leu-Gly-NAP (2), P-Gly-Phe-Ala-NAP (3), P-Gly-Leu-Ala-NAP (4), P-Gly-Val-Ala-NAP (5), P-Gly-Val-Phe-NAP (6), P-Gly-Phe-Leu-Gly-Phe-NAP (7), and P-Gly-Phe-Gly-Phe-NAP (8) (from Reference [144]).

The release of NAP was linear for the first 30 minutes in all cases. The release, however, slowed down after a couple of hours of incubation. When the NAP-PHPMA systems were incubated in rat plasma or rat serum, the release of free NAP for five hours was less than 5 percent. In contrast, 30 to 50 percent of total bound NAP was released over the same time period on incubation with purified bovine spleen cathepsin B. This study has demonstrated that peptidyl side chains in PHPMA copolymers can be synthesized in such a way that they are resistant to hydrolysis in plasma or serum while degradable by lysosomal enzymes at the target site. An anticancer drug, daunomycin, was also attached to the PHPMA via a tetrapeptide side chain. Daunomycin was released upon incubation of the drug-polymer conjugates in a mixture of lysosomal enzymes [149].

A similar study was done by Ringsdorf et al. [150]. Six different oligopeptides were used as a spacer to attach either NAP or bis(2-chloroethyl)-amine, an anticancer drug, to PHPMA. The *in vitro* release study showed that the nature of the oligopeptides influenced the formation of the enzyme-substrate complex, and thus the release of NAP from the polymer. Approximately 3 percent of bound NAP was released from the polymer when incubated in blood for 24 hours. However, the release was much higher in the presence of lysosomal enzymes. This indicated that drugs may be released only after the drug-polymer complexes have penetrated into cells.

Seymour et al. examined the *in vivo* distribution of adriamycin (ADR) covalently bound to PHPMA via oligopeptide linkages (Gly-Phe-Leu-Gly) [151]. When free ADR was administered intravenously into rats, the initial high levels of free ADR in plasma were followed by high levels of free ADR in other tissues such as the liver and the heart. In the case of polymer-bound ADR, however, the initial concentration of free ADR in both plasma and tissue was significantly reduced. The concentration of free ADR in the heart decreased, probably due to the reduced ability of the ADR-PHPMA conjugates to leave circulation. The circulating half-life of the PHPMA-ADR conjugate was approximately 15 times longer than that of the free drug. Since PHPMA itself is not biodegradable, it is desirable to use short polymer chains which may be eliminated through glomerular filtration [102,145].

Laakso et al. coupled primaquine (PQ) and trimethoprim (TMP), which contain primary amino groups, to starch microparticles via oligopeptide spacers [152]. The drugs were derivatized with oligopeptides and subsequently coupled to carbonyldiimidazole-activated starch microparticles. Both drugs were released from the starch microparticles upon incubation in a lysosome-enriched solution. The release of PQ was faster than that of TMP under the same conditions. As shown in Figure 8.22, the release of PQ from tetrapeptide derivatized microparticles was faster than that from the pentapeptide derivatized ones. This again indicated the importance of the oligopeptide structure in enzyme-induced hydrolysis.

Stjärnkvist et al. investigated the immune response of drugs coupled to biodegradable microspheres in mice [92]. Leu-Ala-Lys-dinitrophenyl (DNP) was covalently bonded to polyacryl starch microparticles. Upon incubation in a lysosomal milieu for 24 hours at 37°C, about 10 percent of Lys-DNP was released from the microparticles. The released Lys-DNP induced a weak immune response.

Bennet et al. examined the release of naltrexone monoacetate from the copolymer of N^5-(3-hydroxypropyl)-L-glutamine and L-leucine (PHGL) [153]. Naltrexone monoacetate, a narcotic antagonist, was covalently linked to PHGL. In the aqueous environment, the linkages were hydrolyzed and subsequently released free drugs to the surrounding medium. *In vitro* studies showed a burst effect in drug release. Up to 40 percent of the total drug was released during the burst phase, which was followed by the relatively constant release phase (Figure 8.23). The burst effect was considered to be due to the release of physically adsorbed drug, edge effects, and hydrophilic impurities acting as swelling enhancers. The release rate was dependent on the size of the particles. Smaller particles released the drug more quickly than larger particles throughout the release experiment. The burst effect was also observed from the plasma-naltrexone concentrations measured after subcutaneous injection of the conjugate particles into rats.

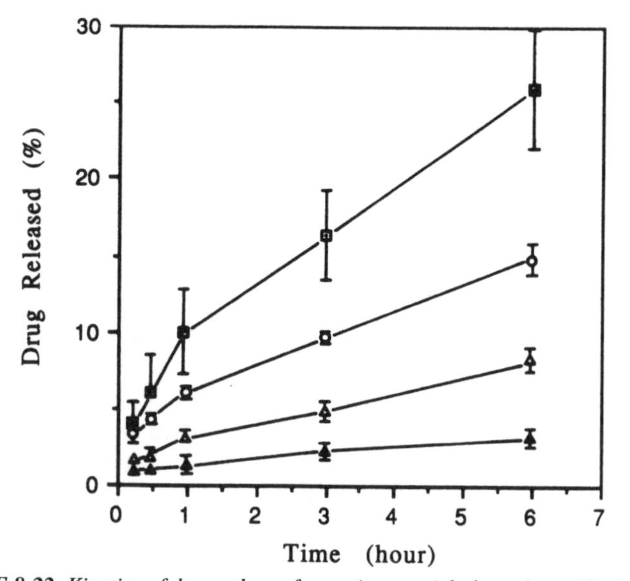

FIGURE 8.22. Kinetics of drug release from microparticle-bound peptide-drug derivatives in a lysosome-enriched fraction. The free drug released is expressed as the percent of the amount initially bound to the microparticles; Leu-Ala-Leu-Ala-Leu-PQ (○), Ala-Leu-Ala-Leu-PQ (⊡), Leu-Ala-Leu-PQ (△), and Leu-Ala-Leu-TMP (▲) (from Reference [152]).

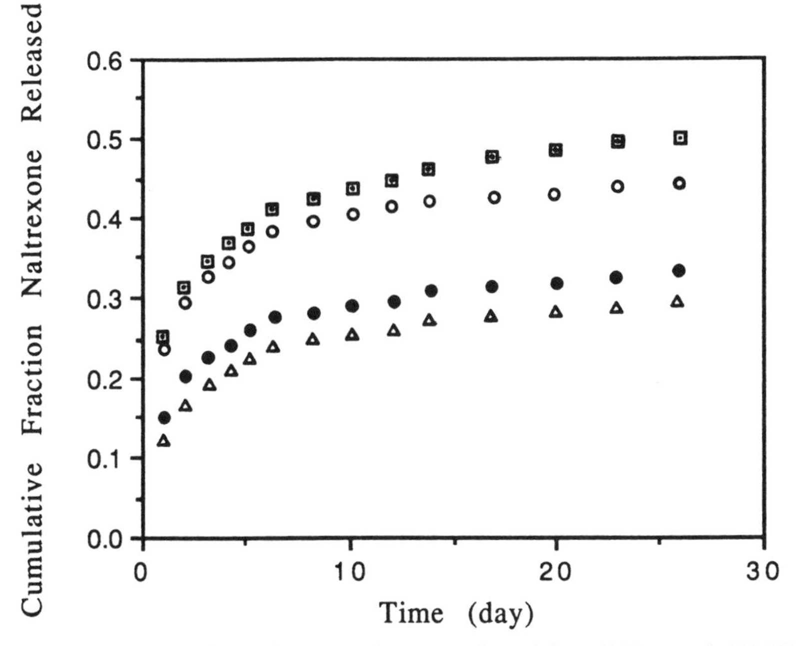

FIGURE 8.23. Cumulative fraction naltrexone released from N14A-copoly(HPG^{50}/ Leu^{50}) conjugate particles of various sizes; 20–50 μm (⊡), 50–100 μm (○), 100–200 μm (●), and 200–350 μm (△) (from Reference [153]).

8.5 REFERENCES

1. Heller, J. 1980. "Controlled Release of Biologically Active Compounds from Bioerodible Polymers," *Biomaterials*, 1:51–57.
2. Heller, J., R. W. Baker, R. M. Gale and J. O. Rodin. 1978. "Controlled Drug Release by Polymer Dissolution. I. Partial Esters of Maleic Anhydride Copolymers–Properties and Theory," *J. Appl. Polym. Sci.*, 22:1191–2009.
3. Chang, T. M. S. 1976. "Biodegradable Semipermeable Microcapsules Containing Enzymes, Hormones, Vaccines, and Other Biologicals," *J. Bioeng.*, 1:25–32.
4. Heller, J. 1985. "Control of Polymer Surface Erosion by the Use of Excipients," in *Polymers in Medicine II. Biomedical and Pharmaceutical Applications*, E. Chiellini, P. Giusti, C. Migliaresi and L. Nicolais, eds., New York, NY: Plenum Press, pp. 357–368.
5. Baker, R. W. and H. K. Lonsdale. 1974. "Controlled Release; Mechanisms and Rates," in *Controlled Release of Biologically Active Agents*, A. C. Tanquary and R. E. Lacey, eds., New York, NY: Plenum Press, pp. 15–22.
6. Crank, J. 1975. *The Mathematics of Diffusion, 2nd Ed.* London: Oxford University Press.
7. Baker, R. W. 1987. *Controlled Release of Biologically Active Agents*. New York, NY: John Wiley & Sons, pp. 1–83.
8. Jost, W. 1960. *Diffusion in Solids, Liquids, Gases*. New York, NY: Academic Press.

9. Zeoli, L. T. and A. F. Kydonieus. 1983. "Physical Methods of Controlled Release," in *Controlled Release Technology. Bioengineering Aspect*, K. G. Das, ed., New York, NY: John Wiley & Sons, pp. 61–120.
10. Kost, J. and R. Langer. 1987. "Equilibrium Swollen Hydrogels in Controlled Release Applications," in *Hydrogels in Medicine and Pharmacy, Volume III. Properties and Applications*, N. A. Peppas, ed., Boca Raton, FL: CRC Press Inc., pp. 95–108.
11. Garrett, E. R. and P. B. Chemburkar. 1968. "Evaluation, Control and Prediction of Drug Diffusion through Polymeric Membranes II," *J. Pharm. Sci.*, 57(6):949–959.
12. Baker, R. W., M. E. Tuttle, H. K. Lonsdale and J. W. Ayres. 1979. "Development of an Estriol-Releasing Intrauterine Device," *J. Pharm. Sci.*, 68(1):20–26.
13. Zhang, G., J. B. Schwartz and R. L. Schnaare. 1991. "Bead Coating I. Change in Release Kinetics (and Mechanism) due to Coating Levels," *Pharm. Res.*, 8(3):331–335.
14. Flynn, G. L., S. H. Yallkowsky and T. J. Roseman. 1974. "Mass Transport Phenomena and Models: Theoretical Concepts," *J. Pharm. Sci.*, 63(4):479–510.
15. Higuchi, T. 1963. "Mechanism of Sustained-Action Medication: Theoretical Analysis of Rate of Release of Solid Drugs Dispersed in Solid Matrices," *J. Pharm. Sci.*, 52(12):1145–1149.
16. Roseman, T. J. 1972. "Release of Steroids from a Silicone Polymer," *J. Pharm. Sci.*, 61(1):46–50.
17. Roseman, T. J. and W. I. Higuchi. 1970. "Release of Medroxyprogesterone Acetate from a Silicone Polymer," *J. Pharm. Sci.*, 59(3):353–357.
18. Luzzi, L. A., M. A. Zoglio and H. V. Maulding. 1970. "Preparation of Evaluation of the Prolonged Release Properties of Nylon Microcapsules," *J. Pharm. Sci.*, 59: 338–341.
19. Crank, J. 1975. "Diffusion in a Plane Sheet," in *The Mathematics of Diffusion, 2nd Ed.* London: Oxford University Press, pp. 44–68.
20. Ritger, P. L. and N. A. Peppas. 1987. "A Simple Equation for Description of Solute Release I. Fickian and Non-Fickian Release from Non-Swellable Devices in the Form of Slabs, Spheres, Cylinders or Discs," *J. Contr. Rel.*, 5:23–36.
21. Higuchi, T. 1961. "Rate of Release of Medicaments from Ointment Bases Containing Drugs in Suspension," *J. Pharm. Sci.*, 50(10):874–875.
22. Higuchi, W. 1962. "Analysis of Data on the Medicament Release from Ointments," *J. Pharm. Sci.*, 51(8):802–804.
23. Schwartz, J. B., A. P. Simonelli and W. I. Higuchi. 1968. "Drug Release from Wax Matrices I. Analysis of Data with First-Order Kinetics and with the Diffusion-Controlled Model," *J. Pharm. Sci.*, 57(2):274–277.
24. Alfrey, T., E. F. Gurnee and W. G. Lloyd. 1966. "Diffusion in Glassy Polymers," *J. Polymer Sci.*, Part C, 12:249–261.
25. Peppas, N. A. and R. W. Korsmeyer. 1987. "Dynamically Swelling Hydrogels in Controlled Release Applications," in *Hydrogels in Medicine and Pharmacy, Volume III. Properties and Applications*, N. A. Peppas, ed., Boca Raton, FL: CRC Press Inc., pp. 118–121.
26. Ritger, P. L. and N. A. Peppas. 1987. "A Simple Equation for Description of Solute Release I. Fickian and Anomalous Release from Swellable Devices," *J. Contr. Rel.*, 5:37–42.
27. Bamba, M., F. Puisieus, J. P. Marty and J. T. Carstensen. 1979. "Mathematical

Model of Drug Release from Gel-Forming Sustained Release Preparations,'' *Int. J. Pharmacol.*, 3:87–92.

28. Peppas, N. A., R. Gurny, E. Doelker and P. Buri. 1980. ''Modelling of Drug Diffusion through Swellable Systems,'' *J. Membr. Sci.*, 7:241–253.
29. Korsmeyer, R., E. Von Meerwall and N. A. Peppas. 1986. ''Solute and Penetrant Diffusion in Swellable Polymers. II. Verification of Theoretical Models,'' *J. Polym. Sci. Polym. Phys.*, 24:409–434.
30. Lee, P. I. 1980. ''Diffusional Release of a Solute from a Polymeric Matrix. Approximate Analytical Solutions,'' *J. Membr. Sci.*, 7:255–275.
31. Brannon-Peppas, L. and N. A. Peppas. 1989. ''Solute and Penetrant Diffusion in Swellable Polymers. IX. The Mechanisms of Drug Release from pH-Sensitive Swelling-Controlled Systems,'' *J. Contr. Rel.*, 8:267–274.
32. Kim, C. J. and P. Lee. 1991. ''Swelling-Controlled Drug Release from Anionic Polyelectrolyte Gel Beads,'' *Proceed. Intern. Symp. Control. Rel. Bioact. Mater.*, 18:451–452.
33. Crank, J. 1975. ''Non-Fickian Diffusion,'' in *The Mathematics of Diffusion, 2nd Ed.* London: Oxford University Press, pp. 254–265.
34. Hopfenberg, H. B. and K. C. Hsu. 1978. ''Swelling-Controlled, Constant Rate Delivery Systems,'' *Polym. Eng. Sci.*, 18(15):1186–1191.
35. Hopfenberg, H. B. 1970. ''Anomalous Transport of Penetrants in Polymeric Membranes,'' in *Membrane Science and Technology: Industrial, Biological, and Waste Treatment Processes*, J. E. Flinn, ed., New York, NY: Plenum Press, pp. 16–32.
36. Windle, A. H. 1985. ''Case II Sorption,'' in *Polymer Permeability*, J. Comyn, ed., London and New York, NY: Elsevier Applied Science Publishers, pp. 75–118.
37. Heller, J., S. Y. Ng, D. W. Penhale, B. K. Fritzinger, L. M. Sanders, R. A. Burns, M. G. Gaynon and S. S. Bhosale. 1987. ''Use of Poly(ortho Esters) for the Controlled Release of 5-Fluorouracyl and a LHRH Analogue,'' *J. Contr. Rel.*, 6:217–224.
38. Heller, J. 1985. ''Controlled Drug Release from Poly(ortho Esters)–A Surface Eroding Polymer,'' *J. Cont. Rel.*, 2:167–177.
39. Linhardt, R. J., H. B. Rosen and R. Langer. 1983. ''Bioerodable Polyanhydrides for Controlled Drug Delivery,'' *Polymer Preprints*, 24:47–48.
40. Leong, K. W., B. C. Brott and R. Langer. 1985. ''Bioerodible Polyanhydrides as Drug-Carrier Matrices. I. Characterization, Degradation, and Release Characteristics,'' *J. Biomed. Mat. Res.*, 19:941–955.
41. Mathiowitz, E. and R. Langer. 1987. ''Polyanhydride Microspheres as Drug Carriers. I. Hot-Melt Microencapsulation,'' *J. Contr. Rel.*, 5:13–22.
42. Rippe, E. G. and J. R. Johnson. 1969. ''Regulation of Dissolution Rate by Pellet Geometry,'' *J. Pharm. Sci.*, 58:428–431.
43. Cooney, D. O. 1972. ''Effect of Geometry on the Dissolution of Pharmaceutical Tablets and Other Solids: Surface Detachment Kinetics Controlling,'' *Am. Inst. Chem. Engrs. J.*, 18(2):446–449.
44. Brooke, D. and R. J. Washkuhn. 1977. ''Zero-Order Drug Delivery System: Theory and Preliminary Testing,'' *J. Pharm. Sci.*, 66(2):159–162.
45. Joshi, A. and K. J. Himmelstein. 1991. ''Dynamics of Controlled Release from Bioerodible Matrices,'' *J. Cont. Rel.*, 15:95–104.

46. Hopfenberg, H. B. 1976. "Controlled Release from Erodible Slabs, Cylinders, and Spheres," in *Controlled Release Polymeric Formulations, ACS Symposium Series,* D. R. Paul and F. W. Harris, eds., Washington DC:American Chemical Society.

47. Baker, R. W. 1987. "Biodegradable Systems," in *Controlled Release of Biologically Active Agents*. New York, NY: John Wiley & Sons, pp. 120–121.

48. Pitt, C. G., M. M. Gratzl, A. R. Jeffcoat, R. Zweidinger and A. Schindler. 1979. "Sustained Drug Delivery Systems II: Factors Affecting Release Rates from Poly(e-caprolactone) and Related Biodegradable Polyesters," *J. Pharm. Sci.*, 68(12): 1534–1538.

49. Heller, J. and R. W. Baker. 1980. "Theory and Practice of Controlled Drug Delivery from Bioerodible Polymers," in *Controlled Release of Bioactive Materials*, R. W. Baker, ed., New York, NY: Academic Press, pp. 1–17.

50. Lee, P. 1991. "Kinetic Considerations of Drug Delivery from Swelling-Controlled and Erosion/Diffusion-Controlled Systems," *Proceed. Intern. Symp. Control. Rel. Bioact. Mater.*, 18:315–316.

51. Baker, R. W. 1987. *Controlled Release of Biologically Active Agents*. New York, NY: John Wiley & Sons, pp. 84–131.

52. Morimoto, Y., K. Sugibayashi, M. Okumura and Y. Kato. 1980. "Biomedical Applications of Magnetic Fluids. I. Magnetic Guidance of Ferro-Colloid-Entrapped Albumin Microsphere for Site Specific Drug Delivery *in Vivo*," *J. Pharm. Dyn.*, 3:264–267.

53. Gupta, P. K. and C. T. Hung. 1990. "Albumin Microspheres. V. Evaluation of Parameters Controlling the Efficacy of Magnetic Microspheres in the Targeted Delivery of Adriamycin in Rats," *Int. J. Pharm.*, 59:57–67.

54. Burger, J. J., E. Tomlinson, E. M. A. Mulder and J. G. McVie. 1985. "Albumin Microspheres for Intra-Arterial Tumor Targeting. I. Pharmaceutical Aspects," *Int. J. Pharm.*, 23:333–344.

55. Tomlinson, E., E. M. Burger, E. M. A. Schoonderwoerd and J. G. McVie. 1984. "Human Serum Albumin Microspheres for Intraarterial Drug Targeting of Cytostatic Compounds. Pharmaceutical Aspects and Release Characteristics," in *Microspheres and Drug Therapy. Pharmaceutical, Immunological and Medical Aspects*, S. S. Davis, L. Illum, J. G. McVie and E. Tomlinson, eds., Amsterdam: Elsevier Science Publishers B. V., pp. 75–89.

56. Gupta, P. K. and C. T. Hung. 1989. "Review. Albumin Microspheres I: Physico-Chemical Characteristics," *J. Microencap.*, 6(4):427–462.

57. Gallo, J. M., C. T. Hung and D. G. Perrier. 1984. "Analysis of Albumin Microsphere Preparation," *Int. J. Pharm.*, 22:63–74.

58. Gupta, P. K., C. T. Hung and D. G. Perrier. 1986. "Albumin Microspheres. II. Effect of Stabilization Temperature on the Release of Adriamycin," *Int. J. Pharm.*, 33:147–153.

59. Gupta, P. K., J. M. Gallo, C. T. Hung and D. G. Perrier. 1987. "Influence of Stabilization Temperature on the Entrapment of Adriamycin in Albumin Microspheres," *Drug Develop. Ind. Pharm.*, 13(8):1471–1482.

60. Kramer, P. A. 1974. "Albumin Microspheres as Vehicles for Achieving Specificity in Drug Delivery," *J. Pharm. Sci.*, 63:1646–1647.

61. Gupta, P. K., C. T. Hung and F. C. Lam. 1989. "Factorial Design Based Optimization of the Formulation of Albumin Microspheres Containing Adriamycin," *J. Microencap.*, 6(2):147–160.

62. Gupta, P. K. and C. T. Hung. 1989. "Albumin Microspheres II: Applications in Drug Delivery," *J. Microencap.*, 6(4):463–472.

63. El-Samaligy, M. and P. Rohdewald. 1982. "Triamcinolone Diacetate Nanoparticles, a Sustained Release Drug Delivery System Suitable for Parenteral Administration," *Pharm. Acta Hevl.*, 57:201–204.

64. Dau-Mauger, A., J. P. Benoit and F. Puisierx. 1986. "Preparation and Characterization of Crosslinked Human Serum Albumin Microspheres Containing 5-Fluorouracil," *Pharm. Acta Hevl.*, 61:119–124.

65. Leucuta, S. E., R. Riska, D. Daicoviviu and D. Porutiu. 1988. "Albumin Microspheres as a Drug Delivery System for Epirubicin: Pharmaceutical, Pharmacokinetic and Biological Aspects," *Int. J. Pharm.*, 41:213–217.

66. Burgess, D. J. and S. S. Davis. 1988. "Potential Use of Albumin Microspheres as a Drug Delivery System. II. *In vivo* Deposition and Release of Steroids," *Int. J. Pharm.*, 46:69–76.

67. Willmott, N. and P. J. Harrison. 1988. "Characterization of Freeze-Dried Albumin Microspheres Containing Anti-Cancer Drug Adriamycin," *Int. J. Pharm.*, 43: 161–166.

68. Gupta, P. K., C. T. Hung and D. G. Perrier. 1986. "Albumin Microspheres. I. Release Characteristics of Adriamycin," *Int. J. Pharm.*, 33:137–146.

69. Royer, G. P., T. K. Lee and T. D. Sokoloski. 1983. "Entrapment of Bioactive Compounds within Albumin Beads," *J. Parentheral Sci. Tech.*, 37:34–37.

70. Lee, T. K., T. D. Sokoloski and G. P. Royer. 1981. "Serum Albumin Beads: An Injectable, Biodegradable System for the Sustained Release of Drugs," *Science*, 213(10):233–235.

71. Willmott, N., J. Cummings and A. T. Florance. 1985. " *In vitro* Release of Adriamycin from Drug-Loaded Albumin and Haemoglobin Microspheres," *J. Microencap.*, 2:293–304.

72. Ishizaka, T. and M. Koishi. 1983. "*In vitro* Drug Release from Egg Albumin Microcapsules," *J. Pharm. Sci.*, 72:1057–1061.

73. Willmott, N., Y. Chen and A. T. Florence. 1988. "Haemoglobin, Transferrin and Albumin/Polyaspartic Acid Microspheres as Carriers for the Cytotoxic Drug Adriamycin. II. *In vitro* Drug Release Rates," *J. Contr. Rel.*, 8:103–109.

74. Gupta, P. K., C. T. Hung and N. S. Rao. 1989. "Ultrastructural Disposition of Adriamycin-Associated Magnetic Albumin Microspheres in Rats," *J. Pharm. Sci.*, 78(4):290–294.

75. Gallo, J. M., P. K. Gupta, C. T. Hung and D. F. Perrier. 1989. "Evaluation of Drug Delivery Following the Administration of Magnetic Albumin Microspheres Containing Adriamycin to the Rat," *J. Pharm. Sci.*, 78(3):190–194.

76. Willmott, N., J. Cummings, J. F. B. Stuart and A. T. Florance. 1985. "Adriamycin-Loaded Albumin Microspheres: Preparation, *in vivo* Distribution and Release in the Rat," *Biopharm. Drug Disp.*, 6:91–104.

77. Goosen, M. F. A., L. Y. F. Leung, S. Chou and A. M. Sun. 1982. "Insulin-Albumin Microbeads: An Implantable, Biodegradable System," *Biomat. Med. Dev. Art. Org.*, 10(3):205–218.

78. Hung, C. T., A. D. McLeod and P. K. Gupta. 1990. "Formulation and Characterization of Magnetic Polyglutaraldehyde Nanoparticles as Carriers for Poly-*l*-lysine-methotrexate," *Drug Dev. Ind. Pharm.*, 16(3):509–521.

79. Kwon, G. S., Y. H. Bae, H. Cremers, J. Feijen and S. W. Kim. 1992. "Release of Macromolecule from Albumin-Heparin Microspheres," *Int. J. Pharm.*, 79:191–198.

80. Madan, P. L., D. K. Madan and J. C. Price. 1976. "Clofibrate Microcapsules: Preparation and Release Rate Studies," *J. Pharm. Sci.*, 65(10):1476–1479.

81. Yoshioka, T., M. Hashida, S. Muranishe and H. Sezaki. 1981. "Specific Delivery of Mitomycin C to the Liver, Spleen and Lung: Nano- and Microspherical Carriers of Gelatin," *Int. J. Pharm.*, 81:131–141.

82. Yan, C., X. Li, X. Chen, D. Wang, D. Zhong, T. Tan and H. Kitano. 1991. "Anticancer Gelatin Microspheres with Multiple Functions." *Biomaterials*, 12: 640–644.

83. Ciao, C. S. L. and J. C. Price. 1989. "The Modification of Gelatin Beadlets for Zero-Order Sustained Release," *Pharm. Research*, 6(6):517–520.

84. Tabata, Y. and Y. Ikada. 1989. "Synthesis of Gelatin Microspheres Containing Interferon," *Pharm. Research*, 6(5):422–427.

85. Raymond, G., M. Degennard and R. Mikeal. 1990. "Preparation of Gelatin: Phenytoin Sodium Microspheres: An *in vitro* and *in vivo* Evaluation," *Drug Devel. Ind. Pharm.*, 16(6):1025–1051.

86. Goto, S., M. Komatsu, K. Tagawa and M. Kawata. 1983. "Preparation and Evaluation of Gelatin Microcapsules of Sulfonamides," *Chem. Pharm. Bull.*, 31(1):256–261.

87. Nixon, J. R. and S. E. Walker. 1971. "The *in vitro* Evaluation of Gelatin Coacervate Microcapsules," *J. Pharm. Pharmacol.*, 23:147S–155S.

88. Laakso, T., P. Artursson and I. Sjöholm. 1986. "Biodegradable Microspheres IV: Factors Affecting the Distrubution and Degradation of Polyacryl Starch Microparticles," *J. Pharm. Sci.*, 75(10):962–967.

89. Laakso, T. and I. Sjöholm. 1987. "Biodegradable Microspheres X: Some Properties of Polyacryl Starch Microparticles Prepared from Acrylic Acid-Esterified Starch," *J. Pharm. Sci.*, 76(12):935–939.

90. Artursson, P., P. Edman, T. Laakso and I. Sjöholm. 1984. "Characterization of Polyacryl Starch Microparticles as Carriers for Proteins and Drugs," *J. Pharm. Sci.*, 73(11):1507–1513.

91. Artursson, P., P. Edman and I. Sjöholm. 1984. "Biodegradable Microspheres. I. Duration of Action of Dextranase Entrapped in Polyacrylstarch Microparticles *in Vivo*," *J. Pharmacol. Exp. Therapeutics*, 231(3):705–712.

92. Stjärnkvist, P., L. Degling and I. Sjöholm. 1991. "Biodegradable Microspheres XIII: Immune Response to the DNP Hapten Conjugated to Polyacryl Starch Microparticles," *J. Pharm. Sci.*, 80(5):436–440.

93. Stjärnkvist, P., T. Laakso and I. Sjöholm. 1989. "Biodegradable Microspheres XII: Properties of the Crosslinking Chains in Polyacryl Starch Microparticles," *J. Pharm. Sci.*, 78(1):52–56.

94. Artursson, P., D. Johansson and I. Sjöholm. 1987. "Receptor-Mediated Uptake of Starch and Mannan Microparticles by Macrophages: Relative Contribution of Receptors for Complement, Immunoglobulins and Carbohydrates," *Biomaterials*, 9:241–246.

95. Stjärnkvist, P., L. Degling and I. Sjöholm. 1991. "Immune Response to DNP-Haptenated Starch Microparticles," *Proceed. Intern. Symp. Control. Rel. Bioact. Mater.*, 18:393–394.

96. Kost, J. and S. Shefer. 1990. "Chemically-Modified Polysaccharides for Enzymatically-Controlled Oral Drug Delivery," *Biomaterials*, 11:695–698.

97. Degling, L., P. Stjärnkvist and I. Sjöholm. 1991. "Antileishmanial Effect of IFN-Gamma Coupled to Starch Microparticles," *Proceed. Intern. Symp. Control. Rel. Bioact. Mater.*, 18:391–392.

98. Wing, R. E., S. Maiti and W. M. Doane. 1987. "Factors Affecting Release of Butylate from Calcium Ion-Modified Starch-Borate Matrices," *J. Cont. Rel.*, 5:79–89.

99. Heller, J., S. H. Pangburn and K. V. Roskos. 1990. "Development of Enzymatically Degradable Protective Coatings for Use in Triggered Drug Delivery Systems: Derivatized Starch Hydrogels," *Biomaterials*, 11:345–350.

100. Larsen, C. 1989. "Dextran Prodrugs–Structure and Stability in Relation to Therapeutic Activity," *Advanced Drug Delivery Reviews*, 3:103–154.

101. Vercauteren, R., D. Bruneel, E. Schacht and R. Duncan. 1990. "Effect of the Chemical Modification of Dextran on the Degradation by Dextranase," *J. Bioact. Comp. Polym.*, 5:4–15.

102. Crepon, B., J. Jozefonvicz, V. Chytrý, B. Říhová and J. Kopeček. 1991. "Enzymatic Degradation and Immunogenic Properties of Derivatized Dextrans," *Biomaterials*, 12:550–554.

103. Edman, P., B. Ekman and I. Sjöholm. 1980. "Immobilization of Proteins in Microspheres of Biodegradable Polyacryldextran," *J. Pharm. Sci.*, 69(7):838–842.

104. Fisher, E. and E. Stein. 1960. "Dextranase," in *The Enzymes, Vol. 4A*, P. D. Bloyer, H. Lardy and K. Myrback, eds., New York, NY: Academic Press, pp. 304–330.

105. Serry, T. W. and E. J. Hehre. 1956. "Degradation of Dextrans by Enzymes of Intestinal Bacteria," *J. Bacteriol.*, 71:373–380.

106. Edman, P. and I. Sjöholm. 1982. "Treatment of Artificially Induced Storage Disease with Lysosomotropic Microparticles," *Life Sciences*, 30:327–330.

107. Edman, P., I. Sjöholm and U. Brunk. 1983. "Acrylic Microspheres *in vivo* VII: Morphological Studies on Mice and Cultured Macrophages," *J. Pharm. Sci.*, 72(6):658–665.

108. Edman, P. and I. Sjöholm. 1983. "Acrylic Microspheres *in vivo* VIII: Distribution and Elimination of Poyacryldextran Particles in Mice," *J. Pharm. Sci.*, 72(7): 796–799.

109. Schacht, E., F. Vandoorne, J. Vermeersch and R. Duncan. 1987. "Polysaccharides as Drug Carriers. Activation Procedures and Biodegradation Studies," in *Controlled-Release Technology Pharmaceutical Applications*, *ACS Symposium Series*, P. I. Lee and W. R. Good, eds., New York, NY: ACS.

110. Schacht, E., R. Vercauteren and S. Vansteenkiste. 1988. "Some Aspects of the Application of Dextran in Prodrug Design," *J. Bioact. Comp. Polym.*, 3:72–80.

111. Vansteenkiste, S., A. De Marre and E. Schacht. 1992. "Synthesis of Glycosylated Dextrans," *J. Bioact. Compat. Polym.*, 7:4–14.

112. Kojima, T., M. Hashida, S. Muramishi and H. Sezaki. 1980. "Mitomycin C-Dextran Conjugate: A Novel High Molecular Weight Pro-Drug of Mitomycin C," *J. Pharm. Pharmacol.*, 32:30–34.

113. Takakura, Y., M. Kitajima, S. Matsumoto, M. Hashida and H. Sezaki. 1983. "Development of a Novel Polymeric Prodrug of Mitomycin C, Mitomycin C-Dextran Conjugate with Anionic Charge. I. Physicochemical Characteristics and *in vivo* and *in vitro* Antitumor Activities," *Int. J. Pharm.*, 37:135–143.

114. Sezake, H., Y. Takakura and M. Hashida. 1988. "Control of Biopharmaceutical Properties by Dextran Conjugates," *J. Bioact. Comp. Polym.*, 3:81–85.

115. Doelker, E. 1987. "Water-Swollen Cellulose Derivatives in Pharmacy," in *Hydrogels in Medicine and Pharmacy, Vol. II. Polymers*, N. A. Peppas, ed., Boca Raton, FL: CRC Press Inc., pp. 115–160.

116. Huber, H. E., L. B. Dale and G. L. Christenson. 1966. "Utilisation of Hydrophilic Gums for the Control of Drug Release from Tablet Formulations I. Disintegration and Dissolution Behaviour," *J. Pharm. Sci.*, 55:974–976.

117. Ford, J. L., M. H. Rubinstein, F. McCaul, J. E. Hogan and P. Edgar. 1987. "Importance of Drug Type, Tablet Shape and Added Diluents on Drug Release Kinetics from Hydroxypropylmethycellulose Matrix Tablets," *Int. J. Pharm.*, 40:223–234.

118. Feely, L. C. and S. S. Davis. 1988. "Influence of Surfactants on Drug Release from Hydroxypropylmethylcellulose Matrices," *Int. J. Pharm.*, 41:83–90.

119. Ford, J. L., M. H. Rubinstein and J. E. Hogan. 1985. "Propranolol Hydrochloride and Aminophylline Release from Matrix Tablets Containing Hydroxypropylmethylcellulose," *Int. J. Pharm.*, 24:339–350.

120. Hogan, J. E. 1989. "Hydroxypropyl Methycellulose Sustained Release Technology," *Drug Develop. Ind. Pharm.*, 15(6/7):975–999.

121. Feely, L. C. and S. S. Davis. 1988. "The Influence of Polymeric Excipients on Drug Release from Hydroxypropylmethylcellulose Matrices," *Int. J. Pharm.*, 44: 131–139.

122. Ford, J. L., M. H. Rubinstein and J. E. Hogan. 1985. "Formulation of Sustained Release Prometazine Hydrochloride Tablets Using Hydroxypropyl Methycellulose Matrices," *Int. J. Pharm.*, 24:327–338.

123. Ford, J. L., K. Mitchell, D. Sawh, S. Ramdour, D. J. Armstrong, P. N. C. Elliot, C. Rostron and J. E. Hogan. 1991. "Hydroxypropylmethylcellulose Matrix Tablets Containing Propranolol Hydrochloride and Sodium Dodecyl Sulphate," *Int. J. Pharm.*, 71:213–221.

124. Moldenharuer, M. G. and J. G. Nairn. 1991. "The Effect of Rate of Evaporation on the Coat Structure of Ethylcellulose Microcapsules," *Int. J. Pharm.*, 17:49–60.

125. Benita, S. and M. Donbrow. 1982. "Dissolution Rate Control of the Release Kinetics of Water-Soluble Compounds from Ethyl Cellulose Film-Type Microcapsules," *Int. J. Pharm.*, 12:251–264.

126. Donbrow, M. and Y. Samuelov. 1980. "Zero Order Drug Delivery from Double-Layered Porous Films: Release Rate Profiles from Ethyl Cellulose, Hydroxypropyl Cellulose and Polyethylene Glycol Mixtures," *J. Pharm. Pharmacol.*, 32:463–470.

127. Jalsenjak, I., C. F. Nicolaidou and J. R. Nixon. 1976. "The *in vitro* Dissolution of Phenobarbitone Sodium from Ethyl Cellulose Microcapsules," *J. Pharm. Pharmacol.*, 28:912–914.

128. Salib, N. N., M. E. El-Menshawn and A. A. Ismail. 1976. "Ethyl Cellulose as a Potential Sustained Release Coating for Oral Pharmaceuticals," *Pharmazie*, 31:721–723.

129. Deasy, P. B., M. R. Brophy, B. Ecanow and M. M. Joy. 1980. "Effect of Ethylcellulose Grade and Sealant Treatments on the Production and *in vitro* Release of Microencapsulated Sodium Salicylate," *J. Pharm. Pharmacol.*, 32:15–20.

130. John, P. M., H. Minatoya and F. J. Rosenberg. 1978. "Microencapsulation of Bitolterol for Controlled Release and Its Effect on Bronchodilator and Heart Rate Activities in Dogs," *J. Pharm. Sci.*, 68(4):475–481.

131. Benita, S. and M. Donbrow. 1982. "Release Kinetics of Sparingly Soluble Drugs from Ethyl Cellulose-Walled Microcapsules: Theophylline Microcapsules," *J. Pharm. Pharmacol.*, 34:77–82.

132. Heller, J., R. F. Helwing, R. W. Baker and M. E. Tuttle. 1983. "Controlled Release of Water-Soluble Macromolecule from Bioerodible Hydrogels," *Biomaterials*, 4:262–266.

133. Davis, B. K. 1982. "Control of Diabetes with Polyacrylamide Implants Containing Insulin," *Experientia*, 28:348.

134. Torchilin, V. P., E. G. Tischenko, V. N. Smirnov and E. I. Chazov. 1977. "Immobilization of Enzymes on Slowly Soluble Carriers," *J. Biomed. Mater. Res.*, 11:223–235.

135. Saffran, M., G. S. Kumar, C. Savariar, J. C. Burnham, F. Williams and D. C. Neckers. 1986. "A New Approach to the Oral Administration of Insulin and Other Peptide Drugs," *Science*, 233:1081–1084.

136. Shalaby, W. S. W., W. E. Blevins and K. Park. 1992. "*In vitro* and *in vivo* Studies of Enzyme-Digestible Hydrogels for Oral Drug Delivery," *J. Controlled Rel.*, 19:131–144.

137. Juskor, W. and G. Levy. 1967. "Absorption, Metabolism, and Execretion of Riboflavin-5′-Phosphate in Man," *J. Pharm. Sci.*, 56:58–62.

138. Kopeček, J. 1984. "Controlled Biodegradability of Polymers–A Key to Drug Delivery Systems," *Biomaterials*, 5:19–25.

139. Ulbrich, K., J. Strohalm and J. Kopeček. 1982. "Polymers Containing Enzymatically Degradable Bonds. VI. Hydrophilic Gels Cleavable by Chymotrypsin," *Biomaterials*, 3:150–154.

140. Rejmanová, P., B. Obereigner and J. Kopeček. 1981. "Polymers Containing Enzymatically Degradable Bonds, 2. Poly[*N*-(2-hydroxypropyl)methacrylamide] Chains Connected by Oligopeptide Sequences Cleavable by Chymotrypsin," *Makromol. Chem.*, 182:1899–1915.

141. Kopeček, J., P. Rejmanová and V. Chytrý. 1981. "Polymers Containing Enzymatically Degradable Bonds, 1. Chymotrypsin Catalyzed Hydrolysis of *p*-Nitroanilides of Phenylalanine and Tyrosine Attached to Side-Chains of Copolymers of *N*-(2-Hydroxypropyl)methacrylamide," *Makromol. Chem.*, 182:799–809.

142. Ulbrich, K., J. Strohalm and J. Kopeček. 1981. "Polymers Containing Enzymatically Degradable Bonds, 3. Poly[*N*-(2-hydroxypropyl)methacrylamide] Chains Connected by Oligopeptide Sequences Cleavable by Trypsin," *Makromol. Chem.*, 182:1917–1918.

143. Dittert, L. W., G. M. Irwin, C. W. Chong and J. V. Swintosky. 1968. "Acetaminophen Prodrugs II. Effect of Structure and Enzyme Source on Enzymatic and Nonenzymatic Hydrolysis of Carbonate Esters," *J. Pharm. Sci.*, 57(5):780–783.

144. Rejmanová, P., J. Kopeček, J. Pohl, M. Baudys and V. Kostka. 1983. "Polymers Containing Enzymatically Degradable Bonds, 8. Degradation of Oligopeptide Sequences in *N*-(2-Hydroxypropyl)methacrylamide Copolymers by Bovine Spleen Cathepsin B," *Makromol. Chem.*, 184:2009–2020.

145. Duncan, R., H. C. Cable, J. B. Lloyd, P. Rejmanová and J. Kopeček. 1983. "Polymers Containing Enzymatically Degradable Bonds, 7. Design of Oligopeptide Sequences in *N*-(2-Hydroxypropyl)methacrylamide Copolymers to Promote Efficient Degradation by Lysosomal Enzymes," *Makromol. Chem.*, 184:1997–2008.

146. Rejmanová, P., J. Kopeček, R. Duncan and J. B. Lloyd. 1985. "Stability in Rat Plasma and Serum of Lysosomally Degradable Oligopeptide Sequences in *N*-(2-Hydroxypropyl)methacrylamide Copolymers," *Biomaterials*, 6:45–48.

147. Rihová, B., V. Vetricka, J. Strohalm, K. Ulbrich and J. Kopeček. 1989. "Action of Polymeric Prodrugs Based on *N*-(2-Hydroxypropyl)methacrylamide Copolymers. I. Suppression of the Antibody Response and Proliferation of Mouse Splenocytes *in Vitro*," *J. Control. Rel.*, 9: 21–32.

148. Říhová, B. and J. Kopeček. 1985. "Biological Properties of Targetable Poly[*N*-(2-hydroxypropyl)methacrylamide] Antibody Conjugates," *J. Control. Rel.*, 2:289–310.

149. Subr, V., R. Duncan and J. Kopeček. 1990. "Release of Macromolecules and Daunomycin from Hydrophilic Gels Containing Enzymatically Degradable Bonds," *J. Biomater. Sci., Polymer Edn.*, 1:261–278.

150. Ringsdorf, H., B. Schmidt and K. Ulbrich. 1987. "Bis(2-chloroethyl)amine Bound to Copolymers of *N*-(2-Hydroxypropyl)methacrylamide and Methacryloylated Oligopeptides via Biodegradable Bonds," *Makromol. Chem.*, 188:257–264.

151. Seymour, L. W., K. Ulbrich, J. Strohalm, J. Kopeček and R. Duncan. 1990. "The Pharmacokinetics of Polymer-Bound Adriamycin," *Biochem. Pharmacol.*, 39(6): 1125–1131.

152. Laakso, T., P. Stjärnkvist and I. Sjöholm. 1987. "Biodegradable Microspheres VI: Lysosomal Release of Covalently Bound Antiparasitic Drugs from Starch Microparticles," *J. Pharm. Sci.*, 76(2):132–140.

153. Bennett, D. B., X. Li, N. W. Adams, S. W. Kim, C. J. T. Hoes and J. Feijen. 1991. "Biodegradable Polymeric Prodrugs of Naltrexone," *J. Contr. Rel.*, 16:43–52.

CHAPTER 9

Future Development

The development of new controlled drug delivery systems is important for many reasons. New delivery systems revitalize old drugs by reducing their pharmaceutical shortcomings and improving the biopharmaceutical properties of the drugs. New delivery systems can also extend the patent life of those drugs whose patent protection has expired. According to the Drug Price Competition and Patent Term Restoration Act, a new delivery system, which makes a drug product a new or better therapy, can protect the drug from generic competition for up to 17 years [1,2]. This exclusivity provision protects the new dosage form of the drug, provided that it is distinguishable from current therapy [1]. Considering the fact that developing a new drug compound costs on the average more than $100 million, one can easily understand why such emphasis has been placed on the development of new drug delivery devices. New drug delivery systems are also important for the delivery of many genetically engineered protein drugs, such as insulin, growth hormone, interferon, erythropoietin, and tissue plasminogen activator. These protein drugs are effective only by following a therapeutic regimen which requires multiple daily injections [3]. The limiting step in the development and commercial success of protein drugs is the availability of suitable delivery systems. New delivery systems for nontraditional routes of administration with the capacity for self-regulating delivery are necessary for those drugs if they are to be useful in treating chronic disease. As described in earlier chapters, biodegradable hydrogels have been used successfully in the preparation of delivery systems for protein drugs. Although significant advances have been made in the development of innovative drug delivery systems using biodegradable hydrogels, there are still a number of important issues that should be given careful consideration in the design of truly useful dosage forms. These issues are briefly discussed here.

9.1 DRUG LOADING

A drug is usually loaded into hydrogels either by equilibrating hydrogels in a drug-containing solution followed by drying [4–6], or by incorporating the drug during the preparation of hydrogels [7,8]. In the former case, solvent removal from the gel can produce the migration of the drug to the gel surface depending on the properties of the solvent and the drug. When this occurs, a significant portion of the drug is present on the gel surface and the drug distribution in the hydrogel is not homogeneous. This tends to result in the significant burst release of the drug. If a drug is present during polymerization, bioactivity of the drug could be compromised if its chemical structure and/or conformation is modified by a chemical initiator, monomer constituent, or temperature increase during polymerization.

The loading of high molecular weight drugs, such as peptide and protein drugs, is even more difficult. Due to the large size of peptide and protein drugs, they do not readily diffuse into the hydrogel. Thus, the efficiency of drug loading by swelling or equilibration of hydrogels in the drug-containing solution is very low. Unless the pore size of the hydrogel is substantially larger than the size of the protein drugs, drug diffusion into the hydrogel will be significantly inhibited and most drugs will remain on the outer region of the hydrogel. The model protein drugs which are incorporated in this way are always released within an hour or two. The incorporation of protein drugs in the polymerization mixture is also undesirable due to the temperature increase during polymerization and the presence of unreacted monomers, initiators, and crosslinkers in the hydrogel. Furthermore, the release of the incorporated drugs is usually limited, unless the hydrogel is degraded. For protein drugs, drug loading into hydrogels is a tricky business. Apparently, an efficient loading process is needed for the practical application of hydrogels as delivery systems for protein drugs. Protein drugs may be loaded into hydrogels by electrophoresis [9]. In this process, protein molecules are transported into hydrogels by an electrochemical gradient. This process allows loading of large amounts of protein drugs in a short period of time. The loading of a large amount of protein drugs can also be achieved by using water-soluble polymers which form gels, either physical or chemical gels, in the presence of protein drugs. For the formation of chemical gels, water-soluble polymers can be modified and purified before crosslinking in the presence of protein drugs. Since the crosslinking of the modified polymers can be done by either ultraviolet or gamma irradiation, problems associated with purifying the final hydrogel system can be avoided [10].

9.2 DRUG TARGETING

The delivery of a drug to specific cells or organs in the body and the release at target sites present a major challenge in the development of drug delivery systems. The drug targeting, which can enhance the efficacy of drugs dramatically, requires a thorough understanding of the various transport mechanisms in the body [11]. The difficulty (or ease) of drug targeting depends on the target sites and delivery routes. In general, the drug targeting can be classified into passive and active targeting. The passive targeting relies on the natural distribution of drug carriers in the body. For example, when polymeric drug delivery systems are injected intravenously, most of them tend to be transported to and accumulated in the liver, lungs, and spleen [11]. The natural distribution of microspherical drug carriers may be controlled to a certain extent by modifying the particle surface properties [12,13]. In active targeting, dosage forms are delivered to specific organs or cells in the body. Active targeting may be achieved using homing devices such as antibodies or saccharide units which are able to interact with specific receptors on the target cell surfaces [14,15].

Physical targeting is another form of active targeting. Drug delivery systems for physical targeting are sensitive to physical stimuli such as pH, temperature, electric field, or magnetic field. Local application of such stimuli will increase the drug release rate at a specific site in the body. This requires the use of a signal inducer inside or in close vicinity to the target organ [11]. There are various hydrogels which can respond to the physical stimuli. Signal-sensitive, biodegradable hydrogels are particularly useful for physical targeting of drugs.

Targeting of oral drug delivery systems to a certain site in the gastrointestinal (GI) tract is also challenging. The dosage forms pass through the gastrointestinal tract in a relatively short period of time. The gastric residence time is usually less than a few hours and the transit time in the small intestine ranges from three to four hours under both fast and feed conditions [16,17]. Thus, the absorption of most drugs from the GI tract is limited by the short GI transit time irrespective of the controlled release properties of the device. If the target sites are located in the upper GI tract, retention of the drug delivery device in the stomach would be beneficial. Biodegradable hydrogels have been successfully used as an effective gastric retentive device [18]. The delivery of drugs to specific segments in the GI tract has been achieved using homing devices. Site-specific delivery to the colon has been achieved by using glycosidic prodrugs [19], polymeric coatings [20], and hydrogels [21] which are degraded by enzymes secreted from the bacterial flora of the colon [22]. The use of bioadhesive, biodegradable hydrogels will be particularly useful in the development of site-specific delivery systems for the GI tract.

9.3 DRUG RELEASE

The study of drug release is usually focused on examining release kinetics and mathematical modeling. Too much attention has been paid to zero-order release or near zero-order release. Unfortunately, zero-order drug release does not necessarily mean zero-order drug absorption by the body. In many situations drug absorption decreases over time. A particular example is the reduced absorption as the dosage form goes through the gastrointestinal tract. Since most drugs are absorbed from the upper portion of the small intestine, duodenum, and jejunum, even zero-order drug release devices show the transient maximum followed by a decrease in blood concentration. In this case, more drugs have to be released to compensate for reduced drug absorption. More emphasis should be placed on overcoming the physiological barriers that hinder zero-order drug absorption.

While zero-order drug absorption is desirable in most cases, there are many situations where maintaining constant drug concentration in blood is not beneficial. Some drugs, e.g., insulin and naltrexone, need to be released only in response to a physiological need [23,24]. It is necessary to develop self-regulated drug delivery devices which vary drug output in response to physiological needs [25,26]. The self-regulated drug delivery devices require biosensing mechanisms which transfer a physiological event to a chemical or electrical signal. Hydrogels are useful in the development of signal-sensitive drug delivery devices. Signal transduction could lead to changes in gel swelling, structure, or degradation, which would ultimately trigger the release of drugs.

9.4 DEGRADATION AND ELIMINATION

The degradation of a drug delivery system has at least two purposes: control of the drug release and removal of the system after its use. There are many different types of biodegradation, and each type results in degradation products with different molecular weights. For parenteral applications or implants in the body, degradation into water-soluble, low molecular weight products is preferred, since the elimination of high molecular weight products is rather difficult. It would be even better if the degradation products changed into naturally occurring metabolites. In other applications such as oral drug delivery, the degradation into low molecular weight products may not be that critical. In fact, high molecular weight products may be preferable to smaller fragments, since the latter can be absorbed into the blood stream and may cause undesirable side effects.

In cases where the degradation products remain as high molecular weight polymers, it is desirable to prepare hydrogels using biocompatible polymers such as those used as plasma expanders. Human albumin, hydroxyethyl starch (or hetastarch) [27], dextran [27–29], polyvinylpyrrolidone (PVP) [30], and poly(vinyl alcohol) [30] have been used as plasma expanders. Poly(ethylene oxide) (PEO) has not been used as a plasma expander, but it is known to be highly biocompatible [31]. When given intravenously, polymers with a molecular weight of about 70,000 or smaller are in general passively excreted by the kidneys [29,32,33]. The higher molecular weight polymers tend to be distributed widely throughout the body, with the liver, lungs, spleen, and lymph nodes showing the highest concentrations [33]. Thus, it is necessary to use polymers which can be degraded by enzymes until small enough to undergo kidney elimination. For enzymatic degradation in the body, polysaccharides are preferred to synthetic polymers.

Since one of the reasons for using biodegradable drug delivery systems is to avoid removal of the system after its use, degradation in a reasonably short period of time is necessary. Physicochemical factors and environmental conditions that affect the degradation rate must be defined for each system in close relation to the drug release. Unless degradation is related to controlling the drug release rate, degradation should not dramatically alter the mechanical strength and the physical integrity of the system until all the drug is released. Pharmacological and biological issues such as long-term tissue tolerance, potential antigenicity, toxicity of polymer and degradation products, and elimination from the body must be studied [34].

9.5 THE USE OF BIOACTIVE POLYMERS

Biodegradable hydrogels degrade to release individual polymer molecules or smaller fragments during or after the drug is released. It is well known that a large number of natural polyanions possess innate physiological properties and various synthetic polymers elicit significant bioactivities, such as antitumor, antiviral, and antibacterial effects [35–39]. Most of the bioactive polymers are polyelectrolytes and they are excellent candidates for making hydrogels. It would be beneficial if biodegradable hydrogel systems were made of polymers which upon release exerted bioactivity or produced smaller fragments possessing bioactivity. At this point, the exact relationship between polymer structure and bioactivity is not well established. Thus, it may be difficult to find polymers with specific bioactivity. Nevertheless, the idea of using bioactive polymers as a part of biodegradable hydrogels presents new opportunities in the development of future drug delivery devices.

The issues discussed in this chapter seem to be difficult enough to keep all of us busy for the next decade. Considering the rate of progress made in the last decade in the areas of pharmaceutics, polymer chemistry, protein chemistry, and pharmacology, however, we are optimistic that these issues can be solved in the very near future. With the wisdom of established researchers and the new ideas of young investigators, we can overcome the hurdles on the way to development of biodegradable, self-regulating drug delivery devices.

9.6 REFERENCES

1. 1987. *The Sourcebook for Innovative Drug Delivery.* Santa Monica, CA: A Canon Communications, Inc., p. 9.
2. Flynn, J., J. O'C. Hamilton and J. Weber. 1992. "A Shot in the Arm for Drugmakers," *Business Week* (September 21):86–93.
3. Chien, Y. W. 1992. *Novel Drug Delivery Systems, 2nd Ed.* New York, NY: Marcel Dekker, Inc., Chapter 11.
4. Robert, C. C. R., P. A. Buri and N. A. Peppas. 1987. "Influence of the Drug Solubility and Dissolution Medium on the Release from Poly(2-hydroxyethyl Methacrylate) Microspheres," *J. Controlled Release*, 5:151–157.
5. Siegel, R. A., M. Falamarzian, B. A. Firestone and B. C. Moxley. 1988. "pH-Controlled Dependent Release from Hydrophobic/Polyelectrolyte Copolymer Hydrogels," *J. Controlled Release*, 8:179–182.
6. Kim, S. W., Y. H. Bae and T. Okano. 1992. "Hydrogels: Swelling, Drug Loading, and Release," *Pharm. Res.*, 9(3):283–290.
7. Bernfield, P. and J. Wan. 1963. "Antigens and Enzymes Made Insoluble by Entrapping Them into Lattices of Synthetic Polymers," *Science*, 142:678–679.
8. Heller, J., R. F. Helwing, R. W. Baker and M. E. Tuttle. 1983. "Controlled Release of Water-Soluble Macromolecules from Bioerodible Hydrogels," *Biomaterials*, 4:262–266.
9. Shalaby, W. S. W., A. A. Abdallah, H. Park and K. Park. "Loading of Bovine Serum Albumin into Hydrogels by an Electrophoretic Process and Its Potential Application to Protein Drugs," *Pharm. Res.* (submitted for publication).
10. Kamath, K. R. and K. Park. 1992. "Use of Gamma-Irradiation for the Preparation of Hydrogels from Natural Polymers," *Proceed. Intern. Symp. Control. Rel. Bioact. Mater.*, 19:42–43.
11. Arshady, R. 1990. "Biodegradable Microcapsular Drug Delivery Systems: Manufacturing Methodology, Release Control and Targeting Prospects," *J. Bioact. Compat. Polymers*, 5:315–342.
12. Harper, G. R., M. C. Davies, S. S. Davis, Th. F. Tadros, D. C. Taylor, M. P. Irving and J. A. Waters. 1991. "Steric Stabilization of Microspheres with Grafted Polyethylene Oxide Reduces Phagocytosis by Rat Kupffer Cells *in Vitro*," *Biomaterials*, 12: 695–700.
13. Kissel, T. and M. Roser. 1991. "Influence of Chemical Surface-Modifications on the Phagocytic Properties of Albumin-Nanoparticles," *Proceed. Intern. Symp. Control. Rel. Bioact. Mater.*, 18:275–276.
14. Gregoriadis, G., ed. 1979. *Drug Carriers in Biology and Medicine*. London: Academic Press.
15. Říhová, B., K. Veres, L. Fornusek, K. Ulbrich, J. Strohalm, V. Větrička, M. Bilej and J. Kopĕcek. 1989. "Action of Polymeric Prodrugs Based on *N*-(2-Hydroxypropyl)-methacrylamide Çopolymers. II. Body Distribution and T-Cell Accumulation of Free and Polymer-Bound [125I]Daunomycin," *J. Controlled Rel.*, 10:37–49.

16. Khosla, R. and S. S. Davis. 1989. "Gastric Emptying and Small and Large Bowel Transit of Non-Disintegrating Tablets in Fasted Subjects," *Int. J. Pharm.*, 52:1–10.
17. Davis, S. S, J. G. Hardy and J. W. Fara. 1986. "Transit of Pharmaceutical Dosage Forms through the Small Intestine," *Gut*, 27:886–892.
18. Shalaby, W. S. W., W. E. Blevins and K. Park. 1992. "*In vitro* and *in vivo* Studies of Enzyme-Digestible Hydrogels for Oral Drug Delivery," *J. Controlled Rel.*, 19: 131–144.
19. Tozer, T. N., J. Rigod, A. D. McLeod, R. Gungon, M. K. Hoag and D. R. Friend. 1991. "Colon-Specific Delivery of Dexamethasone from a Glucoside Prodrug in the Guinea Pig," *Pharm. Res.*, 8:445–454.
20. Saffran, M., G. S. Kumar, C. Savarian, J. C. Burnham, F. Williams and D. C. Necker. 1986. "A New Approach to the Oral Administration of Insulin and Other Peptide Drugs," *Science*, 233:1081–1084.
21. Kopeček, J., P. Kopeckova, H. Brøndsted, R. Rathi, B. Říhová, P.-Y. Yeh and K. Ikesue. 1992. "Polymers for Colon-Specific Drug Delivery," *J. Controlled Rel.*, 19:121–130.
22. Friend, D. R., ed. 1992. *Oral Colon-Specific Drug Delivery*. Boca Raton, FL: CRC Press Inc.
23. Heller, J. 1988. "Chemically Self-Regulated Drug Delivery Systems," *J. Controlled Rel.*, 8:111–125.
24. Reckos, K. V., J. A. Tefft, B. K. Tritzinger and J. Heller. 1992. "Development of a Morphine-Triggered Naltrexone Delivery System," *J. Controlled Rel.*, 19:145–160.
25. Hrushesky, W. J. M., R. Langer and F. Theeuwes, eds. 1991. *Temporal Control of Drug Delivery, Annals of the New York Academy of Sciences, Vol. 618.*
26. Kost, J., ed. 1990. *Pulsed and Self-Regulated Drug Delivery*. Boca Raton, FL: CRC Press Inc.
27. Schuerch, C. 1985. "Biomedical Applications of Polysaccharides," in *Bioactive Polymeric Systems. An Overview*, C. G. Gebelein and C. E. Carraher, Jr., eds., New York, NY: Plenum Press, pp. 365–386.
28. Schuerch, C. 1972. "The Chemical Synthesis and Properties of Polysaccharides of Biomedical Interest," *Adv. Polymer Sci.*, 10:173–194.
29. Turco, S. and R. E. King. 1987. *Sterile Dosage Forms. Their Preparation and Clinical Application*. Philadelphia, PA: Lea & Febiger, Chapter 9.
30. Braybrook, J. H. and L. D. Hall. 1990. "Organic Polymer Surfaces for Use in Medicine: Their Formation, Modification, Characterization and Application," *Prog. Polym. Sci.*, 15:715–734.
31. Amiji, M. and K. Park. In press. "Surface Modification of Polymeric Biomaterials with Poly(ethylene Oxide), Albumin, and Heparin for Reduced Thrombogenicity," *J. Biomater. Sci. Polymer Edn.*
32. Larsen, C. 1989. "Dextran Prodrugs–Structure and Stability in Relation to Therapeutic Activity," *Adv. Drug Delivery Rev.*, 3:103–154.
33. Robinson, B. V., F. M. Sullivan, J. F. Borzelleca and S. L. Schwartz. 1990. *PVP. A Critical Review of the Kinetics and Toxicology of Polyvinylpyrrolidone (Povidone)*. Chelsa, MI: Lews Publishers, Chapter 5.
34. Kopeček, J. 1984. "Controlled Biodegradability of Polymers–A Key to Drug Delivery Systems," *Biomaterials*, 5:19–25.
35. Batz, H.-G. 1977. "Polymeric Drugs," *Adv. Polymer Sci.*, 23:25–53.

36. Donaruma, L. G. 1975. "Synthetic Biologically Active Polymers," in *Progress in Polymer Science*, *Vol. 4,* A. D. Jenkins, ed., New York, NY: Pergamon Press, pp. 1–25.

37. Ottenbrite, R. M., K. Takemoto and M. Miyata. 1987. "Biomedical Polymers and Polycarboxylic Acid Polymer Drugs," in *Functional Monomers and Polymers*, K. Takemoto, Y. Inaki and R. M. Ottenbrite, eds., New York, NY: Marcel Dekker, Inc., pp. 423–459.

38. Klotz, I. M. 1987. "Enzyme Models–Synthetic Polymers," Chapter 2 in *Enzyme Mechanisms*, M. I. Page and A. Williams, eds., London, England: The Royal Society of Chemistry.

39. Carraher, C. E., Jr. and C. G. Gebelein, eds. 1982. *Biological Activities of Polymers, ACS Symposium Series 186*. Washington, DC: American Chemical Society.

INDEX